Reluctant Genius

Reluctant Genius

ALEXANDER GRAHAM BELL
AND THE PASSION FOR INVENTION

Charlotte Gray

ARCADE PUBLISHING • NEW YORK

Arcade Publishing books may be purchased in bulk at special discounts for
sales promotion, corporate gifts, fund-raising, or educational purposes. Special
editions can also be created to specifications. For details, contact the Special Sales
Department, Arcade Publishing, 307 West 36th Street, 11th Floor, New York, NY
10018 or info@skyhorsepublishing.com.

Arcade Publishing® is a registered trademark of Skyhorse Publishing, Inc.®,
a Delaware corporation.

Visit our website at www.arcadepub.com.

10 9 8 7 6 5 4 3 2 1

Library of Congress Cataloging-in-Publication Data is available on file.

ISBN: 978-1-61145-060-6

Printed in the United States of America

Contents

**PART 3   MONSTER KITES AND FLYING MACHINES
1889–1923**

MAPS

For my family on both sides of the Atlantic

Part 1

CIRCUITS AND CONNECTIONS
1847–1876

It is a neck and neck race between Mr. Gray and myself who shall complete our apparatus first. He has the advantage over me in being a practical electrician—but I have reason to believe that I am better acquainted with the phenomena of sound than he is—so that I have an advantage there. . . . The very opposition seems to nerve me to work and I feel with the facilities I have now I may succeed. . . . I shall be seriously ill should I fail in this now I am so thoroughly wrought up.

Alexander Graham Bell to his parents, November 1874

Chapter 1

THE GREAT WHITE PLAGUE
1847–1870

The plaintive wail echoed through the stone stairwell of the tall, stately Edinburgh house: "Mama, Mama." There was a pause, then the childish voice started up again, in a monotonous rhythm: "Mama, Mama, Mama." Finally, a door opened on the floor below and a woman emerged, asking petulantly, "Good gracious, what can be the matter with that baby?" The wail stopped abruptly, and a door slammed above. Perplexed, the woman leaned over the stair rail, looking up for an infant in distress.

There was no infant. Instead, there were two dark-haired, mischievous teenage boys and an intriguing invention. The boys were Melville (Melly) and Alexander (Alec) Bell, sons of the woman's landlord, Professor Melville Bell. Melville Bell was a bombastic character with a leonine head, an untrimmed black beard, and a shock of untidy hair who lived with his family on the upper floors of 13 South Charlotte Street. A former actor, he was fascinated by the production of sound. He had challenged his two older sons to build a "speaking machine," and the boys had thrown themselves into the project with gusto. They had pored over their father's anatomy books. They had begged the local butcher to let them watch him carve up a lamb's head. They had scoured the garbage behind their neighbors' stables for bits of wood, metal, and wire that they could use in their construction. Alec took impressions of the upper and lower jaws of the human skull that sat on a shelf in his father's study. Then, in the secrecy of their shared bedroom, the brothers set to work. Alec laboriously shaped a jaw, upper gum, hard palate, teeth, tongue, and pair of lips out of gutta-percha (a primitive form of rubber), wood, and wire. Meanwhile, Melville, who was more nimble-fingered, created a larynx of tin and rubber and experimented with providing "breath" by blowing into a tin tube "throat." It was difficult to manipulate all the working parts, but after a few rehearsals Alec and Melly's machine could wail "Mama." Eager to escape their mother's notice, the boys tiptoed out of their bedroom and carried their contraption into the shared stone stairwell of the house to test it. Years later, Alec recalled with glee the moment when their neighbor came out to rescue the bawling baby: "We quietly slipped into our house, and closed the door, leaving our neighbour to pursue [her] fruitless quest for the baby. Our triumph and happiness were complete." The speaking machine then became a family toy—an entertainment for social gatherings, the subject of further experimentation for Alec and Melly, and a trophy for Melville, who bragged about having given his clever sons the challenge in the first place. Alec's cousin Mary Symonds never forgot the contraption, later recalling, "I think I was somewhat afraid of it, it gave such uncanny sounds."

Sounds. Alec Bell's childhood was full of sounds. Many were typical of a middle-class household in the mid-nineteenth century. There was the bossy boom of the paterfamilias—Melville Bell—whose powerful, well-modulated voice echoed through the house on Charlotte Street. There were the Scottish airs and Presbyterian hymns sung at Bell family get-togethers around the piano in the parlor. Beyond the house's elegant sash windows, there was the hubbub of parades and pipers, dogs and drunks in the streets of the Scottish capital. Horses' hooves clattered on the cobbles; knife sharpeners and meat-pie peddlers shouted out their wares; ragged street boys jeered at passing carriages. Once a week, there would be the rumble and roar of coal being delivered down a chute to the house's basement; coal fires were the only source of heating during Edinburgh's long, chilly winters, and sooty smoke spewed from the city's forest of brick chimneys. But there were also sounds peculiar to the Bell household—sounds designed to improve human communication. During the daytime, Alec would hear the rhythmic chants of his father's pupils, who had come to get help from the professor with their lisps, stutters, and enunciation. "Da, da, da . . ." would echo through the house, followed by "Pa, pa, pa . . ." In the evenings, strange grunts and whistles often emanated from his father's study. Alec himself would communicate in an unconventional way with his mother, Eliza Bell, who was deaf. Alone among her three sons, Alec had found a way to talk with her, by speaking in a deep voice close to her forehead so that she could pick up the vibrations.

Alec absorbed all of these distinctive sounds, and developed an unusually discriminating ear. At night he would lie in bed and identify each church bell that rang out over the ancient city and which neighbor's dog was barking at a stranger. When he sat down at the grand piano in the high-ceilinged second-floor parlor, he could play anything he had heard by ear, and he would improvise by the hour. His extraordinary ability to distinguish minute variations of pitch and tone would shape his entire career.

Born in Edinburgh on March 3, 1847, Alexander Bell came from a family that had been preoccupied with sound for at least two generations. These days, we would label both his father and his grandfather "speech pathologists." Such a title did not exist then, but since these were men untroubled by modesty, they each cheerfully adopted an even fancier title: "professor of elocution." They took care to speak "proper English" and insisted that their families follow suit, so that

Alec's birthplace, 16 South Charlotte Street, was in Edinburgh's elegant New Town.

there was never any hint of Scottish brogue in Alec's speech. There was plenty of demand for Professor Melville Bell's services, and not simply from those who had speech impediments. In an era when a person's accent reflected his or her position within a rigid social hierarchy, speaking "proper English" was a sign of both education and class.

Alec's grandfather, who was also called Alexander, had started life as a shoemaker in St. Andrews, a few miles from the capital. But he had ambition, and he climbed the social ladder to become first an actor, then a teacher, and finally a "corrector of defective utterance." His son Alexander Melville, Alec's father, was born in Edinburgh, where as a young man he began to cough and gasp in the city's grimy, damp, soot-filled streets—sure signs of the respiratory infections that were the bane of nineteenth-century life. So, in 1838, young Melville took passage across the Atlantic to the British colony of Newfoundland, famous for its bracing winds and clean air. Working as a clerk for the shipping firm Thomas McMurdo and Company, he quickly recovered his health, made many friends, and become quite a figure in St. John's, the colonial capital, as an organizer of amateur dramatics. After four years, he returned to Edinburgh in fine fettle, with an unshakable belief in the New World's healthy climate. Emboldened by his social and dramatic successes in faraway Newfoundland, as well as by his father's example, he hung out

*Eliza Symonds Bell captured the
closeness of her three sons (Alec is
on the right) in an early watercolor.*

his shingle as an elocution teacher. (George Bernard Shaw probably based Professor Henry Higgins, in his play *Pygmalion,* on either Alec's father or his grandfather.) A friend then introduced him to Eliza Grace Symonds, the sweet-tempered daughter of a Royal Navy surgeon who lived with her widowed mother and scraped a living as a painter of miniatures. Eliza was a handsome woman, with a long, solemn face and dark eyes, but she struggled to hear any remarks addressed to her since a childhood infection had damaged her hearing. Within months, Melville Bell had proposed to, and been accepted by, Miss Symonds.

Eliza and Melville's marriage was a happy and long-lasting union, although Eliza was ten years older than her husband and too deaf to hear much of his stirring renditions of passages from great authors. (Melville was said to read Charles Dickens better than Dickens himself. When the elders of his church rebuked him for promoting the "ungodly Mr. Dickens," Melville stalked out of their presence, vowing that he would never darken the church's door again. He kept to his word.) A year after their 1844 wedding, their eldest child, Melville, or "Melly," arrived. Alexander was born two years later, and Edward, the youngest, the following year. Eliza Bell depicted her three sons as curly-haired angels in a watercolor she painted when the boys were still small: Melly is a scholarly young man, Edward is a delightful little boy in skirts, and Alexander is a lively youngster taking aim with an arrow at an invisible target. The boys may have been angelic to their mother, but to their neighbors they were rambunctious, noisy lads, always shouting and banging doors as they rushed in and out of the house. Alec asserted his independence early. Exasperated by being the third Alexander Bell in a row, he decided to add Graham to his own name, becoming "Alexander Graham Bell," after he met a Canadian student of his father called Alexander Graham.

Alec was born in a flat in 16 South Charlotte Street, but soon after his birth his family moved first to a larger apartment, around the corner at 13 Hope Street, and then, when he was six, to 13 South Charlotte Street. This was a spacious four-story house that the Bells were able to purchase thanks to Melville's success as a lecturer and teacher. Melville, Eliza, and their three sons occupied the ten rooms on the top two floors, and the lower two floors were rented to tenants. The house was just off Charlotte Square—an elegant Georgian square in the city's New Town. By 1847, New Town was hardly new (its pale yellow-gray sandstone terraces had been planned nearly a century earlier and largely completed around 1820), but it was far more modern than the dark, cramped medieval buildings of the Old Town, clustered at the foot of Edinburgh Castle. When young Alec threaded his way through the Old Town's twisting alleys and closes, he would see on each side decaying stone tenement buildings, often ten or twelve stories high, crammed with people from every walk of life— from supreme court judges to street vendors. Pigs and dogs ran freely, and sanitation was nonexistent. He had to sidestep the great clots of tubercular spittle flecked with blood that passersby casually spat onto the cobbles, and he had to keep a watchful eye on the windows above him. Edinburgh pedestrians all knew that the cry of "Gardy loo!" (a Scots approximation of the French "Prenez garde à l'eau") meant that a chamber pot was about to be emptied on their heads from an overhead window. The odors of garbage, sewage, and coal fires had earned for the Scottish capital the nickname "Auld Reekie."

Melville Bell was an early enthusiast of photography: Alec (on left) disliked sitting still for family portraits.

In the more salubrious streets of New Town, half a mile from the castle, the Bell household was cheerful and busy, with regular visits from an extended network of relatives and friends. Eliza Bell supervised her children's education, and when her sons were small, she played the piano for family sing-alongs, aided by a special ear trumpet attached to the instrument's sounding board. By the time he was ten, Alec had taken over as the family's pianist. He and his two brothers also excelled at entertaining guests, often with voice tricks. They crowed like cocks, clucked like hens, or performed as ventriloquists, making puppets recite nursery rhymes. Their cousin Mary Symonds recalled how Alec "used to chase an imaginary bee around the room, imitating the buzzing of the bee, and then the muffled sound when it seemed to be caught in the hand."

But Alec was also a typical middle child, sandwiched between a brainy elder brother, who carried off several school prizes and on whom his father doted, and a sickly younger brother whose health dominated his mother's attention. Despite his ready laugh, his face wore a quizzical expression in repose, and his deep-set black eyes were serious and intense. He could ham it up at parties, but he was not by nature gregarious: he often retreated into solitude, particularly when he was preoccupied with a project. "He was a thoughtful boy," in the words of Mary Symonds, "always courteous and polite." He was particularly sensitive to his mother's hearing problems, which threatened to cut her off from everyday communication. In addition to communicating with her by speaking close to her forehead, he had mastered the English double-hand manual alphabet, so that he could silently spell out conversations to her. When relatives and friends gathered at the Bells' dining-room table, all chattering at once and clattering their plates and forks, Alec would sit attentively at Eliza's side, spelling out to her with his fingers what various people were saying so that she never felt left out. Thanks to his close relationship to his mother, Alec was untouched by the assumption, common at the time, that somehow deafness involved intellectual disability.

Alec's relationship with his father was more complicated: throughout his life, he would be torn between a gnawing hunger for Melville's approval and resentment of his domineering manner. Melville Bell was an authoritarian parent, convinced that he knew what was best for his children. When the boys were young, they tiptoed around the house while their father was present. Elocution lessons provided Melville's bread and butter; a steady stream of stutterers and mumblers arrived at his door, looking for help. Professor Bell corrected their speech problems and prepared students for public recitals. In 1860, he outlined his theories in *The Standard Elocutionist*, which also included several literary passages arranged for public performance. The book is said to have run to 168 printings in Britain and to have sold a quarter of a million copies in the United States by 1892. The author frequently boasted of his success, before grumbling that he had never received from his publishers the royalties he was owed. However, *The Standard Elocutionist* did give him the credibility to become a regular lecturer at the University of Edinburgh.

Elocution lessons, however, were not Melville Bell's abiding interest. He was particularly fascinated by phonetics—the way the human voice

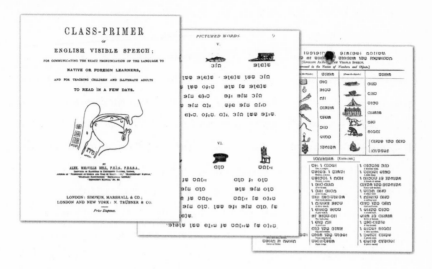

Melville Bell's Visible Speech system made it possible to transcribe, and reproduce, any sound a person could make.

actually produces sounds. When young Alec stood outside his father's study, he often heard the oddest grunts and hisses emanating from it. When he opened the door, he would find bushy-bearded Melville, his stern brow knit in concentration, staring at his reflection in the mirror as he contorted his tongue, jaw, and lips into strange expressions, uttered a sound, then made a rapid sketch of his mouth. Melville's pride and joy was "Visible Speech," a series of symbols he had developed to denote different sounds. The basic symbol for each consonant was a horseshoe curve, and that of a vowel a vertical line. How Melville wrote these symbols depended in each case on the particular action of the tongue, the breath, and the lips—so as he wrote, he would constantly emit different sounds and check his reflection. There were modifying symbols, such as hooks and crossbars, to signify particular vocal positions, and additional symbols for actions like suction and trilling. Melville Bell spent years cataloging every sound a human mouth could make and devising a way to put them on paper. The culmination of his work would be *Visible Speech: The Science of Universal Alphabetics*, which he would publish in 1867. Anyone who mastered his Visible Speech symbols, he claimed, could reproduce any sound exactly, even if he or she had never heard it before and had no clue what it meant.

Melville Bell's efforts to systemize speech, while seemingly abstruse, were typical of the intellectual fervor of the mid-nineteenth century— a time when human knowledge seemed to be expanding exponentially as the world was shrinking. Steam power, which had launched the Industrial Revolution, had speeded up travel within and between continents. Trains traveled five times faster than the fastest stagecoach, and steamships had cut the average duration of a transatlantic crossing from forty to twelve days. Melville Bell was not the only Victorian eager to chart the unknown. Intrepid missionaries and explorers spread out to every corner of the globe, intent on mapping the vast areas still left blank in their atlases and on making contact with the heathens (*always* heathens, in the Victorian view) who inhabited them.

In southern Africa, David Livingstone slogged north from Cape Town, preaching the Gospel despite being maimed by a lion and felled by swamp fever. In northern Africa, Richard Burton rode off into the desert, determined to find the source of the Nile. These explorers met a bewildering array of hitherto unknown peoples and tribes, speaking different languages. But they could rarely communicate with them, as these languages were unknown and in some cases lacked their own alphabets. For nearly a century, voice experts had tried to construct a written phonetic system that could be used to transcribe any language from anywhere in the world. Now, Melville Bell announced proudly, he had achieved a workable system. He was convinced that his book would bring him both fame and fortune.

While his father was perfecting his elaborate system of curves and hooks, Alec attended Edinburgh's Royal High School, the most important school in Scotland. When it was built in 1829, everything Greek was in fashion, and the school was modeled on the Temple of Theseus in Athens. The city could boast as graduates several stars of the Scottish Enlightenment, including philosopher David Hume, political economist Adam Smith, and writers Robert Burns and Sir Walter Scott. This intellectual explosion prompted Edinburgh to style itself "the Athens of the North." But Alec wasn't particularly interested in Athens, and despite Scotland's impressive record of inventors and engineers (the Glaswegian instrument-maker James Watt commercialized the steam engine in the 1770s), the Royal High School neglected the sciences, Alec's favorite subjects. So, to his father's dismay, Alec's school record was unimpressive. Chronically untidy and late for class, Alec often skipped school altogether to go bird-watching on Arthur's Seat, the rise of land just beyond Edinburgh Castle. Instead of learning Greek, he preferred to collect plants, shells, small skeletons, and birds' eggs. For Alec, as for many boys, high school was just a distraction from more exciting pursuits. Years later he wrote rather apologetically, "I passed through the whole curriculum of the Royal High School, from the lowest to the highest class, and graduated, but by no means with honours, when I was about fourteen years of age."

Outside the classroom, however, he demonstrated the ingenuity and single-mindedness that would shape his later career. While still a youngster, he invented not only the speaking machine but also, for a local mill-owner whose son was one of his friends, a machine for removing the husks from grain. He installed his collection of birds' eggs, dried grasses, and fossils in one of the rooms at the top of the Charlotte Street house and announced that this was his "laboratory." His brothers and friends were enrolled in what he grandly termed "The Society for the Promotion of Fine Arts among Boys." Each youthful member was dignified with the title "professor" and invited to give a lecture. Alec adopted the title "professor of anatomy" and took great delight in dissecting the corpses of small creatures, including rabbits and mice, for his fellow professors' benefit. He lost his audience, however, when he thrust a knife into the belly of a dead piglet and the foul gas trapped in its intestines was released with an eerie groan. A posse of young professors scuttled down the stairs of the tall house and out into the street.

Along with his brothers, Alec embraced the exciting new medium of photography. The technology was still in its earliest stages, and the boys laboriously coated glass plates with collodion, dunked them into a pan of silver nitrate, exposed them to the desired image, then developed them. The process took hours and the results were often disappointing, but Alec would always persevere. This was the kind of learning he liked: hands-on discovery. And he could lose himself in music. He would sit at the piano, sight-reading sheet music with extraordinary ease. He would play so intensely, with such concentration, for so long that he would end up with a splitting headache.

A studio photograph of Alec at fifteen, taken in 1862, a few months after he left the Royal High School, shows a skinny, clean-shaven young man with a prominent nose, thick, dark hair swept back from a broad brow, and an intense gaze. The debonair young gentleman in this portrait looks closer to twenty than his real age, particularly as he is holding a top hat so shiny it could have belonged to an undertaker. This young cosmopolite is a dramatic contrast to the unkempt youngster who appears in the blurred family photos of earlier years. The studio photograph was taken during the year Alec lived with his widowed grandfather in London. The older Alexander Bell had taken his grandson in hand in no uncertain terms, insisting that the young man spend more time on his studies, practice the piano more regularly, and dress like an English gentleman rather than a Scottish farm boy. He had made Alec exchange his comfortable, shabby tweeds for an Eton jacket. However, the photograph suggests that Alec didn't enjoy this enforced transformation: he holds his top hat awkwardly, and looks distinctly uncomfortable. Throughout his life, he resisted attempts to dress him up or pin him down for studio photos—Alexander Graham Bell always hated "ceremony." Nevertheless, his year in London, away from home, gave him a taste for freedom from paternal pressure. His year with his grandfather, he would write, "converted me from a boy somewhat prematurely into a man."

By now, young Alec had sensed that he needed more education, even if his high school record was dismal. Because money was tight in the Bell household, he and Melly made a deal: one would stay home in Edinburgh and attend university for a year or two while the other earned the money to pay for it, then they would trade places. So when Alec was sixteen, he left home to teach for a year at Weston House, a boys' boarding school in the handsome old market town of Elgin on the Moray coast, in the northeast corner of Scotland. In return for teaching piano and elocution, he received board, ten pounds, and further instruction in Latin and Greek. In the meantime, Melly attended the University of Edinburgh. Between August 1863 and early 1866, everything went roughly according to plan. Alec spent a total of eighteen months in this period in Elgin, where his students were so impressed by his serious demeanor and London manners that they never realized that several of them were older than he was. He reveled in his independence, and in the dramatic landscape: "I spent many happy hours lying among the heather on the Scottish hills, breathing in the scenery around me with a quiet delight that is even now pleasant for me to remember."

But two developments ensured that Melly and Alec's plan would never be completed.

The first was their father's obsession with his universal phonetic system. By now, Melville Bell had nearly perfected Visible Speech, and he decided that before he published it he should drum up public support in the hope of getting a government subsidy for the book. After Alec's return from Elgin in the summer of 1864, Melville drilled the three brothers in his elaborate system of notation. Then he took his boys on the road. In Edinburgh, Glasgow, and London, various speech experts dictated words to Professor Bell in different accents, dialects, and languages, including Hindi, Persian, and Urdu. While his sons remained outside the room, Professor Bell transcribed them onto a blackboard in Visible Speech symbols. Then he invited one of his sons into the room. Reading the carefully transcribed lines of horseshoes, lines, hooks, and crossbars, Alec or one of his brothers would reproduce the sounds perfectly. In one packed theater, Alec was summoned on stage by his father to reproduce sounds that, it seemed to him, could have no meaning to anybody. The symbol that Melville had written on the blackboard required him to blow a puff of air while the tip of his tongue touched the roof of his mouth. When he performed this act of lingual gymnastics, the audience burst into applause. Alec had reproduced one of the most obscure sounds in the universe: the Sanskrit cerebral T, described by a linguistics professor in the audience as almost impossible for an English-speaking person to pronounce. "The professor expressed surprise," Alec later noted, "that Mr. Bell's son should have given it correctly at the very first trial, without ever having heard the sound at all."

Alec enjoyed doing the demonstrations—they weren't so different from the parlor games at family parties—but they diverted him from his goal of a university education. They didn't even elicit the hoped-for government subsidy. However, the demonstrations did spur Alec toward experiments of his own into the production of sounds. He started playing around with a set of tuning forks, to investigate the composition of vowel sounds. His keen ear had allowed him to detect that certain vowel sequences were composed of both ascending and descending musical scales. He also devised instruments with stretched membranes, to measure the vibrations in air created by human speech—instruments that were a crucial step toward what would become his greatest invention.

The second development that sabotaged Alec's hopes for a decent education was his brother Edward's deteriorating health. By now, Alec's grandfather had died and Melville and Eliza Bell had moved to London, where they established themselves in his house in Harrington Square, just behind Euston Station. Melville was convinced that he would be able to lobby officials for recognition of the Visible Speech system more easily from here. But his youngest son, Edward, known as Ted, had begun to lose weight, to cough, and to struggle for breath. The Bells watched with sinking hearts: these were the warning symptoms of tuberculosis, the contagious lung disease that flourished in the dank, sooty atmosphere of nineteenth-century cities. Tuberculosis—variously known as TB, consumption, or the Great White Plague—was as much a scourge and a stigma as AIDS today. A diagnosis of tuberculosis spelled the end of a young man's hopes for a glittering career, or a young woman's hopes of a good marriage. There was no cure. The only remedy that sometimes worked was prolonged rest in a sanatorium somewhere far from polluted industrial cities like Edinburgh and London. But the Bells could not afford to send Ted away. Instead, the young man lay on the couch in the drawing room, in the largest, dirtiest city in the world, growing paler and weaker by the day.

With money even tighter now, Alec's parents wanted their boys close in this family crisis, so Alec moved to England. But he firmly resisted his father's demand that he move back into the family home. Instead, he struggled to make a living from teaching in a school in Bath, on the edge of the English Cotswolds, and tried to enroll in a university degree course. For a young man sensitive to family duty but eager to carve out his own path, this was a grim time. Alec was now taller than his father, which only seemed to make Melville even more pugnacious in argument. Jutting out his bushy beard as if it were a weapon, Melville would vociferously oppose his son's plans to combine teaching and studies, insisting that Alec's health would suffer. Alec's health certainly seems to have been unreliable: he complained of headaches, depression, and sleeplessness. Perhaps this wasn't surprising considering the undisciplined intensity of his work habits. In a pattern that would last a lifetime, he would sit up all night reading by the yellow glow of gas lamps or working obsessively on sound experiments. He stumbled toward his bed only when he saw the pale light of dawn seeping through the curtains. His health problems and his careless disregard for them drove his mother into a fever of overprotectiveness ("Keep clear of pickles . . . take a dose of Cammomilla on going to bed, and one of Belladonna on getting up . . . wear your flannel shirts . . . surrender yourself to Papa's judgment").

The pressure to live at home was relentless. His parents yearned for his company: "We would not like to be without one of you at home," his mother wrote. Then, in 1867, the dreaded blow fell. Edward, aged only eighteen, passed away. "Sorrow we can never cease to feel for the loss of our darling Edward," his heartbroken mother wrote, "but it must be a sorrow chastened by submission to the will of God. He was a dear good boy and our way will be dark without him." With Ted gone, twenty-year-old Alec could no longer keep his distance from his father. He reluctantly sacrificed his hard-won autonomy and joined his bereaved parents in London. Melville and Eliza were immensely relieved, because, like most parents, they firmly believed that they knew better than their son what was good for him. "Young birds are very prone to try the strength of their wings too soon," Eliza added. "The parent birds know best the proper time for independent flying."

Residence in Harrington Square had its advantages. Alec had more opportunity to play around with the kind of sound experiments he loved, and Trouve, the family's good-natured Skye terrier, was a willing accomplice. Alec would fill a bowl with scraps of meat, then set to work on the family pet. With a judicious mix of treats and jaw adjustments, he taught Trouve to talk. The terrier was more versatile than the speaking machine of Edinburgh days. "The fame of the dog spread," Alec later recalled. "Many were the visitors who came to the house to see this dog sit up on his hind legs, and with a little assistance from my hand growl forth the words, 'How are you Grandmama?'" (Trouve's repertoire consisted of "Ow," "ah," "oo," "ga," and "ma.")

Alec also had the chance to mingle with his father's colleagues and former students from Edinburgh days. Among the latter was James Murray, who had moved from Edinburgh to London after the death of his wife and child, to work as a bank clerk. Alec's father decided to take this enthusiastic amateur dialectician along to a meeting of the Philological Society, at which scholars stroked their beards and discussed with enthusiastic pedantry the origin, pronunciation, and meaning of words. Murray would devote his life to the society's most famous project: the definitive English dictionary, eventually published as the *Oxford English Dictionary*. But in the late 1860s, both James Murray and Alec Bell had other pursuits in mind. In the summer of 1867, Alec was best man at Murray's second marriage. Alec himself was courting Marie Eccleston, a young trainee teacher from Lancashire who often called at the Bells' Harrington Square home. Eliza Bell liked this cheerful, rather forceful young woman: "He will not find many like Marie in strength of character," she wrote to a friend.

Alec was busy these days: he attended physiology and anatomy classes at the University of London and taught deaf children at a private school in Kensington. He and his father were successfully using Visible Speech to teach deaf children to pronounce words correctly, by learning how to shape their mouths. But when Melville Bell traveled to Canada and the United States in 1868 to promote Visible Speech, Alec also had to keep an eye on his father's pupils and publications in London. He barely slept as he tried to continue his sound experiments despite the added teaching load. When Melville Bell returned from his travels, he found Alec looking skinny, unkempt, and exhausted. This, of course, confirmed Melville's self-serving view that his son was far too irresponsible to be independent. Alec was "a perfect baby," Melville wrote to a friend in Canada, "and needs to be told when to wrap up in going out, when to change boots or wet clothes etc. etc."

Yet the Bells had every reason to worry about their children's health. In 1868, their eldest son, Melly, who had married a pretty young woman called Carrie Ottoway and had set up a speech therapy practice in Edinburgh, became the proud father of a son, Edward. Before the child reached his second birthday, he died of an unidentified disease that was probably tuberculosis. Alec knew that Melly's health was also uncertain. Then one day James Murray and his new wife called at the house in Harrington Square. A troubled Alec met them at the door and drew them conspiratorially into the drawing room. Running an agitated hand through his long, lank black hair, he confided to them, "I don't know how to tell Father, but Melville is bringing up blood again." Within a few weeks, in late May 1870, the Great White Plague had taken the life of twenty-five-year-old

Melly. He was buried next to his grandfather and youngest brother in London's Highgate Cemetery.

It is hard to imagine the heartbreak in the Bell household. Black crepe was draped over the Harrington Square windows; Eliza retired weeping to bed for days; Melville could barely bring himself to mention his dead sons. For twenty-three-year-old Alec, both the sorrow and the pressure were intense. A few years earlier, he and his brothers had been lively young mischief-makers, tearing up and down the stairs of their Charlotte Street house, terrorizing neighbors and cousins. Death had been treated as a distant joke—as boys, he and Melly had even made a pact that whoever died first would try to contact the other from the afterlife. Now Alec found himself the only surviving brother, missing the easy companionship he had once taken for granted and bearing the impossible weight of his parents' grief. His home felt like a morgue, and yet he knew that now he could never leave.

Eliza Bell worried incessantly about Alec's health, and what she called his "head-achey fits." Melville Bell paced around his London study for hours, his brow knit with frustration. Pain at his loss and anxiety about his remaining son were compounded by the dismay he felt that his precious Visible Speech was going nowhere, despite praise in learned journals. In an effort to make the system more accessible, he had published a simplified version in 1868, in a sixteen-page pamphlet entitled *English Visible Speech for the Million*. But "the Million" chose to ignore it. Only the most ardent philologists understood the point of an extremely complicated system that recorded strange sounds without regard to their meaning.

Abruptly, Melville Bell took a decision that, for a well-established professional in his fifties, was extraordinarily brave. He recalled how the bracing climate of Newfoundland had restored his own health thirty years earlier, and he remembered his visit only two years before to several Scottish friends who had settled in Canada. Melville had spent a few days with the Reverend Thomas Henderson in Paris, a small farming community set in the lush countryside of southern

Ontario, fifty miles west of Toronto, the provincial capital. He had marveled, he wrote home to Eliza, that the Hendersons all looked *"as young* as when I saw them last. The climate, trying as it is in its extremes of heat & cold, evidently agrees well with our fellow countrymen." Now he decided that he, Eliza, Alec, and their son Melville's widow, Carrie Ottoway Bell, for whom they felt responsible, should immediately uproot themselves from their comfortable London life and move to Canada.

Melville was a great one for just barging ahead, on the assumption that all around him would bend to his will. Alec was not so sure. He still nursed a passion for Marie Eccleston; he still hoped to earn a degree at the University of London; he enjoyed teaching at the school in Kensington; he couldn't imagine leaving the most important city in the world for a distant colony. But Melville had made up his mind, and Eliza was so desperate to leave behind the soot and sad memories of London that Alec knew he had no choice. Moreover, Marie Eccleston did not seem as heartbroken by his impending departure as he wished. There was an almost cavalier tone to her farewell note: "Don't grieve about your examinations etc. . . . [A]ll the degrees in the world would not make up for ill-health." Marie's final piece of advice suggests she knew her beau better than he knew himself: "Don't get absorbed in yourself—it is one of your great failings," she suggested. "Mix freely with your fellow [men]."

In June, Alec took the train to Edinburgh to dispose of Melly's piano and household effects. Slumped in the corner of the carriage, staring wistfully at fields and moors steeped in history, he felt trapped, by his own poor health as much as by his father's firm decision. The following evening, after he had packed up Melly's meager belongings, his mood was even gloomier. Perched on an uncomfortable wooden chair in his dead brother's dusty, abandoned study, he listened to the sounds of the city he had loved since childhood and tried to imagine a new life in the New World. All Alec knew about the colonies in British North America was that they had swallowed up large numbers of Scots emigrants during the previous half-century and had formed a self-governing "Dominion" only three years earlier.

His bleak stereotype of Canada seems to have been based on popular books like Catharine Parr Traill's *The Backwoods of Canada,* first published in 1836 and reprinted frequently, which described the struggle to survive in a pitiless, thinly populated landscape. "[I] tried to imagine myself in the Backwoods of Canada," he wrote to his parents. "It was not very hard to imagine . . . sitting on borrowed chairs, in the empty classroom." But he could see his parents' desolation, and he reluctantly admitted that they had cause to be anxious about their last remaining son. Despite his own repeated denials, all the signs of TB were there. His headaches had intensified, and he coughed repeatedly. Although he was well over six feet tall, his weight was down to about 130 pounds and his face was gaunt. Eliza reassured Alec, "You don't really think you are going into the backwoods, do you? You are merely going into a country house, and will have civilised society there, just as much as you have here." The Bell family's passage was booked on a steamer that left London on July 21, 1870. In later years Alec would tell friends, "I went to Canada to die."

Chapter 2

THE BACKWOODS OF CANADA
1870

Alec was withdrawn and gloomy during the ten-day voyage across the Atlantic; his parents saw him only at mealtimes, when he barely touched his food. Most days, he stayed in his cabin, studying a new book he had bought in London by the German scientist Hermann von Helmholtz, entitled *On the Sensations of Tone*. He also scribbled in a small notebook (on the cover of which he wrote "Thought Book of A. Graham Bell") his lingering resentment against his father's decision to emigrate. "A man's *own* judgement should be the final appeal in all that *relates to himself*. Many men . . . do this or that *because someone else* has thought it right." But once the Bells had landed at Quebec City and had transferred themselves and their baggage to the steamer that would take them up to the St. Lawrence River to Montreal, his spirits began to lift. A sense of eager anticipation rippled through the party as they contemplated the spectacular scenery in the early August sunshine. Thick green forests interspersed with pretty little villages crowded down to the edge of the mighty river, and the sound of church bells rang across the water from the silvery spires of sturdy stone churches.

Alec took deep breaths of the cool, clean air, fragrant with the smells of pine resin and wood smoke. He allowed himself to address a few civil words to Melville and to accompany his sister-in-law as she took a turn around the deck.

If Alec's expectations of the Great Dominion of the North were based on *The Backwoods of Canada,* his first glimpses of his new country must have been as reassuring as his mother's words before they left England. The Bells disembarked from the St. Lawrence steamer at Montreal, the largest city in Canada—a bustling seaport of nearly 100,000 people, the hub of a rapidly expanding railroad system, and the center of the Canadian banking system. There was a comfortable familiarity to the city: although the language of the warehousemen and porters was an incomprehensible French dialect, once the Bells left the docks Alec heard more English spoken than French, often with a Scottish accent. Many of the streets were named after Scotsmen: McTavish, Drummond, Mackay. There was nothing strange about the architecture, either. Scottish merchants who earlier in the century had amassed fortunes in the fur trade had built themselves splendid stone mansions along Sherbrooke Street that were reminiscent of the granite mansions built in the Scottish countryside by wealthy lairds. When the four Bells breakfasted in their hotel, they were happy to be offered porridge, even if they were surprised to see fellow guests eating it with maple syrup. As they strolled around the cobbled streets, only the muggy heat of an eastern Canadian summer was utterly foreign.

From Montreal, the family took the Grand Trunk Railway train west, to their friends the Hendersons in southern Ontario. As the iron engine steamed through the sunlit countryside, Alec stared out of the window at the prosperous little towns, well-established farms, and snake fences weaving their way around flat fields of wheat. Within a few days of unpacking his bag at the Hendersons' rectory in Paris, Melville Bell had seen a ten-and-a-half-acre farm called Tutelo Heights, eight miles away. With customary brio, he snapped it up for $2,600. The main house, a pleasant two-story white building, boasted four bedrooms, a

Tutelo Heights, the Bell home in Canada, was an Italianate villa near the friendly little manufacturing town of Brantford.

large dining room, a study and parlor, a good-sized kitchen, and an attractive front porch with gingerbread trim. There was a glassed-in plant conservatory, and behind it a small workroom. There were also an orchard filled with apple, plum, and cherry trees, a stable, a carriage shed, a henhouse, a pigsty, an icehouse, a well, and rainwater cisterns. At the back of the house, beyond a high bluff, wound the Grand River on its way to Lake Erie. And on the other side of the river were the church spires and factory chimneys of Brantford, one of the busy little manufacturing towns (Elora, Waterloo, Berlin, Galt, and Guelph were its friendly rivals) that had sprung up in the Grand River Valley in the mid-nineteenth century.

In 1818, only twelve Europeans lived by the ford across the Grand River that was named after Mohawk chief Joseph Brant. When the Bells arrived fifty-two years later, there were close to ten thousand. Although Tutelo Heights was four miles from the center of town, just before seven each morning they would hear the whistles and bells of a hundred factories, summoning mechanics, foremen, timekeepers, bookkeepers, and laborers. Massey mowers and reapers, Tisdale's iron

stoves, Lily White Gloss Starch, T. J. Fair cigars—"Made in Brantford"
was stamped on goods that were dispatched to every corner of British
North America. The burghers of Brantford were intensely patriotic,
raising their glasses to Queen Victoria at celebratory dinners as if they
lived in Kent or Cumberland back in England. The town boasted two
daily papers (the *Brantford Courier* and the *Daily Expositor*) and,
of particular interest to Melville Bell, its own rail link to the United
States—the Buffalo and Brantford Railway had opened in 1854.

Tutelo Heights fulfilled Eliza Bell's promise that, in Canada, the
Bells would live in a comfortable "country house." And Melville didn't
have far to look for "civilised society." Between the town and Tutelo
Heights lay the property of Ignatius Cockshutt, one of Brantford's
leading citizens and owner of the best dry goods store in town. Eliza
and Carrie soon discovered that Mr. Cockshutt's bolts of silks and wool
cloth were as good as anything on offer in Edinburgh stores. A short
drive away from the Bells' new home was Bow Park, a large estate
belonging to George Brown. By 1870, Brown was a leading Liberal
MP in the Canadian parliament in Ottawa, as well as the wealthy editor of the influential Liberal newspaper the *Toronto Globe*. He was
also a fellow Scotsman—born in Edinburgh, he had been educated,
like Alec, at the Royal High School. Thanks to Brown, along with their
other friends from Edinburgh days, the Bells easily slipped into the
Scots-dominated society of late-nineteenth-century Canada.

Yet Brantford had little to offer Alec intellectually. The nearest
school for the deaf was in Belleville, 180 miles to the east. The nearest university was sixty-eight miles away, in Toronto. All Alec could do
was acknowledge his poor health and take things easy. He spent the
rest of that summer at Tutelo Heights quietly sitting in the garden
with a rug, a pillow, and a book, feeling the sun on his face and brooding about the life he had left behind. Memories of his dead brothers
were never far from his mind, and he often caught his mother staring
sadly at a photograph she kept tucked in a drawer, of Edward lying in
his coffin. Melville busied himself with organizing his new estate, and

Carrie helped Eliza master such rural obligations as planting a garden and keeping hens. But Alec felt bored and frustrated. He never really warmed to the flat landscape of southern Ontario. "Go where you will in Canada," he wrote a few years later, "your horizon is bounded by trees! You seem to be perpetually in the midst of a clearing! . . . The fields are full of stumps. Here and there some solitary giant tree—a relic of the primeval forest—stands mourning over the remains of its companions—to show how recently the land has been reclaimed from the wilderness."

However, enforced idleness did give Alec time to notice what was happening in the wider world. In particular, he began to see how his little experiments back in Britain with tuning forks and the speaking machine might fit into the explosion of inventions and scientific advances in North America and Europe.

The mid-nineteenth century was a wonderful time to be alive for a young man with a quick brain and endless curiosity. Today, we can understand the technological revolution of Bell's day only if we compare it with the impact that microprocessors have had on our own lives. The same coal-driven steam power that had accelerated travel and shrunk the world had also transformed a way of life that had existed virtually unchanged for over a century in the United States, and far longer in Britain's rural areas. The Industrial Revolution picked up speed in the early nineteenth century, as steam-driven machinery vastly increased productivity and spawned factories in the rapidly expanding cities of northern England and the United States. Working conditions in the new industrial cities were brutal, but ordinary people's lives on both sides of the Atlantic were immeasurably improved by such mechanical inventions as the sewing machine, the rotary printing press, the mechanical reaper, and the steam train. The entrepreneurial spirit thrived in North America. It soon appeared as though every young go-getter had a blueprint for a new gadget in his back pocket: better bobbins, music stands, valves, kitchen implements. "There is no clinging to old ways," a

German visitor to the United States noted in the 1820s. "The moment an American hears the word 'invention' he pricks up his ears." A town like Brantford, the Bells' new home, owed its prosperity to technological ingenuity coupled with entrepreneurial push.

By the time that Alec had moved to Brantford, the second wave of the Industrial Revolution had begun, as scientists and inventors scrambled to produce technological, electrical, and chemical inventions rather than purely mechanical innovations. The big breakthrough in electricity had come at the end of the eighteenth century, when Italian physicist Alessandro Volta had built the first electric battery. Although weak and primitive compared to modern batteries, Volta's invention was a major advance since it could produce a continuous flow of electricity along a wire. Before this breakthrough, researchers could produce only short bursts of current with little control over the voltage. Electrical pioneers in Germany, England, and America quickly grasped that if electricity could be carried through a wire, the same current might be used to carry a message. Ten years before Alec was born, in 1837, American artist Samuel F. B. Morse demonstrated an electrical telegraph that used pulses of electric current to send messages over a wire. Morse and his collaborator Alfred Vail devised the famous code of dots and dashes to represent letters, numbers, and basic punctuation. In their receiver, an electromagnet moved a stylus so that it marked long and short strokes onto a moving paper tape. Over the next decade, Morse and Vail made numerous improvements to their original design. They developed relays (a switch operated by an electromagnet) so that messages could be sent over great distances, and they invented the telegraph key, a switch that could be operated at the touch of a finger, so that telegraph operators could send messages more quickly. In 1843, President John Tyler signed a congressional appropriation to Morse to build the first telegraph line, from Baltimore to Washington. The first message was sent over the line on May 24, 1844. Morse, son of a Congregational minister, tapped out the exclamation of the prophet Balaam in Numbers 23:23: "What hath God wrought?"

Either Morse or God had wrought a revolution. Previously, the fastest way a message could travel was along lines of signal poles or towers. These semaphore telegraphs required an operator at each station and could work only in clear weather. Most messages, however, were transmitted by pony express over land, or by schooner over water. When Lucrece Morse, Samuel Morse's first wife, had died in New Haven in 1825, her husband had been busy painting portraits in Washington. He did not hear the tragic news of her death until several days later—too late even to attend her funeral. Now, thanks to his invention, the message could outrun the messenger and he might have heard the same day. The potential of this extraordinary invention was immediately recognized by American power brokers. Within a week of Morse's inaugural message, the telegraph was used to transmit to Washington the results of the vote at the Democratic Convention in Baltimore on the next presidential candidate. The convention had been deadlocked between Martin Van Buren and Lewis Cass. After the ninth ballot, a staccato message came down the wire: "Polk is unanimously nom." The convention had picked a dark-horse candidate: James K. Polk of Tennessee.

By the time Alec arrived in Canada, Samuel Morse's "electro-magnetic telegraph" had been in use for more than twenty years and Morse was enjoying fame as the Lightning Man. (Lightning was the only electrical manifestation that most people understood. When Morse had tried to explain to members of the U.S. Congress how his invention actually worked, they looked, according to one witness, as blank as if he "had spoken in Hebrew." Only a successful demonstration convinced them of the telegraph's potential.) In 1870, the length of the telegraph wire, strung between an army of wooden telegraph poles and thrumming across the continent, approached 250,000 miles. Massachusetts was linked to California, Ottawa to Montreal, and every city and town in between either was or hoped to be connected to the expanding web of wire. The telegraph was part of an emerging commercial infrastructure that was producing and distributing wealth at record speeds. Newspapers, insurance companies, and political parties had all leaped to exploit it. Businesses in New York City could communicate directly with their offices in Boston, Toronto, New Orleans, or San Francisco. The Western Union Telegraph Company, incorporated in 1857, was busy buying up local telegraph companies and was well on its way to the goal of every nineteenth-century American industry— a complete monopoly. Outside the United States, Morse's telegraph wires linked Berlin to Aachen and Hamburg in Prussia, Stockholm to Uppsala in Sweden, London to every major industrial city in England. In India, Calcutta was connected to Bombay and Madras; Sydney and Adelaide were linked in Australia.

Even more remarkable, the New World was now connected to the Old World by thousands of miles of undersea cables. When the first attempt to lay a 1,700-mile-long transatlantic telegraph cable was made in 1856, from the island of Valentia off the southwest corner of Ireland to Newfoundland, the American press had exploded with excitement. "THE GREAT WORK OF THE AGE" read one newspaper headline, and the *New York Tribune* asked, "Where in the annals of the world have we the evidence of a stride the one-millionth part as sublime as this in its immensity?" A decade later, when the cable-laying was successfully completed by the 700-foot-long steamship the *Great Eastern,* euphoria rolled across the ocean. Queen Victoria and President James Buchanan exchanged telegraphed congratulations, and toasts to "England and America United" were heard at banquets on both continents. New York City, noted its chronicler George Templeton Strong, threw itself into an "orgasm of glorification."

It is hardly surprising, then, that the electrical telegraph caught Alec Bell's imagination, given his lifelong interest in communications. While he was still teaching in Bath, he had hooked up a primitive telegraph system, powered by homemade electrical batteries, between neighboring houses. Now, in Brantford, temporarily freed from the obligation to earn his living, he allowed his mind to grapple with the complexity of telegraph technology. As his health improved, he decided to convert the workroom behind the conservatory into a makeshift laboratory. He returned to the book that had absorbed him during his transatlantic voyage—Helmholtz's *On the Sensations of Tone.* Helmholtz had managed to synthesize vowel sounds by keeping several tuning forks in simultaneous and continuous vibration. Soon Alec had put together a construction that held upright a series of tuning forks, with which he tried to replicate Helmholtz's work. But it wasn't just sound that fascinated Alec: he was also intrigued by electricity and its new uses for communication.

A single issue now preoccupied him: Could Helmholtz's theories on the nature of sound have any application for the electric telegraph? In particular, could they solve a puzzle with which amateur engineers all over the United States were grappling? Nearly thirty years after its first commercial application, the telegraph system was still limited to sending one message at a time. The race was on to increase its capacity, by figuring out a way first to send two messages over one wire simultaneously in opposite directions and then to transmit several messages in the same direction simultaneously. Alec was determined to join this race. He had only a sketchy understanding of the nature of electricity, but he did understand how human beings produced sound—a knowledge that none of the other competitors in the race could boast. As the skinny, pale young man lay in the garden at Tutelo Heights, feeling the late summer sunshine bring new color to his face, he dreamed about a multiple telegraph that would marry Helmholtz's acoustical theories with Morse's system of electrical impulses. His goal was a "harmonic telegraph," which would employ sympathetic vibration to send several messages on a single wire at the same time.

In 1844, Samuel Morse had catapulted the world into a new era when he vastly accelerated the speed of communications. Alexander Graham Bell would invent something far more sophisticated than the Lightning Man's electrical telegraph. The young Scotsman would make instant communication accessible to everybody, and change the world forever. But he would achieve this only after several false starts, and only because he found the right people to help him.

Chapter 3

BOSTON BOUND
1871–1874

By the end of August 1870, Alec's health was starting to improve and his energies began to return. His gaunt cheeks and lanky frame filled out; his black eyes, deep-set and intense, lost their feverish look; he grew impatient with the inactivity of his routine. When the apples in the orchard ripened during the warm fall days, he helped the hired man pick them. Once the temperature dropped and the leaves began to turn, he started playing the piano again in the parlor while Carrie and his mother sewed thick woolen petticoats in preparation for their first harsh Canadian winter. He explored the countryside around Tutelo Heights and discovered the Six Nations Reserve a few miles down "a most *awful* road."

This was home to the Iroquois Loyalists who had left the Mohawk Valley of upper New York State after the American Revolution and had traveled north into what was then British North America. The six different Indian bands living there still spoke their own languages: Mohawk, Tuscarora, Oneida, Onondaga, Cayuga, and Seneca. The Six Nations Reserve was an exotic little subculture amid the sea of stifling conformity and materialism that was English Canada. Alec struck up a particular friendship with George Johnson, an attractive young Mohawk chief with an English-born wife. Alec was fascinated to meet a genuine North American Indian and to learn about wampum belts, peace pipes, and beadwork. George allowed him to record the Mohawk language in Visible Speech and to dress up in the fur hat and fringed jacket that he wore for ceremonial occasions on the reserve. The most permanent memento of this friendship was the Mohawk war dance that the chief taught the immigrant. In the coming years, at moments of triumph in his life, Alec would raise his arms and, with a gleeful whoop, break into the dance. It never failed to startle observers.

Otherwise, life in small-town Canada was stultifying for the young Scotsman, eager to pursue his own interests in education and scientific experiments. He dreaded being a financial burden on his parents. The proceeds from the sale of their London effects and the income from *The Standard Elocutionist* were not enough to keep all the Bells in style at Tutelo Heights. Moreover, Brantford's numerous factories were not doing so well during these years: the Canadian economy was in a slump, and the American market was largely closed to Canadian goods. Alec's father was frustrated by both the depression of his neighbors and the shortage of openings in Brantford for a highly qualified elocution teacher. His only local sources of income were public elocutionary performances, of a type he had come to regard as below him in Scotland. In August 1871, the *Brantford Courier* reported that at a concert commemorating Sir Walter Scott, "Prof. Melville Bell's great elocutionary powers were brought into requisition, and all who heard his reading will bear us out in the assertion that he literally caused many of the *dramatis personae* of Scott's works to pass in review before the admiring eyes of his audience." This would have been fine praise for an actor, but "Professor" Melville Bell considered himself a cut above that kind of thing. It made for a difficult atmosphere in the Bell home: Alec knew that they were in Canada only for his sake, and nobody was really enjoying the experience. Only Eliza, relieved by the improvement in her beloved son's health, maintained a sunny optimism. Every sigh and grunt that Melville uttered sounded, to Alec, like a reproach.

Melville Bell looked south for business to boost his battered ego and strained finances. Once his family was settled, he caught the train to Buffalo and then on to New England, on a money-making lecture tour that featured readings from Shakespeare interspersed with promotional talks about Visible Speech. His portly figure, resplendent in frock coat and muttonchop whiskers, took to the stage in venues across the northeastern United States. In Boston he renewed the acquaintance he had struck up two years earlier with Miss Sarah Fuller, principal of the newly established Boston School for Deaf Mutes, where lip-reading and speech were taught. He was invited to give a further series of lectures on using Visible Speech in the education of deaf children. Eager to get home, he declined the offer but mentioned that his son Alec might be interested in traveling down to Boston. When Alec heard what his father had done, he saw his chance to leave home. "I should not personally object to teaching Visible Speech in some well-known institution," he wrote his father, "if you would get an appointment—even if it was not remunerative." Melville spoke to the Massachusetts Board of Education, then telegraphed his son that there was a possibility of various teaching stints, including a month at Miss Fuller's school in Boston and another month at Miss Rogers's Clarke Institution for Deaf Mutes at Northampton, which also taught the "oral method" (lip-reading) of communication to deaf children. Miss Fuller and Miss Rogers, two formidable pioneer educators, were keen to know more about Visible Speech.

A couple of times a week, Eliza would send Alec into Brantford to collect the mail. Eliza, like so many immigrants, longed for letters that would connect her with the Old Country. She missed the network of cousins scattered through the British Isles; she would inquire wistfully why they didn't follow the Bells' example and move to the colony. So Alec would dutifully ride off in the ponycart, down Mount Pleasant

Road to Brantford, often accompanied by his widowed sister-in-law. Like Alec, Carrie was blossoming in the Canadian air; her cheeks glowed with health, and she enjoyed Brantford's church socials and tea parties. The two attractive newcomers waved at neighbors and acquaintances as their vehicle bounced along the rutted track.

Each time Alec reached the post office, he searched eagerly for an envelope bearing a U.S. rather than a British stamp. In March 1871, the letter for which he longed finally arrived. The Massachusetts Board of Education had passed a $500 appropriation to finance additional teaching for deaf children. Alec was invited to arrive in Boston as soon as possible, on a short-term contract to teach with Miss Fuller.

After only eight months in Canada, Alec was Boston-bound. It nearly broke Eliza's heart to say goodbye to her only remaining son. Would she ever see him again? She channeled her distress into her usual fever of concern about his health, and he was barely out of the house before she began firing off instructions. "Remember Boston is famed for a prevalence of East wind in the Spring. Don't throw aside your warm clothing till the warmer weather has set in," she wrote in an early letter. "The Americans keep their rooms too hot, and this would debilitate you." Just in case Alec missed his mother's message, she was even blunter in her next letter: "Our only comfort and stay now is in you."

On a blustery April day, Alec arrived in Boston and carried his battered leather suitcase from the train station to his lodgings in a Beacon Hill boarding house. He walked past elegant churches, thriving markets, and eight-story office buildings that were far larger and older than anything Brantford had to offer. He eavesdropped shamelessly on passersby, fascinated by the New England drawl that grated painfully on his ear. With its teeming streetlife and mingled odors of saltwater, commerce, and congested tenements, Boston reminded him of Edinburgh or London. His spirits lifted with every step he took on the cobbled, crowded streets as he realized that he had escaped from the backwoods into a metropolis.

The splendid Massachusetts State House, designed by Charles Bullfinch and completed in 1798, dominated Boston and gave Alec a sense of the city's importance.

He would soon discover that Boston's "Brahmins," as members of the city's elite were always known, were, like the grandees of his hometown of Edinburgh, so convinced of their city's intellectual superiority that they had given it a Greek flourish: they dubbed it "the Athens of America." But there was some truth to the title. New England was the cradle of American industry, and Boston was its capital. Inventors, electricians, engineers, machinists, educators, and skilled artisans congregated there. The republic's political heart might beat in Washington, and New York City might be the financial center, but Boston housed the technological elite of the nation.

Alec's excitement at the Boston bustle intensified an even greater thrill: enthusiasm for the career he believed lay ahead. By using Visible Speech alongside other techniques, he was going to help hearing-impaired children and adults assimilate into the speaking world. His resolve hardened after an early encounter in Boston with a small child from New York City who was trapped in silence. The four-year-old girl was a "deaf-mute," as children who could neither hear nor speak were labeled in the nineteenth century. It was a label with dreadful connotations: "deaf-mutes" were assumed to lack most intellectual faculties. Their families, unable to communicate with them, consigned them to residential asylums for life. The stigma attached itself to the handicapped child's entire family: the suspicion that the condition might be genetic could ruin the marital prospects of the child's siblings. This particular four-year-old was causing her parents endless stress because of her violent temper tantrums; her two little brothers were afraid of her. So her unhappy parents had already chosen an institution for her, and once she had been shipped off they would likely never mention her in polite company. Warm-hearted Alec was appalled at her plight. He didn't believe she was ineducable, and he was convinced that frustration triggered her tantrums. As he wrote to his parents, "The little thing gives one a most painful idea of what an uneducated deaf mute may be."

Yet Alec still had no formal training in deaf education, or appreciation of the fierce controversies that already raged in the United States over how best to educate the deaf. There was an endearing guilelessness about the lanky young man with a wispy beard who nervously brushed his threadbare coat before his first day of work. After combing back his unruly dark hair, he strode up the hill to Pemberton Square and the large brick building that housed the Boston School for Deaf Mutes. Alec's first concern was to make a good impression on Sarah Fuller, the principal, who was ten years his senior. Fuller had a welcoming manner and an air of quiet authority as she dealt with her thirty charges. "I never saw *Love, Goodness,* and *Firmness* so blended in one face before," Alec wrote home.

Alec's gifts as an instructor were soon evident. During the first Visible Speech lesson in his Boston classroom, he drew a face on the blackboard and showed the children how to point to their own features as he pointed to the face's features. His engaging mix of clear instruction and funny expressions delighted his audience. By the time he erased from the blackboard face all but the elements that were represented in Visible Speech symbols, the children were totally enthralled. Within half an hour, each student had learned to identify four classes of symbols. Each of them could, when invited by Alec, move their lips, tongues, and mouths to utter particular sounds. "It was a triumph to see even the youngest toddler catching the idea so quickly," Alec reported to his parents.

The class's progress impressed Miss Fuller, the school governors, and a reporter from the *Boston Journal*. (Alec proudly sent the clipping to his parents in Brantford.) A couple of weeks later, Alec established a class for deaf adults. He advertised private elocution lessons, and, he reported triumphantly to his parents, he had been "obliged to decline no less than *three private pupils*"—two deaf girls and a stammerer. "I find myself making headway every day. New ideas are being constantly suggested [to me] by the defects arising in the school." He took on more and more private pupils for the afternoons, started an evening class, and worked feverishly to prepare new teaching materials. His black eyes sparkled with enthusiasm for his work, but the dark circles under them soon returned. To keep up with his responsibilities, he allowed himself less and less sleep.

Alec wrote to his parents at least once a week, often enclosing items from the Boston newspapers. ("I have sent you a paper containing a Yankee poem I think Papa could manufacture into something good.") When his first three months in Boston were up, he caught the train north to spend the summer in Brantford. There he caught up on his sleep and listened to Melville's ambitious plans to promote Visible Speech in North America, most of which involved Alec dutifully following his father around the continent. On the rare occasion he managed to escape from his parents' fond embrace, he walked down the cart track to the Six Nations Reserve to renew his friendship with George Johnson. The two men celebrated their reunions with some foot-stomping war-dance displays.

By September 1871, Alec was back in Boston, where his reputation as a teacher of the deaf, using Visible Speech, was spreading. He wrote an article on his techniques for the *American Annals of the Deaf and Dumb,* and read a paper at the annual convention of American deaf-school teachers in Flint, Michigan. His methods were effective, but it was his empathy and patience that made him such a superb teacher. Those early children's parties in Charlotte Street, featuring buzzing bees and crowing cocks, had trained the inner actor. Although shy and awkward outside the classroom, Alec had the ability to captivate a crowd, whether it was composed of deaf children, interested laypeople, or skeptical scientists. When he was with a deaf child, the youngster felt enveloped and secure in Alec's passion to help him or her communicate. The superintendent of Boston schools thought his results "more than satisfactory; they are wonderful," and suggested that the Visible Speech system "must speedily revolutionize the teaching in all articulating deaf-mute schools."

In April 1872, Alec moved to Northampton, Massachusetts, to teach at the Clarke Institution—the other New England school that used the oral method. However, he started to have misgivings about whether the oral method was appropriate for all deaf children, especially those born deaf. "I have been studying the subject of the education of the deaf and dumb very deeply since I have been here," he explained in a letter home. "It makes my very heart ache to see the difficulties the little children have to contend with on account of the prejudice of their teachers. You know that here all communication is strictly with the mouth . . . and just fancy little children who have no idea of speech being made dependent on lip-reading for almost every idea that enters their heads. Of course their mental development is slow. It is a wonder to me that they progress at all."

Alec had stumbled into the controversy that has now characterized the debate on deaf education for close to two centuries. He was familiar with the oral method, which taught deaf children to read lips and use a spoken language. This is where Visible Speech had some potential as a teacher's aid. The oral method thrived in Britain and Germany; one early proponent of lip-reading was the Scotsman Thomas Braidwood. But Miss Fuller's and Miss Rogers's schools—the two schools at which Alec had taught since he had come to the United States—were aberrations within North America. The most popular communication technique for the deaf in the United States was a different system: sign language, which allowed deaf people to communicate with each other by silent gesture.

Sign language had been developed by the Abbé de L'Epée, a French cleric, in the eighteenth century. Not far from Boston, in Hartford, Connecticut, a pupil of the Abbé de L'Epée called the Reverend Thomas Gallaudet had founded the Hartford School for Deaf Mutes in 1817. This was a much bigger operation than either Miss Fuller's or the Clarke school, and it was the model for nearly all the schools for deaf children in the United States and Canada.

The debate between deaf educators who were partisans of the spoken word, like Miss Fuller and Alec himself, and advocates of visible gestures such as Gallaudet was ferocious. In the nineteenth century it had a philosophical and religious dimension that was as hard for Alec Bell, no great churchman himself, to understand as it is for us today. European "oralists" appealed to the notion of speech as God's special gift to mankind, which it was cruel to withhold from deaf children. Since the Renaissance, the human voice had been regarded as an image of the divine soul, and language as the source of civilization. Oralists argued that unless deaf children learned to speak, they could never understand God's word or receive absolution. These arguments, however, were meaningless to supporters of sign language, who pointed out that vocal training for the deaf could be a long and arduous process—almost a cruelty in itself—whereas gestural signs (which are not based on spoken words) offered a naturalness, universality, and lucidity not available in spoken language, which is arbitrary and obscure. Those who championed sign language believed that it was a system of communication that developed naturally among the deaf and that it was as legitimate a language as French, English, or Hindi.

Alec was still trying to come to grips with this debate when he received an unexpected invitation in May 1872. Edward Gallaudet, principal of the prestigious Hartford School and son of its founder, wanted to see this talented young teacher in action. Alec agreed to give a demonstration of the oral method, using Visible Speech, in the bastion of sign language. He caught the train to Hartford, eighty miles southwest of Boston, and found his way through the state capital's leafy streets to the Gallaudet headquarters. The size of the Hartford school took him aback: the four-story building, with its rows of neatly shuttered windows, its Corinthian pilasters, and its graceful portico, all surrounded by trim lawns, was far grander than any institution he had taught in. But he refused to be overawed by the wealth of this establishment. His first initiative was to introduce the Hartford students to an unfamiliar sensation—the use of their vocal chords.

A correspondent for *Silent World,* a magazine for the deaf, described the exercises that this energetic new teacher took students through each morning. Alec stood at the front of the Hartford classroom like a conductor, his arms outstretched and his face aglow with enthusiasm for what he was doing. The students stood up, stretched their arms to "open their lungs," then, "in a low tone and all together, they say what may sound like i . . . i . . . i . . . i . . . i . . . i, as long as Mr. Bell wishes. He moves his hand, with thumb and forefinger close together, slowly from left to right, for this, and spreads out his fingers quickly when he wants them to stop. Then he begins again but with his thumb and forefinger wide apart, and such a roar comes up as makes the floor tremble, the windows rattle, and the hall resound again. With these simple motions . . . he has the whole two hundred and fifty voices, from deep bass to shrill treble, under sufficient control to make them roar in concert or die away softly." In the warmth of a summer's day, these extraordinary sounds floated out of open windows into the Hartford streets, startling horses and bringing passersby to an astonished stop. The students reveled in both their unheard chorus and their young teacher's encouragement.

At Hartford, Alec began to see the potential of sign language, alongside lip-reading, as a teaching tool. He mentioned to his father, in a letter, that he was learning sign language. Melville was furious, and the next letter that Alec received from Brantford included a scolding. Melville wanted Alec to promote the Visible Speech gospel, and its use in deaf education was, in its inventor's view, only one of its many applications. He was exasperated that his son was allowing himself to be distracted by sign language. But Visible Speech was not really Alec's first love. For Alec, "rescuing" the hearing-impaired was an end in itself. The disagreement between father and son created a rift between them—a rift that widened as Alec refused to be cowed. Alec was good at what he was doing, and much as he loved his father, he was not going to be bullied from a distance into abandoning his plans. However, Alec never embraced sign language with the Gallaudets' fervor; he never accepted the argument that sign language was "natural," and he continued to insist that most deaf children were better off if they remained in the speaking world.

As a young male teacher, Alec was doubly a rarity in the New England school system. Between 1825 and 1860, a quarter of all native-born New England women were schoolteachers for at least a portion of their lives. Few of them, however, had any specialist training with deaf children, and many school boards still preferred to hire male teachers (although their salary was two or three times that of female teachers). So Alec, a male teacher who could teach deaf children, found himself in demand. Offers of employment reached him from all over the eastern seaboard, including the Columbia Institution for the Instruction of the Deaf and Dumb in Washington. But the offers came too late. The Scots immigrant had decided to settle in Boston and establish himself as a teacher of the deaf there. He had adopted the city's own view of itself. "Though Washington is the political centre," he acknowledged to Miss Fuller, "Boston is the intellectual centre of the States." And he wanted to stay connected to the intellectual center so that he could pursue his scientific interests.

Like Edinburgh, Boston was fueled by an Enlightenment belief in scientific progress and human perfectibility. Between 1810 and 1865, while its population jumped from 30,000 to 140,000, Boston was far ahead of other American cities in social reforms: movements for prison reform, women's suffrage, and temperance; campaigns against slavery and warfare—all caught fire here. The 1776 American Revolution was launched from Boston, and heated debates within the city's churches, taverns, and newspaper columns over the abolition of slavery had helped to precipitate the Civil War in 1861. *The Liberator,* published by William Lloyd Garrison out of a hole-in-the-wall office on Washington Street, was one of the elements that had reshaped the antislavery campaign in the northern states into a powerful moral crusade.

Foremost among Bostonian preoccupations was education. The Massachusetts State Board of Education had laid the foundations for the country's first public school system, with properly trained and paid teachers, books, and decent schoolhouses. On Boylston Street, a beautiful Romanesque Revival building housed the Boston Public Library, founded in 1848 as the first large municipal public library in America. Even the poorest city resident could settle between the marble columns of its reading room to peruse the latest publications. And within a stone's throw of Alec's Beacon Hill lodgings were some of the finest academic institutions in the country. America's oldest university, Harvard, was just across the Charles River in Cambridge, while the Massachusetts Institute of Technology had recently been incorporated (its first building was being constructed near the museum on Boylston Street). The city also claimed leadership in the manufacture of scientific apparatus and in medical science. The first issue of *The New England Journal of Medicine* had been published here in 1812, and Massachusetts General Hospital claimed that its physicians were the first to use ether as an anesthetic during surgery, in 1846. If Alec strolled out of Pemberton Square and into the Public Garden—the best laid-out public garden in the country—he could stop and admire a florid Victorian concoction of granite and red marble: the Ether Monument.

The string of civic achievements had given Bostonians a robust belief that there was a solution to every problem. Accustomed to Scots

Thomas Edison, a trained telegraph operator, applied for his first patent in 1868 and seemed bound to win the race to invent a multiple-message telegraph.

pessimism and Canadian self-doubt, Alec embraced this confident new attitude. He was excited to see that the city had emerged as the hub of the closely knit telegraph fraternity, thanks to its concentration of mechanical expertise and capital. He realized that, in Boston, he could support himself by teaching, and in his spare time return to the telegraph experiments he had begun years ago in England and had continued to ponder in his parents' garden in Brantford.

Of course, there were other eager young inventors with the same goal in mind. One was Michigan-born Thomas Edison, a short, jaunty, hard-of-hearing hustler whom Alec would rival in accomplishment and fame. Edison had recently spent twelve months in Boston before moving on to Newark, New Jersey. Like Alec, he wanted to be the person who discovered how to send several messages at once over a single telegraph wire. Edison appeared more likely to succeed in the quest because he had the technical background that Alec lacked. While Alec was still collecting birds' eggs in Edinburgh, the cocky Midwestern youth, his exact contemporary, had got his first job as a "Knight of the Key," or telegraph operator, on the Grand Trunk Line at Stratford, Ontario. He moved rapidly up the telegraph tree, and qualified as a first-class operator while spending evenings and weekends reading

Michael Faraday's *Experimental Researches in Electricity,* and tinkering with telegraph repeaters and sealing wax. Hungry to make his mark on the world, Edison had even rigged up a primitive private-line telegraph system in Boston on which to experiment. With the reckless panache that would characterize his career, he had strung his lines over the roofs of houses next to his customers' offices without bothering to ask permission from their owners.

Edison was determined to solve the problem of multiple messages by manipulating the electric current. Alec, still preoccupied with Helmholtz's experiments with vibrating sounds, wanted to explore the idea that the same wire might carry different tones. At the newly built Massachusetts Institute of Technology in Back Bay, there was a complete set of Helmholtz's acoustic apparatus that he was invited to examine. Perhaps because Helmholtz's book on sound was in German, Alec's understanding of his work was flawed: he thought the German scientist had not only created synthetic vowel sounds but also had sent them from one point to another over an electric wire, in the same way that the telegraph worked. It was this misunderstanding that led him to believe that the transmission of the spoken word was possible. He dreamed of transmitting not just electrical impulses but the sound of the human voice. Soon he was tinkering with electromagnets and tuning forks until dawn crept across the sky. Despite his mother's entreaties, he had again become a night owl, allowing himself only three or four hours of sleep.

Alec's only source of income, regrettably, remained teaching. After another summer with his parents in Brantford, he returned to Boston in September 1872, intending to build up a practice of private students and to teach at Miss Fuller's school. He took a second-floor apartment in 35 West Newton Street in the South End. He could afford rooms in the bow-fronted brownstone house because this once-solid residential district was rapidly going downhill, as fashionable Bostonians moved across town to the newly developed Back Bay area. His two-room suite consisted of a parlor and a second room that served as his schoolroom by day and his bedroom by night. ("If you use your school room for a bedroom," fussed Eliza, "take care it is thoroughly aired.") He put an advertisement offering speech lessons in the *Boston Transcript* and soon had a dozen students.

One, a five-year-old boy named George Sanders who had been born deaf, moved into another suite in 35 West Newton Street along with his nurse. "Little George progressing splendidly," Alec wrote home in November. He was a "loving and lovable little fellow. . . . Expect great results with him if I can have him with me for two or three years." Eliza reluctantly accepted that she had lost her son to Boston: "You are now, my dear Alec, upon your own legs, and must have all your wits about you." Keen to show his mother that she still mattered to him, Alec sent her the latest model of a hearing tube. "Well my dear boy the tube arrived safely," Eliza wrote. "I must use it for company and strangers, it being too powerful for everyday use."

As soon as the last student of the day left, Alec would eagerly return to his telegraph experiments. He set up a table in his parlor and began to gather the components with which to replicate Helmholtz's device for producing an electric current of sustained frequency. He made most of his own equipment, and his table had a cover that he could lock to protect his work from competitors' eyes. Flicking his dark hair out of his eyes, he would crouch over his equipment in the flickering gaslight, laboriously winding copper wire around iron cores to make electromagnets and cursing his own clumsiness. Maybe, he hoped, a telegraph breakthrough would make him some money, but he couldn't afford to apply for a patent, so he had to operate in total secrecy.

The pressure of teaching by day and experimentation by night soon triggered the return of his headaches. Back in Canada, his parents fretted about their weak-chested son's susceptibility to the kinds of infections and sicknesses that were as common in Boston's overcrowded tenements and smoky back streets as they had been in Edinburgh. Melville, still irritated that Alec was straying from the Visible Speech campaign, huffed and puffed. "I value your health above all things," he said, adding, "I prescribe a steamboat trip on the upper lakes." In June, Alec returned to Canada and did as his father suggested. He spent July on a steamer, traveling through Georgian Bay to Lake Superior, then back through Lake Huron. He stopped coughing and his head cleared.

Alec's third winter in Boston was easier than the previous one. The most important change was his appointment as professor of "Vocal Physiology and Elocution" at the newly established School of Oratory in Boston University. The university had begun in 1839 as the Newbury Biblical Institute in Vermont but had moved to Boston and now boasted several new faculties, including schools of law, music, and oratory. Isaac Rich, a successful fish merchant, had bequeathed it his $1.5 million estate. (The university's directors wisely decided not to name the university after its benefactor; after all, who would donate funds to "Rich University"?) A significant part of the new university was housed at numbers 18 and 20 Beacon Street, including the College of Liberal Arts, the School of All Sciences, and the School of Oratory, as well as a room used for daily chapel meetings. Alec was given an office at 18 Beacon Street, where he could receive his private pupils, and was paid five dollars an hour for his university lectures.

For a twenty-six-year-old without a university degree, a professorship was a tremendous boost to both ego and credibility. He couldn't help grinning enthusiastically at the young men and the sprinkling of young women (the university was among the first to admit women) who clattered up and down the steep, narrow staircase outside his second-floor office. It is easy to imagine that Melville Bell's pride in his son's progress was diluted by a spurt of envy: a steady appointment at a prestigious university was exactly the kind of position he thought *he* should hold. He was soon nagging Alec about allowing university duties to distract him from the important job of promoting Visible Speech. When was Alec going to get a promotional pamphlet for the VS system printed, and start talking it up? *That* was the way to make some money. Alec assured Melville that his university duties would occupy only about five hours a week. His timetable consisted of lectures at the university on Mondays and practical classes on Wednesdays. Within days of his arrival, he had twenty students in his university class, and several private pupils to teach on Tuesdays, Thursdays, and Fridays.

Then George Sanders's grandmother offered Alec free board and lodging in her home in Salem, an eighteen-mile train ride from Boston, in exchange for continuing George's education. George adored the teacher who, the previous year, had taught him the rudiments of lip-reading and articulation. Now Alec gave the six-year-old his morning assignments before he caught the train into Boston. In the evenings, as Alec walked back from Salem station toward the Sanderses' clapboard mansion on Essex Street, Georgie would wait at the parlor window, in his smart little suit and with his hair neatly combed. As soon as the child saw his teacher, his gray eyes would light up. After a little more instruction, George would go to bed and Alec would turn his attention to his experiments. One day George took an illicit peek into his tutor's bedroom. "I found the floor, the chairs, the table, and even the dresser covered with wires, batteries, coils, cigar boxes, and an indescribable mass of miscellaneous equipment," he would recall fifty years later. "The overflow was already in the basement, and it wasn't many months before he had expanded into the carriage house." Mrs. Sanders turned a blind eye to the mess—her grandson's progress was too impressive to allow a tangle of wires or Alec's nocturnal work habits to disrupt the relationship. (She did worry about the young Scotsman's lack of sleep, however, so she got into the habit of cutting a few inches off each of his candles, forcing him to go to bed earlier.) By December, George could read. By spring, Alec had developed a way that they could spell out words and communicate easily in both public and private.

Alec described in a newspaper article how he and George "talked" to each other. It was a method that relied on neither Visible Speech nor sign language, nor on the two-handed manual alphabet by which Alec communicated with his mother. For George Sanders, Alec had developed a special glove. He had written on this white glove letters, in spots that could easily be touched, then taught George to use it.

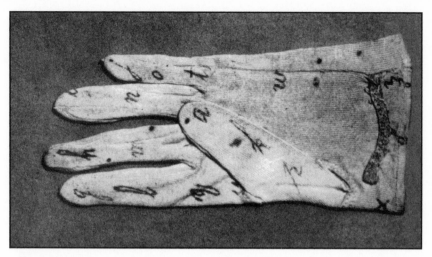

The alphabet glove, with which Alec taught George Sanders to communicate.

"George presented me with the word for some object he desired; we shall suppose the word 'doll.' I then covered up the word with the exception of the first letter 'd,' and directed his attention to the glove. After a little searching he discovered the corresponding letter upon the glove . . . and so with the other letters." Once George learned the single-glove system, his fingers could work as fast as fingers on a keyboard. Soon Alec didn't need to wear the glove: George knew which spot on the hand denoted which letter. "The use of the glove alphabet was so little noticeable," Alec continued, "that [soon] I could talk to him very freely in a crowd without attracting the attention of others. I took him to Barnum's museum and talked to him all the time the lions were being fed, and I am sure that no one among the spectators had the slightest suspicion that the boy was deaf."

Alec marveled at the contrast between the little deaf girl from New York, locked away in an institution because she was "ineducable," and George Sanders, released from his silent prison by the ability to communicate. It reinforced his conviction that institutionalizing deaf children together was counterproductive, and unfair to those who wanted to live in a larger world. He always insisted that George Sanders should not associate with other deaf children. "I was seventeen years old," George told a newspaper reporter in 1929, "before I even knew there were any other deaf people in the world."

In late October 1873, a new student enrolled with Boston University's professor of vocal physiology and elocution. Mabel Hubbard was the fifteen-year-old daughter of Gardiner Hubbard, a Boston lawyer who was president of the Clarke Institution for Deaf Mutes at Northampton, run by Miss Rogers. A tall, slim girl with thick golden brown hair and a broad brow, Mabel had a wonderfully expressive face. A mischievous smile constantly pulled at the corners of her mouth, and her large gray eyes sparkled with joie de vivre. The previous year her father had seen Alec teach a class at the Clarke Institution, and decided that his own deaf daughter should take instruction from this brilliant young teacher. Mabel arrived at Alec's Beacon Hill office to discuss a course of private lessons to improve her articulation. She was accompanied by Mary True, a chatty young woman who had been Mabel's teacher when she was a little girl. Mary now taught at Miss Fuller's school in Boston and, like Gardiner Hubbard, had met Alec Bell the previous year and been dazzled by his success with the deaf.

When Alec met his new student, he was astounded to see how well she functioned in a speaking world. Mabel had never functioned in any other world, or learned sign language. Despite her youth and hearing loss, she exuded some of the high spirits and confidence that characterized good-looking daughters of wealthy men. The key to her self-assurance was her brilliant ability to read lips: strangers were often unaware she had no hearing. Only the intensity with which she watched her companion's mouth and her distorted vowels when she herself spoke gave her away. For Alec, she was living proof that if a deaf person was completely integrated into speaking society, she need not be regarded as "abnormal."

Mabel was less impressed with her new teacher, whom she already considered "a quack doctor." She had been reluctant to meet the Scotsman, and did so only to keep Mary and her father happy. She did not know what to make of this disheveled figure who looked a lot older than twenty-six. "I both did not, and did like him," she later recalled. "He was so interesting that I was forced to listen to him, but he himself I disliked. He dressed carelessly and in a horrible, shiny broadcloth, which made his jet-black hair look shiny. Altogether, I did not think him exactly a gentleman." A gentleman could not have been more polite and respectful than this gauche teacher, but at least his collar would have been clean and his shoes shined. "There was a lack of assurance in the habit he had of passing his hand over the thick high curl over his forehead, a nervousness in smoothing the slight moustache. One felt he desired to put you at your ease while he was hardly at his." Mabel was even less impressed by Alec's gloomy office, with its dingy green walls. She gazed out of the window at the gravestones in the Granary Burying Ground, where revolutionary leaders such as Paul Revere lay. Then she glanced around at the bare room, the empty grate, and the hideous flowered rug. It looked like a dentist's office.

On that gusty fall afternoon, there was no spark of mutual sympathy, let alone attraction, between the self-possessed fifteen-year-old and the preoccupied twenty-six-year-old. When arrangements had been made for Mabel to study with Dr. Bell and the two women had left, Alec's overwhelming emotion was relief that he had another source of income. He wrote home to Brantford that Miss Hubbard plus his other private pupils from well-established families were "just those calculated to do me good." As Mabel Hubbard took the streetcar back to Cambridge, she told Mary True that she would take a few lessons with Professor Bell because, if she married within Boston's wealthy elite, improved diction would assist her to keep up a position in society.

Nothing suggested that this businesslike first encounter would be the start of an extraordinary lifelong partnership.

BELL'S BOSTON

Chapter 4

A BRAHMIN CHILDHOOD
1857–1873

Until her fifth birthday, Mabel Hubbard had enjoyed normal hearing. But then she caught scarlet fever. The bacterial infection made its way from her throat up through the Eustacean tubes to her ears and triggered an intense inflammation for which there was no treatment. It destroyed her inner ear, so that she lost both her hearing and her sense of balance. Her plight was agonizingly common in her day. Before the invention of antibiotics in the twentieth century, common childhood infections like diphtheria, chicken pox, scarlet fever, and meningitis frequently left their victims without hearing, if they survived at all. In Mabel's day, there were proportionately thousands more deaf people in the population than there are today. While there are no reliable statistics for the total number of hearing-impaired North Americans, according to one estimate in *American Annals of the Deaf*, infections accounted for half the hearing-impaired population of the United States in 1848. The next most common causes of deafness were aging and accidents; only a small percentage of the population was born deaf.

Mabel, however, had a huge advantage over most of the children who lost their hearing in early childhood. Her parents, Gardiner and Gertrude Hubbard, were well aware of the stigma attached to deafness in their circles and were determined not to ship their cheerful little girl off to an institution or allow her to dwindle into a silent recluse. They shared an unassailable belief in *progress*, and the confidence that, in the second half of the nineteenth century, a solution could be found for any problem. Mabel would continue to be a vital member of their family, their world. They did not even assume that sign language was the only form of communication available to her.

Until faced with the challenge of Mabel's deafness, the Hubbards were just another wealthy, successful, and utterly conventional New England family. Gardiner Hubbard, a tall, thin-faced, wiry man with deep-set eyes and an Abraham Lincoln beard, mixed with the best of Boston's rather haughty Brahmins. A graduate in law from Dartmouth College, he belonged to a posh Boston law firm; he was detached in manner and rarely lost his temper. But Gardiner's aloof exterior was deceptive. As the grandson of an Irish immigrant who had made a fortune in Boston real estate, he was a shrewd operator. He reflected the heady optimism and get-up-and-go spirit of an open-ended era of finance capitalism, when there were no securities regulations, much abuse of the stock market, and a mania for "progress." Gardiner was always looking for opportunities to improve things, and to make a penny in the process. It was no accident that, of all the branches of law that Mabel's father might have chosen, he had become a patent lawyer, with a special interest in mechanical and electrical inventions. He had noticed that it wasn't only the inventors responsible for technological breakthroughs who were making fortunes: their backers and lawyers were also getting rich.

Early in his career, Gardiner had shown that when he set his mind to something, he went after his goal like a terrier. First he pursued Gertrude McCurdy, daughter of a successful New York dry goods merchant, until she agreed to marry him in 1846. Next, he bought a

Gertrude Hubbard treasured this photograph of herself and her second daughter, taken before scarlet fever robbed Mabel of her hearing.

forty-five-acre meadow between the Charles River and Brattle Street, in Cambridge, and built a splendid home on the edge of it. Then he set about upgrading the public utilities. In the 1850s he founded the Cambridge Gas Light Company, promoted a new waterworks, and established a horse-drawn railroad service between Cambridge and Boston. Once these civic improvements were in place—and the value of his property had quadrupled—Gardiner Hubbard started building houses on his estate to sell. At the same time, he kept an eye on the rapidly expanding field of telegraph technology.

Gardiner Hubbard's brusque ambition was softened by the gentle manner of his wife, Gertrude McCurdy Hubbard. Gertrude Hubbard was a tall, square-jawed woman with a melodious voice and eyes as determined as they were gentle. Throughout her life, she emanated an aura of goodness and serene optimism. Her tact and helpfulness represented the womanly ideal of her period: service crowned with self-effacement. The Hubbard marriage was a harmonious partnership, and their large two-story clapboard mansion at Number 146 Brattle Street teemed with relatives and children. Gertrude's aunts, uncles, and cousins often visited. Tragedy struck the Hubbards in 1848, when their first child, a son, died in infancy, but the following

Mabel (right), with sisters Gertrude (left) and Grace (center), was always a bookish, serious child.

year Gertrude gave birth to a healthy daughter, named Gertrude after her mother and always known as "Sister." After Mabel's birth in 1857 came two more girls, Grace in 1859 and Roberta in 1861. The four little girls were raised, dressed, and taught according to the upper-middle-class pattern of their era. This meant starched white pinafores, fastidious manners, regular attendance at the local Presbyterian church (and dark suspicions of "papism"), and the prospect of following their mother into lives of refined domesticity. His eldest daughter was Gardiner's favorite, but Mabel was an attractive child, with her apple cheeks, sparkling gray-blue eyes, and serious manner. She was also the most bookish: she would sit for hours with a picture book, for example, or converse gravely with her father about her toys.

The crisis in Mabel's life erupted in January 1863, when she accompanied her mother to New York to visit her McCurdy grandparents in their tall brick and brownstone house near Washington Square. The visit had begun happily—as soon as Mabel walked up the steps and under the elegant fanlight of Number 10 East 14th Street, she was embraced by her doting grandfather, Robert McCurdy. She showed him her latest treasure, a postcard picture of Mr. and Mrs. Tom Thumb, two midgets currently on display at Barnum's American Museum of Curiosities. A day or two later, though, Mabel fell ill.

"Our dear little lovely and loved one has been and still is alarmingly sick," Robert McCurdy wrote to a niece on February 6. "She was taken 12 days since with scarlet fever. . . . [A]fter a few days it assumed a malignant type and her whole system seemed poisoned, her throat swelling very large, her mouth, eyes and nose almost closed and her breathing difficult." Death hovered over the McCurdy mansion, and the servants went about their business red-eyed and scared. A quarantine notice was pinned to the big oak front door announcing that scarlet fever had struck the household and discouraging all visitors. Straw was scattered over the cobblestones to muffle the sound of carriage wheels. Mabel's father, Gardiner, was summoned from Massachusetts as her mother and grandparents watched the child "bravely struggling for life, with alternate hopes and fears." At first Mabel kept trying to speak, even asking her mother if she thought Jesus wanted little girls to say their prayers when they were so sick they couldn't. But soon the effort of breathing and swallowing made talk impossible. Gertrude and Gardiner Hubbard began reliving the grief they had felt fifteen years earlier, when their infant son had succumbed to fever. They knew that even if Mabel pulled through, she could be left blind, hearing-impaired, or brain-damaged. All the physician could prescribe was "nourishment, watching and faith."

By mid-February the crisis had passed, but the little girl remained inert and unresponsive. When Gertrude Hubbard stroked her daughter's damp brow, she was almost inclined to accept the doctor's gloomy prediction of permanent brain damage. A few days later, however, Gertrude noticed that Mabel's eyes now focused on her and followed her movements. Six weeks after the onset of fever, there was a breakthrough. Gertrude showed Mabel her treasured picture of Mr. and Mrs. Thumb. Mabel murmured, "Little lady." Gertrude realized that the disease had not affected her daughter's intellect, but it had left her completely deaf. Mabel was soon asking, "Mamma, why don't you talk to me?" With tears in her eyes, Gertrude put her face close to her daughter's so that Mabel could see that the problem was not with her mother's speech but with her own hearing.

Mabel and her mother remained in New York for several more weeks that spring, as the child slowly regained her strength. Outside the bedroom windows, the city stirred as the days lengthened. Passersby hailed each other across the street. The chatter of children chalking games onto the paved walks of Washington Square, one block away, drifted through the warm April air. As Gertrude gazed out of the window of her parents' house, she could hear newsboys yell out details of President Lincoln's latest speech and news that the brilliant Confederate general Robert E. Lee was preparing to march north with 75,000 troops. The Civil War between the Confederate states and the Union had been raging for nearly two years. Outraged that the antislavery Republican candidate for president, Abraham Lincoln, had been elected in 1860, the slave-owning southern states had seceded from the Union in February 1861. They adopted a provisional constitution for the newly formed Confederate States of America and attacked Fort Sumter in Charleston, South Carolina. This was the trigger for a far more brutal war than anybody had imagined, as each side violently defended its position on issues of constitutional principle, human rights, and economic self-interest. In 1863, the war was not going well for the Union side. But it all seemed irrelevant to a mother worried sick about her child.

When Mabel was finally well enough to join her mother at the window, she looked out at a silent world. She watched neighbors open their mouths in hushed pantomime; she saw carriages bounce along noiseless streets. She could see shapes, colors, and familiar faces, but she could not hear the laughter, birdsong, music, or chatter that, for most of us, brings our three-dimensional world to life. The damage to her middle ear left her balance uncertain; she would always need a steadying hand in the twilight or at night, or on a moving vehicle.

Eventually Mabel was well enough for her mother to take her home, and Gertrude made her way to New York's train station, holding her daughter's hand tightly. Back in Cambridge, Gertrude's three other daughters—Sister, now thirteen, Roberta, age three, and Grace, age one—had been eagerly awaiting their mother's return. In the homecoming tumble of embraces and laughter, Mabel's sisters were at first oblivious to her inability to hear their voices. Gertrude Hubbard watched the interchanges between her children carefully. She observed how Sister and Roberta hugged Mabel and started chattering to her. She saw how Mabel clung to her sisters as she stared silently into their faces, then stuttered out their names awkwardly. She watched Sister, older and more aware of the problem, speak more and more emphatically to Mabel as she willed her to understand her words. Mabel backed away from Sister, clutching at her mother's hand.

Gertrude set some rules in her Brattle Street home that ensured that Mabel had to keep talking and didn't lose what spoken language she had despite the fact that she could no longer hear her own voice. Gertrude told everybody that if Mabel tried to communicate by signs, they were to ignore them. When Gertrude herself was with Mabel, she made the little girl look carefully at her lips so she could see what words were being formed. Before her illness, Mabel had learned the words of the 23rd Psalm, "The Lord is my shepherd." Now Gertrude insisted she repeat it regularly, then built on the words to improve her vocabulary. She encouraged family and servants to stand face to face with Mabel so that she could see their lips, then talk to her as though she could hear. Gradually Mabel started following the speech of others through their lip movements. She learned to intuit what the conversation was about so that she would know which of similar words ("bat" and "pat" for instance) the speaker meant. And her family learned to say things in different ways if Mabel missed the first way they had phrased a remark.

While Gertrude pushed to keep Mabel in the speaking world, she and Gardiner explored educational possibilities for deaf children. They refused to be discouraged by teachers who told them that Mabel would soon be incomprehensible (one physician insisted that her voice would be "worse than the whistle of a steam engine"). In Boston there was already a fascinating example of the way new teaching methods could help blind children operate in the seeing world. The Perkins Institution for the Blind had been established by the Christian reformer Dr. Samuel Gridley Howe, husband of Julia Ward Howe, in 1839. Dr. Howe's techniques for educating blind children, involving musical instruments and raised alphabets, had attracted educators from all over Europe and North America. His most celebrated success was a young woman named Laura Bridgman, who had been born deaf as well as blind and who could now speak and read raised letters. Charles Dickens himself had gone quite overboard with the Laura Bridgman story in *American Notes*, ranking her, alongside Niagara Falls, as one of the two most impressive phenomena he had witnessed on his trip to the United States. Gardiner Hubbard determined to ask Howe what he recommended for Mabel.

Yet Mabel's speech began to deteriorate as, locked in her silent world, she failed to model the volume of her speech and her pronunciation on the voices of those around her. She sounded increasingly weird. Berta and Grace started excluding her from their games, and she became less talkative, withdrawing into her own silent play. Gertrude herself could not give Mabel as much of her undivided attention as she wanted—besides her four daughters, the Hubbard household now also included Gertrude's invalid mother, and in April 1865, Gertrude Hubbard gave birth to a fifth daughter, Marian.

Go-ahead Gardiner promptly decided that the Brattle Street house needed extensive renovations to accommodate the growing household. Mabel never forgot the innovations: "a wonderful new flooring of inch thick diamond shaped blocks of white and brown wood" in the dining room; crystal gas chandeliers "that caught the morning sunshine and broke it into a thousand lovely rainbows"; rich red velvet wallpaper in the hallway, and soft silver-gray paint on the library walls. Gardiner decreed that while the work was being done, his family should spend the summer in the country. He engaged rooms for his wife and daughters in the little Maine village of Bethel, on the Androscoggin River, in an old white-walled house with green shutters, belonging to the local doctor, Dr. Nathaniel P. True. When the family moved into the house in July, the stolid New England villagers were amused at the sight of the little blonde girls in their dainty white pinafores and black boots. The children captivated Dr. True's twenty-one-year-old daughter, Mary, who had just finished training to be a teacher. Mary was particularly drawn to the child who played by herself, spoke strangely, and called her "Miss Rue." By the time the Hubbards left Bethel, Mary had been hired as the new governess. She would remain with the family for three years.

Mary True proved to be a brilliant teacher for Mabel. Yet she had no instruction manual or training in deaf education, and unlike Alec Bell, there was no "professor of elocution" in her family to whom she could turn for advice. Both she and Mabel were in unknown territory, as Mary tried to enter into Mabel's mind to see how it functioned and then build on its potential to improve Mabel's articulation, vocabulary, reading, writing, and lip-reading skills. Day by day, she made Mabel watch her lips as she sounded out words, then made Mabel copy the positions of her lips, tongue, and mouth. She would not let Mabel revert to baby talk, communicate by gestures, or avoid hard words. Since Mabel was an avid reader, Mary worked hard to find books her student enjoyed, then made her read out loud to learn how words sounded as well as looked: "We read pages upon pages in a School Reader using the words of her own vocabulary to explain new ones, or a phrase to define a word. Sometimes the language was queer, I admit, and the definitions crude, but somehow we got on," she later recalled. Mary was a thorough, old-fashioned teacher. "I taught her grammar after the old style of parsing. She learned the parts of the speech—noun, pronoun, verb, adverb, adjective, conjunction, preposition." Mary expected Mabel to learn anything that her sisters could learn, including the pronunciation, during geography classes, of such complicated names as Chattanooga, Susquehanna, and Chesapeake. But with Mabel, Mary used different teaching methods. She invented word games and took Mabel on walks round Cambridge, showing her how to pay her fare on the streetcar ("Two tickets to Brattle Street, please") or purchase material for the next day's lesson in a shop ("We need some coloring pencils"). When Mary was explaining how government worked, she took Mabel to the Massachusetts State House in Boston to watch legislators in action, with all their ritual and rhetorical flourishes. On their return home, they traced each day's outing on a map.

Mary's job was made easier by her student's intelligence and hunger to learn. Mabel soaked up new information. Her reading skills soon surpassed those of her hearing sisters, and her taste in books was extraordinarily sophisticated. In later years she described how, as a child, "I did not care to romp and play out of doors, all I wanted was to curl up in some quiet corner and read—all day if allowed. My father's library was well-stocked and I had almost free range. When eleven years old I delighted in reading such books as Jane Porter's *Scottish Chiefs,* and before I was thirteen I had read through, with intense interest, Motley's *Rise of the Dutch Republic,* most of Prescott's histories, several large volumes of the Civil War, books of travel, as well as all the stories and novels I could get hold of."

In 1867, Mary took Mabel, now a solemn nine-year-old, back to Bethel with her for the summer. "She was very proud of new words," the teacher later recalled. "'Worthless' and 'valuable' were rolled as sweet morsels under her tongue, and applied with great frequency to my brother Alfred when he teased her. 'You are a very, very, very *worthless* man' settled all scores." Mary True had become such an important figure in Mabel's life that her pupil's parents trusted her with a difficult job: breaking the news to Mabel of the death back in Cambridge of her baby sister, Marian (probably from diphtheria). Mary led the sobbing child to the far end of the Trues' orchard and tried to explain to her the significance of death and heaven.

By the time she was eleven, Mabel's proficiency in lip-reading and all her school subjects was remarkable—and so was the self-esteem that her success gave her. Her parents decided she should join her younger sisters Berta and Grace at Miss Songer's school, a private school for girls in Cambridge. Miss Songer and her colleagues were astonished at Mabel's skills. One teacher confessed herself "surprised at the readiness with which she reads from the lips, as I have never talked with her before, and she understood me without difficulty." After testing, Mabel was assigned to classes where she was the youngest by several years. Mary True had done her job so well that Mabel was the equal in intellectual development and achievement of girls three to five years older. Fifty-four years after Mary had come into Mabel's life, the latter paid her governess this tribute: "She opened my mind and gave me a mental training and grasp of things which has formed a broad and firm foundation on which could be built all I have since acquired."

Meanwhile, Gardiner Hubbard had finally managed to consult Dr. Samuel Gridley Howe about deaf education, and had learned that lip-reading was *not* the technique of choice in the United States. Howe told Hubbard that if he wanted Mabel to rely on the oral method rather than on signing, he should take her to Europe and enroll her in one of the well-established schools there that taught lip-reading abilities. Hubbard decided to follow this advice, but he also determined to make the oral method more available in his own part of the world. Mabel's father lobbied the Massachusetts state legislature to grant a charter for the establishment of a school where very young deaf children might be taught to speak and lip-read. He even enlisted his own daughter in the campaign: when she was nine, Mabel appeared before the Massachusetts legislature as an example of a deaf child who thrived in the mainstream thanks to her lip-reading skills.

Perched on a large chair, dressed in her best silk gown and linen cuffs, and with Mary True at her side, Mabel must have been an appealing figure in front of all those bewhiskered, black-coated men. Urged on by Gardiner Hubbard, Mabel's audience plied her with questions

Mabel Hubbard, aged twelve, was cheerful and self-possessed, apparently unaffected by her disability.

in history, geography, and mathematics as though she were a doctoral candidate instead of a small, hearing-impaired girl. She understood their questions with ease and, with the unselfconscious enthusiasm of a child who had never known discrimination, answered each question promptly and correctly. Only one question posed by a legislator perplexed Mabel: "Are you a deaf child?" Mabel hesitated, and looked questioningly at her father. Watching her troubled little face, Mary True realized that Mabel had not realized she was "different." A few minutes later, Mary saw Mabel staring with an expression of blank incomprehension at some students from Gallaudet's American Asylum at Hartford for the Education and Instruction of the Deaf and Dumb. The students were communicating by sign with each other and, through an interpreter, with the legislators. Mabel Hubbard had no idea what this was all about.

As a result of Gardiner Hubbard's aggressive lobbying (and over Gallaudet's furious objections), the state government approved legislation for a Massachusetts school for the deaf in which children as young as five would be taught to speak and lip-read. The Clarke

Institution for Deaf Mutes opened in Northampton in 1867, with Miss Harriet B. Rogers as principal and Hubbard as president. This was the school in which, five years later, Alec Bell would demonstrate Visible Speech. Gardiner Hubbard would play a lifelong role in this new school, but right now he had to acknowledge it would not answer Mabel's needs. His daughter was already so good at lip-reading that she needed the advanced instruction available only in the kind of well-established schools in Europe that Dr. Howe had mentioned. And Hubbard had an additional motive for shipping his whole family off to Europe. The bitter and bloody Civil War had finally drawn to an end in 1865, after a horrific total of 600,000 Americans had lost their lives, and the republic had then plunged into depression in the subsequent squabbles and scandals. Hubbard's business interests had suffered. It would be cheaper to shut up the Brattle Street house and travel through France and Italy than to try to keep up appearances in Cambridge. In the spring of 1871, the Hubbards sailed out of New York harbor.

Until this point in her life, Mabel's life had been spent within the loving circle of friends and family. Brattle Street was its own cozy little world: Mabel's Blatchford cousins (children of her father's sister) lived next door, and the families all knew the conductors and drivers of the Brattle Street horsecars by name. "In fact, they were little short of household retainers," recalled Mary Blatchford, a particularly haughty cousin. Everybody in Mabel's life was so accustomed to facing her and enunciating slowly and clearly when they addressed her that it had become second nature to them; they barely recalled her disability. Once the Hubbard family set foot in Europe, however, Mabel was on an unfamiliar continent, surrounded by people who spoke a medley of unfamiliar languages and who often made no eye contact. She could lip-read, but she didn't know what the words meant. Her own diction was still odd, and she was suddenly self-conscious: "I am almost afraid of the many strange faces and the stares people give me when I speak," she noted in her journal. Now thirteen, Mabel had to cope not only with the complicated emotions of adolescence, but also with the dawning recognition that she was different from her bubbly, careless sisters. She began to understand that people outside the family regarded her disability as a handicap.

When Gertrude first introduced Mabel to strangers, they saw a solemn, cautious girl, neatly dressed in the full-skirted gowns and buttoned boots of the period. She stayed closer to her mother than her sisters did, but her eyes sparkled and she watched the faces around her with eager curiosity. She laughed easily, and had endless energy for sight-seeing adventures. When she started to talk, however, acquaintances struggled to understand the strange sounds she made. They unconsciously addressed all their remarks to Mabel's sisters and mother, avoiding looking at her, as though there were something wrong with her brain as well as her hearing. Mabel brooded on her limitations. During a family expedition to the Austrian town of Innsbruck that summer, she recorded in her journal how she knelt by the sofa on which her mother lay, weeping softly to herself "over some private grievance of my own—my not being able to hear which I feel more keenly now I am thrown so much on strangers."

In September, while the Hubbards were in Geneva, Gertrude broke the news to her thirteen-year-old daughter that she had been enrolled in a school for the deaf in Vienna. Mabel's response was instantly fearful: "I won't go." Gertrude explained that no one would force Mabel to go anywhere. But as Mabel wrote in her journal, "They thought that if I went there, I would learn German more thoroughly and that it would be much to my improvement; it would help me to speak and understand much better and faster." Mabel would not be alone, her mother assured her. Gardiner Hubbard had invited Miss Rogers, principal of the new Clarke Institution, to join Mabel in Vienna for the winter to see the European approach to the oral method of teaching. Mabel listened to her mother dolefully. Still, she already had not only a Bostonian sense of duty, but also a psychological stamina unusual in a child her age. She consented to follow her parents' wishes: "On the whole I think I had better go to Vienna for Mama says it concerns my future welfare."

Mabel made phenomenal progress at the Austrian school. Within months, she could read, lip-read, understand, and speak as fluently in German as in English, and she had a smattering of French and Italian. Her diction improved. She also made new friends. A young Hungarian Jewish boy, nicknamed Buba, became so fond of her that he wouldn't let any of the other children play with her. Mabel (or "Mapel," as her German-speaking fellow students called her) had a healthy sense of her own dignity and value and didn't appreciate Buba's familiarity: "I consider it something short of a personal insult." She admitted in a letter to her father, however, that she enjoyed competing academically with Buba—and winning. On weekends, she and Miss Rogers were conscientious tourists, visiting the city's monuments, churches, and museums.

Gertrude Hubbard and Sister spent that winter in Paris, while the two youngest girls were packed off to a private school in Switzerland. The following spring, the Hubbard women were reunited in Rome. Their existence in Europe might have come straight from the pages of a Henry James novel: it was a life of luxury hotels and first-class berths in the wagons-lit. While in Rome, despite their sturdy New England Presbyterianism, they had an audience with the pope, which involved close attention to Vatican protocol. Mabel, now fourteen, had shot up the previous winter and had lost the puppy fat from her face and waist. European sophistication had rubbed off on her: she dressed her waist-length brown hair in elaborate styles and never ventured outside without a parasol or umbrella. "I am so tall and look so old that Mama was afraid they would think me too old to be allowed to wear a colored dress," she confided to her journal, "so as I had no black dress of my own, she loaned me her dress. Didn't I feel fine! They say I looked twenty years old."

Life among strangers had matured Mabel to the point where, by her mid-teens, she was old beyond her years. She knew how to mask her deafness; she was unobtrusive in company, and had mastered all the tricks the deaf use to manage in a hearing environment. She was adept at picking up unspoken clues to people's intentions—subtle movements, facial expressions, exchanges of glances. But for all her self-possession and linguistic skill, she recognized that her disability, once acknowledged, would make her "odd" in many people's eyes. In her final weeks in Italy, she wrote in her journal, "As I grow older I feel my loss much more severely. At home I don't remember the idea ever entering my head to wish to hear, but now when I am thrown into the society of strangers, I feel somewhat discontented. Only somewhat for thank God I am getting to understand mere strangers without help."

Mabel's family was oblivious to her strange tones, bizarre intonations, and frequently quizzical expression. For her parents and sisters, it was now perfectly natural to face her when they spoke to her, exaggerate their mouth movements, or tug at her sleeve and stamp their feet to catch her attention. They knew how intelligent she was, and also how, to compensate for her loss of hearing, she had developed extraordinary inner strength. She created an island of welcoming calm in the bustle of their busy family life. Resilience and bravery lay beneath Mabel's gentle manner. They loved her deeply and she, in turn, loved them. Throughout her life, Mabel would always be more comfortable among family than with anyone else.

While Gertrude Hubbard and their four daughters had been in Europe, Gardiner Hubbard had returned to his Boston law practice: there were too many business opportunities in North America to allow him to stay away for long. But he had also kept in close touch with the development of the Clarke Institution in Northampton, for which he had campaigned and where the oral method of deaf education was now well established. On a visit to the Clarke Institution in 1872, he met the tall, dark-haired Scotsman who was spending a month there. Hubbard didn't have much time for Alexander Graham Bell's Visible Speech techniques, but he was impressed by his teaching skills. He determined that, as soon as his family returned from Europe, Mabel would start taking instruction from Mr. Bell.

In the fall of 1873, Mabel began her twice-weekly visits to 18 Beacon Street, and despite her early misgivings about Mr. Bell, she was soon enjoying herself. She received most of her instruction from Abby Locke, a former pupil who now acted as Alec's assistant, but it didn't take long before she found herself hoping, as she made her way from Cambridge, that it was the disheveled professor rather than his assistant who would teach her that day. He was "so quick, so enthusiastic, so compelling, I had whether I would or no to follow all he said, and tax my brains to respond as he desired." Abby and Alec began the lessons in the same way that Alec had begun his first class at Miss Fuller's school when he first arrived in Boston. They drew a baby's face on the blackboard, with all its features, then rubbed out everything except the elements that were represented in the Visible Speech symbols that told the student how to shape her mouth and tongue. "I like my lessons with Prof. Bell very much, but find the baby's head very hard," Mabel reported to her mother during the early weeks.

As autumn gave way to winter, Alec's study became steadily more untidy. Half-written papers piled up on his desk, next to dog-eared library books and sketches of both telegraph equipment and human speaking organs. As usual, Alec had undertaken too much, and he was now finding himself under pressure of deadlines and student demands. But his new star pupil had mastered the system of Visible Speech notation in record time, and her enunciation was transformed. Alec realized that this remarkable young woman possessed an intellectual strength, a cosmopolitan background, and social skills that were rare in the circles he moved in. Mabel Hubbard could be much more than simply a student. Soon Mabel was writing to her mother, who was on an extended visit to Mabel's grandparents in New York, "What do you think they want me to do? Nothing less than to write something in symbols for a periodical on Visible Speech!"

Chapter 5

GOOD VIBRATIONS
1873–1875

Alec Bell and his assistant Abby Locke were stimulating teachers. Mabel's trips to Beacon Street soon became the highlights of her week, especially since the rest of her life was dreary. Her father was trying to break into the business end of the telegraph industry, and his legal activities kept him in Washington. Her mother and sisters were staying with Gertrude Hubbard's parents in New York City—Gardiner Hubbard had a cash-flow problem, so he had put his Cambridge home on the market. The only people rattling around in 146 Brattle Street were Mabel, her cousins Carrie and Mary Blatchford, and a couple of servants.

Throughout the fall of 1873, fifteen-year-old Mabel pined for her family. "It is so sad," she wrote, ". . . that we should go on living like this with no settled home, our family divided, part in New York, Papa in Washington and I here." Mabel hated seeing most of the windows at 146 Brattle Street unlit in the evenings, the pears in the pear orchard lying rotting in the grass, and the herbaceous borders in the garden going to seed. She frequently wandered into the library or the sitting room and, suffused with nostalgia for happier years, quietly lifted the dustcovers in the hope of catching the lingering scent of her mother's eau de toilette or her father's hair pomade. She told her mother in a letter, "Our house, that sweet home where we were born and passed our childhoods' first years still remains to us, but a large placard distinctly visible announces that 'this estate is for sale.' . . . It seems sometimes as if it were almost better to have it sold than to see it looking so lonely & desolate & uninhabited."

However, Mabel was not by nature prone to depression. After the months in Austria, she was used to separation from her family, and she kept herself busy. She visited neighbors, including Edith and Annie Longfellow, daughters of the great poet Henry Longfellow. She learned to navigate around Boston without an escort, despite her deafness and her balance problems. "I expect to go to church this afternoon, I must go alone and it's dreadful to me, I am always frightened at going out alone," she wrote. "However I'm too big to be so babyish and I will go to try and conquer it." She read voraciously and wrote long letters to "My precious Mamma." As the weeks passed, the lessons with Miss Locke and Mr. Bell occupied an increasing amount of space in the lengthy scrawls to her mother. In December she wrote, "Mr. Bell informed me the other day he never had a pupil who improved so— not slowly but—*rapidly*. There's for you Ma Chère!"

In February of the following year, she was so eager not to miss a lesson that she braved a huge snowstorm in order to catch the streetcar into Boston—"Both Miss Locke and Mr. Bell were surprised to see me." After her lesson, Alec insisted on escorting her back to the streetcar stop, since the snow was by then knee-high. As they stepped into the street, they were almost blinded by the whirling snowflakes and found themselves slipping and sliding into each other. Alec held a protective hand out to his pupil since she could not hear his warnings about patches of ice; she grasped it gratefully and gave him a wide smile. Soon the two of them were running downhill, arm in arm and shrieking with laughter. For Alec, it was a sudden reminder that he was, after all, still a young man—and that it wasn't so long since he had enjoyed such adventures in Edinburgh winters with his long-lost brothers. For Mabel, who described their "grand time" in a letter to her mother, it put her fascinating but usually unsmiling teacher in a new light. Mabel's pleas to be reunited with her family in New York petered out: "I enjoy my lessons very much and am glad you want me to stay. Everyone says it would be such a pity to go away just as I am really trying to improve." In this particular letter, Mabel signed her name in Visible Speech symbols.

It remained a teacher-pupil relationship. Mabel went on thinking that Professor Bell was nearly twenty years older than she was. But she enjoyed his company and the way he treated her as an adult. His opinions were so thrillingly different from those that held sway on Brattle Street; she didn't know whether to take him seriously when he insisted, for example, "that America was the refuge for the ruffians of the whole world." For his part, Alec was gratified by the rapid improvement in her speech: he had almost managed to smooth the strange "honk" out of Mabel's voice and to teach her how to modulate its volume. Close attention to her enunciation had made him notice other features—the uncommon clarity of her eyes, the sweetness of her smile, and the infectious nature of her laugh. In a postscript to one letter to her mother, Mabel noted, "What do you think, I have been told I am beautiful!" But Mabel was still very young, and Alec's intellect was engaged elsewhere. As Mabel would recall in later years, "His mind was too full of telegraph ideas for much else."

Alec continued to live at Salem, helping George Sanders each evening, then working far into the night on his own ideas for a multiple telegraph. The mountain of electromagnets, vibrating steel rods, and wire circuits on his worktable grew alongside his conviction that two messages could be sent simultaneously on the same wire if they used two different tones. But every time he solved one puzzle, another emerged. His keen "ear" helped detect vibrations in the rods, but he was hampered by his ignorance of electricity. He filled notebook after notebook and his letters home with sketches of harmonic telegraph devices. He read in the *New York Times* and heard from colleagues at Boston University that other inventors were on the same track as he was. As early as 1861, a German physicist called Philipp Reis had transmitted various sounds (but nothing so complex as speech) on an interrupted electric current over a distance, and called his invention a "telefon." Within the previous couple of years, a professional electrician and inventor from Ohio named Elisha Gray, employed by the Western Electric Company in Chicago, had successfully transmitted music over telegraph wires. In Newark, New Jersey, Thomas Edison was already bragging that he was close to introducing the "quadruplex telegraph," which he claimed might carry two Morse code messages simultaneously in each direction. The competition unnerved Alec, and once again his health began to suffer. By late May, Mabel told her mother that her instructor was "quite sick, having overworked himself."

In June 1874, the discouraged twenty-seven-year-old packed up his instruments and his notebooks, said goodbye to his students, and wearily climbed on the train for another summer in Brantford with his parents. His obsession with telegraphy now overshadowed his commitment to deaf education. As the locomotive steamed toward Buffalo and Canada, Alec sat on the scratchy horsehair seat of the second-class train car staring out at the waving cornfields and rolling meadows of New England. By late afternoon, the train was passing through upper New York State. Slanting light gilded the thick forests and painted the ponds and marshes with yellow gleams—but Alec was too preoccupied to notice. He was thinking about his summer plans, and a new acquisition. A few weeks earlier, he had mentioned to his friend Dr. Clarence Blake, a Boston ear specialist, that he was trying to replicate mechanically the action of the delicate bones and membrane of the human ear. Dr. Blake had asked him why he was trying to reinvent the wheel: why didn't he just use a human ear? Days later, Alec received a little box in the mail from his friend; in it

Alec developed the phonautograph, which incorporated human ear bones, to help the deaf "see" how words should sound.

were the anvil, hammer, and stirrup bones and the membrane of an ear taken from a corpse at the medical school where Dr. Blake taught. Now the macabre package was in his travel bag, on the luggage rack above his head.

Once in Brantford, Alec walked the three miles from the station and appeared unannounced through the French doors of his parents' parlor, with a cry of, "Hoy Hoy! So what's happening?"—his usual style of entry at Tutelo Heights. Eliza jumped up from her seat with delight, sent a servant to find Melville, and immediately began fussing about her son's pallor and thinness. Melville appeared, his face wreathed in smiles as he looked up at his son. Soon the two men were deep in conversation about Alec's telegraph experiments. The following day, Alec and Melville were busy in the laboratory that Alec had established behind the conservatory, Alec working on an apparatus incorporating the human ear that Dr. Blake had sent him. He called this contraption his "ear phonautograph," or sound writer. When Alec spoke into the ear, a pen that was attached to one of the bones behind the membrane moved in sympathy with the membrane's vibrations and traced a wavy line on a piece of smoked glass. Initially, Alec's intention was to produce an instrument that would allow the hearing-impaired to "see" how words should sound. He hoped that his deaf students would be able to compare the lines their own voices made with those of correctly pronounced words so that they could adjust their own pronunciation. But the phonautograph soon took Alec in a different direction. He was struck by the way that sound vibrations acting on a tiny membrane could move relatively heavy bones. Melville's terse diary entry for July 26, 1874, read, "Electric speech (?)"

On August 10, after a morning spent with the phonautograph, Alec strolled out of the house toward the bluff overlooking the Grand River. A large tree had blown down here, creating a natural and completely private belvedere, which Alec had dubbed his "dreaming place." Slouched on a wicker chair, his hands in his pockets, he stared unseeing at the swiftly flowing river below him. Far from the bustle of Boston and the pressure of competition from other eager inventors, he mulled over everything he had discovered about sound. His experiments had demonstrated that vibrating tuning forks produced sound waves, a pattern of compressions and rarefactions that moved at a steady speed away from the source of vibration. Alec understood that sound waves travel through air in the same way that energy moves through the coils of a spring when a section of the spring is compressed and released. He now began to wonder whether electric currents could be made to mimic a sound's pattern of compressions and rarefactions. If they could, any sound could be transmitted electrically.

Alec's knowledge of sound, and his acute ear, meant that he was aware of two other aspects of sound waves. First, he knew that rapidly vibrating objects have a higher frequency than less rapidly vibrating ones—they go through their cycle of vibration more times each second. A sound's frequency determines its pitch: how high or low it seems to a listener. Second, he understood the concept of sympathetic vibration. Such vibrations are set up when sound waves from one vibrating object cause another object of the same natural frequency to vibrate.

The phonautograph apparatus mimicked the action of the human ear. In the process of hearing, sound waves pass through the ear canal and hit the eardrum or membrane, which begins to vibrate. The vibrations move the tiny bones connecting the membrane to the inner ear, and these movements cause waves in a fluid in the inner ear. The fluid then pushes against another membrane, which is covered with thousands of hair cells. These in turn are attached to nerve fibers, which, as the hair-covered membrane moves, send signals to the brain. These signals are interpreted as sounds.

Alec had understood how sound was produced in the human larynx ever since he and Melly had built their speaking machine, all those years ago in Edinburgh. Now he understood how sound was received in the human ear. The next step would be to reproduce the action of the ear membrane and design an instrument to translate the vibrations into sounds. Suddenly the idea struck him that it might be possible to create an undulating electric current that could carry sound along a telegraph wire in the same way that air carried sound waves from the speaker to the hearer. The telephone receiver, pressed to a human ear, could act like an electrical mouth. Current flowing through an electromagnet would cause the receiver's membrane to vibrate. The vibrations, reasoned Alec, would then hit the listener's eardrum, making it vibrate too. The listener's ear would interpret these vibrations as the sounds spoken by the person at the other end of the wire.

This was the "eureka" moment for Alexander Graham Bell—his flash of genius. In his dreaming place overlooking the swirling waters of the Grand River, he had grasped the principle on which the telephone would operate.

Years later, when Alec was defending his patent application, he would describe the moment when "the conception of a membrane speaking telephone became complete in my mind; for I saw that a similar instrument to that used as a transmitter could also be employed as a receiver." This extraordinary leap of the imagination is what put Alec ahead of the pack of inventors scrambling to advance the technology of communication. No one else knew enough about the human ear. His imaginative breakthrough is also what makes Alexander Graham Bell the quintessential inventor of the nineteenth century—the era when an untrained individual working alone could dream up such a crucial scientific advance.

But in that summer of 1874, Alec ran headlong into his own limitations. He was an intuitive inventor, not a trained scientist. He lacked the technical expertise and the knowledge of electricity, let alone the funds, to construct the apparatus required. That would have to wait until he returned to Boston. Moreover, the telephone and the phonautograph were diversions from his main preoccupation: the multiple telegraph, which would simultaneously transmit several Morse code messages rather than a single human voice, and for which there was already a market. This seemed the best bet for making him some money. So for the time being, he abandoned his idea of inducing continuous undulatory currents and returned to the challenge of creating a battery-induced intermittent current with simple, steady frequency. As he fiddled with his electromagnets in the Tutelo Heights laboratory, Melville Bell looked on with impatient disapproval. What did this have to do with Visible Speech? "Alec full of schemes," Melville jotted in his diary on September 7.

The cherries in the Bells' orchard ripened, and summer slipped by. Eliza redoubled her efforts to ensure that her son was well fed and well rested before he returned to a life that she knew was full of "nervous work" and headaches. All too soon, however, Alec was climbing into the Bells' buggy to catch the train. Eliza waved a sad goodbye. She was now in her sixties—gray-haired, losing what little hearing she had, and acutely conscious of the passing years. She wished Alec were more settled, preferably somewhere she could see him more often.

Back in Boston, Alec resumed where he had left off—living in Salem and lecturing at the School of Oratory. Mabel Hubbard re-enrolled as a private student. Her home life was happier now: her mother and sisters had returned to the house on Brattle Street at the start of summer. Once again, the parlor was filled with flowers, afternoon tea was served in a pretty imported tea service, and, on cool evenings, a log fire crackled in the hearth.

But there was a shadow over the Hubbard household: Gardiner Hubbard's business dealings were not going well. Mabel Hubbard's father had been trying to play Jack the Giant Killer, challenging the Western Union Telegraph Company, recently renamed the Western Union Corporation. The Western Union Corporation was one of the first truly national companies in the United States. It had gobbled up most of its competitors in the telegraph field and now administered over two thousand telegraph stations. Hubbard and his allies denounced this near-monopoly and charged that Western Union rates were far too high. They pointed out that telegraph rates were lower in Britain, where the service was provided by the post office. Arguing that cheaper rates would give business a big boost, Hubbard had proposed a competitor for Western Union: a private corporation, the United States Postal Telegraph Company, which would be responsible for the transmission of telegrams. Reception and distribution, argued Hubbard, could then be handled by the U.S. Post Office, which could treat telegrams like letters and charge them the same rates.

Gardiner Hubbard had spent the previous two years in Washington, lobbying Congress to charter his proposed private corporation. If the "Hubbard bill," as the legislation was known, were to pass, the Western Union monopoly would splinter and Hubbard would make a great deal of money. But he faced a formidable opponent in William Orton, the tall, distinguished, and powerful president of Western Union, and a man who fit the plutocratic, cigar-chomping robber-baron stereotype to a tee. Up and down the country, Orton thumped the table and made speeches defending free enterprise and attacking Hubbard's scheme as "government interference." Orton was determined to protect his monopoly, crush Gardiner Hubbard's scheme, and secure his company's technical advantage over any competitors, present or future. He had already taken the precaution of acquiring rights to any multiple telegraph that Thomas Edison, the pushiest young inventor on the scene, might complete.

While Orton ranted, Gardiner Hubbard's other business interests suffered. He was forced to borrow capital from his father-in-law, Robert McCurdy, and to fend off his creditors. His wife struggled to keep up appearances: "When am I to get the wherewithal to pay for bread and butter," she wrote to her husband, "not to mention cake and ice creams . . . butchers and school and carpet women and dressmakers and butter and eggs and servants' wages and all daily expenses?" But Hubbard was damned if he was going to buckle before Western Union and its bully president. He recognized that the entrepreneur or corporation that could market the first multiple telegraph would have the edge in the marketplace. One reason for his return to Boston that summer was the hope that he might find, within Boston's cadre of brainy young telegraph enthusiasts, an inventor who was closer to a breakthrough than William Orton's protégé. If Hubbard could be the first to unveil a way of sending several messages along telegraph wires simultaneously, Congress might be more inclined to pass his bill.

This was the background to a pivotal tea party in Cambridge in the fall of 1874. Gertrude Hubbard had invited her daughter's elocution teacher to visit 146 Brattle Street on a Sunday afternoon. Alec, dressed in his shiny black coat and worn dress shoes, looked distinctly uncomfortable as he was introduced to Mabel's sisters. The drawing room was so opulent, with its crimson damask curtains and heavy gilt valances, and the Hubbard girls were so dainty and gracious in their muslin tea gowns and carefully curled hair. They were unfamiliar company for a young man who spent most of his evenings with a small boy, an elderly lady, and a tangle of copper wiring. His large hands, with their thick fingers and their nails bitten to the quick, seemed particularly clumsy as they held a fragile bone china cup of tea. Unpracticed in small talk, Alec was tongue-tied. Mabel's sisters found the shabbily dressed, nervous Scotsman rather a bore.

But then Gertrude Hubbard invited the visitor to play the grand piano. The chatter in the rest of the room soon died down as the Hubbards all realized that they had a gifted musician in their midst. At the same time, the rich notes of a well-tuned piano reawakened the Edinburgh boy who had entertained his cousins with sound games—and who understood the principle of sympathetic vibration. Alec could not resist demonstrating one of his party tricks to his hosts. Did they know, he asked, that the strings of a piano would repeat a note sung into them? He lifted the lid of the Hubbards' piano, and in a resonant baritone sang into its strings. The strings began to vibrate and to echo Alec's note. Alec explained to his host that the piano would also respond to a telegraphic impulse having the same frequency. Gardiner Hubbard began to get very interested. Did this phenomenon have any value for communication? Alec gave an abrupt laugh as he blurted out his theory: the phenomenon *should* mean that a single telegraph wire could carry several messages at once, if they were all vibrating at a different frequency.

Inventor and Yankee entrepreneur had found one another. Alec was unaware that Mabel's father was on the hunt for a multiple telegraph device; Gardiner Hubbard had no idea that his daughter's teacher spent his nights crouched over a table covered with electromagnets and lengths of wire, or that George Sanders's father, owner of a successful leather business, had already offered to underwrite some of his expenses. Alec had the ideas that Hubbard needed; Hubbard, along with Sanders, had the access to capital to finance them. The Boston lawyer also had the professional qualification to secure patents for any breakthroughs that Alec made.

Soon Gardiner Hubbard was drafting a partnership agreement between Alec, Thomas Sanders, and himself. Gardiner was well aware of the exhaustive and competitive process involved in patent applications, so on November 24 he made a vital suggestion to Alec: "Whenever you recall any fact connected with your invention jot it down on paper as time will be essential to us, and the more things actually performed by you at an earlier date the better for our case. You must not neglect an instant in your work, so that we may file the application for a patent as soon as possible."

The October tea party marked the start of a tumultuous year for Alec. He was determined to fulfill his commitments to the university and his students while pushing ahead with his invention. This meant working all hours of the day and night, causing endless worry to his parents. Melville Bell continued to demand that Alec focus on Visible Speech. Eliza fretted that her son would drive himself to financial ruin: "Does your partnership with these gentlemen bind you to share their expenses in case of failure?" Eliza didn't understand the lure of telegraphy and tried to redirect her son's attention back to what he had always insisted was his vocation: "If the theory of vibration should lead to any clue as to the cause and proper treatment of deafness, it would indeed be a valuable discovery."

When Alec told his parents that he had applied for American citizenship so he could patent his work, he unleashed further anxiety in Tutelo Heights. The Civil War was still fresh in most people's memories, and Eliza fretted that Alec might be summoned for military service. "We do not mind your adopting as your own a country that has sprung from your own," she wrote, "but we shrink from your being compelled to obey the arbitrary laws of that country, even to the sacrifice of your life." Melville Bell asked yet again what this had to do with his own life's work. Alec tried to assuage him: "Should I be able to make any money out of the idea, we shall have Visible Speech put before the world on a more permanent form."

But Alec was now obsessed with winning the race to find a way to send several messages at once by telegraph. The partnership agreement that he had signed with Gardiner Hubbard and Thomas Sanders specified that the two businessmen would put up the money to pay Alec's expenses, including wages for an assistant, and Alec would provide the inventions. They made no provision for additional income for Alec, who was expected to continue supporting himself on his university and teaching income. Alec was too naive to negotiate a better deal. He knew that both Elisha Gray and Thomas Edison were close to unveiling their multiple telegraph prototypes, and that Gray was already playing around with a gadget that employed a metal diaphragm to transmit simple chords. Alec was in a frenzy. "It is a neck and neck race between Mr. Gray and myself who shall complete our apparatus first," he wrote to his parents in November. "He has the advantage over me in being a practical electrician—but I have reason to believe that I am better acquainted with the phenomena of sound than he is—so that I have an advantage there. . . . The very opposition seems to nerve me to work and I feel with the facilities I have now I may succeed. . . . I shall be seriously ill should I fail in this now I am so thoroughly wrought up." Eliza, knowing her son's fragile health, was terrified: "My dear boy, let me . . . caution you to fortify your mind against possible difficulty and disappointment."

Alec frequently caught the horse-drawn tram to Cambridge to visit his benefactor Gardiner Hubbard—and his pupil Mabel. Gertrude Hubbard liked her husband's protégé, and often invited him to join them for midday Sunday dinner, with its unvaried menu of roast beef followed by floating islands—meringue in almond-flavored custard. After the meal, Alec would play the piano. From the piano stool, as his fingers danced over the keys, he would look around at the black walnut furniture, the marble-topped safe, the walls covered in red velvet, all glowing in the light from the gas chandelier, and quietly measure the gulf between this mansion and his own threadbare office. "How well I remember it all," he later admitted to Mabel, "the blazing fire, the comfortable meal, the luxury and love. . . . I looked at [the] happy home much as a friendless beggar looks into the windows of a cheerful room." But he seldom let his mind dwell on the contrast, or on his developing feelings toward Mabel Hubbard. The telegraph race consumed him. After Christmas, he was able to redouble his efforts. He acquired the skilled assistant his partners had promised him: Thomas Watson, a twenty-two-year-old who was employed in a Boston metalworking shop.

Chapter 6

THE FATEFUL "TWANG"
1875

A continual roaring, clanging, and whining echoed through the dingy machine shop at 109 Court Street, Boston. About twenty-five men worked here, under the supervision of the owner, Charles Williams. Bent over metalworking lathes (mostly worked by hand, but Mr. Williams had a couple of engine-driven machines), they manufactured keys, relays, sounders, registers, switchboards, gongs, galvanometers, and a few printing telegraph instruments. These were the simple components required for the meager list of electrical appliances then in use: telegraph equipment, fire alarms, call bells. The loft's brick walls had once been whitewashed, but now only patches of dull white showed here and there through the dirt. Overhead, a tangle of racing leather belts and whirring pulleys hid most of the grimy beams and ceiling. At the center of the room were wooden racks holding steel, iron, and brass sheets and rods. Piles of rough castings littered the floor. In one corner of the room there was a small forge, and a sink where the men washed up. Another corner was partitioned off for the office and salesroom, where finished machines were displayed in glass cases. However, flimsy partitions could not shield customers from the soot that hung thick in the air, the stench of viscous machine oil, or the noise of the metalworking equipment.

There were workshops like this scattered through Boston, where young men could apprentice as electricians and where inventors could get their ideas translated into equipment. But Mr. Williams's shop was known as one of the best. And the most skilled machinist at 109 Court Street was a clean-shaven, taciturn Salem livery stable manager's son called Thomas Watson.

Machine-shop protocol dictated that customers conferred with Mr. Williams about their needs, and once they had placed their orders, Mr. Williams allocated the work among his employees. One day, however, a stranger rushed out of the front office and through the shop to Thomas Watson's workbench. He was, recalled Watson in his autobiography, "a tall, slender, quick-motioned young man with a pale face, black side-whiskers and drooping mustache, big nose and high, sloping forehead crowned with bushy jet-black hair. It was Alexander Graham Bell. . . . He was bringing me two little instruments I had made. . . . They had not been made in accordance with his directions and he had impatiently broken down the rudimentary discipline of the shop by coming directly to me to have them altered."

The instruments in Alec Bell's hands were a practical application of the ideas he had explained to Gardiner Hubbard in the fall of 1874—the transmitter and receiver of his harmonic telegraph. He explained to Watson how the instruments worked on the principle of sympathetic vibrations and, when perfected, would send six or eight telegraph messages over a single wire simultaneously. The transmitter had an electromagnet attached to a reed made of a steel spring, and an adjustable contact screw. The receiver had a similar magnet and reed, but no contact screw. Alec described how the reed of the transmitter, kept in vibration by a battery connected through the contact screw, interrupted the battery current the number of times per second that corresponded to the pitch of the reed. This intermittent current,

Thomas Watson, Alec's first assistant,
was amazed by his new boss's "buzzing ideas."

passing through the telegraph wire to the distant receiver, would set the reed of that receiver into vibration as long as it was tuned to the same pitch as the transmitter reed. Alec theorized that if six transmitters with their reeds tuned to six different pitches were all sending their intermittent currents through the magnets of six distant receivers with reeds tuned to the same pitches, each receiver would select from the mix-up along a single wire its own set of vibrations, and would ignore the rest. The operator at the receiving end could then read the dots and dashes yielded by the receiver. But the instruments had to be tuned with extraordinary accuracy to ensure the separation of each set of vibrations.

Thomas Watson had enough experience of inventors to know that many of their breakthroughs turned out to be utter failures. Most of the self-taught inventors he had met were wild-eyed eccentrics who shouted orders at everybody and treated metal-shop workers like servants. Thomas Edison, for example, had established a reputation within the Court Street shop as a bad-tempered little bully who bilked his backers and annoyed his fellow workers by chewing and spitting tobacco. Watson wasn't surprised to find that Alec's theories were difficult to translate into working prototypes, but he was intrigued by the intense Scotsman. He liked his soft-spoken new boss's "punctilious courtesy to everyone" and his "clear, crisp articulation." As the two men got to know each other better, they often spent evenings together, and Thomas began to appreciate that Alec was very different from the other customers he had met. He was deeply impressed by Alec's table manners (he used his fork, not his knife, to convey food to his mouth) and by the way he could sit down at the piano and immediately begin to play Scottish melodies or Moody and Sankey hymns. "No finer influence than Mr. Bell ever came into my life," he wrote. "The books he carried in his bag lifted my reading to a higher plane. He introduced me to Tyndall, Helmholtz, Huxley and other scientists."

In early 1875, Watson began to work almost exclusively on Alec's harmonic telegraph, usually in a cramped attic above the Court Street machine shop. Alec, for his part, was thrilled to find someone who had the manual dexterity he himself lacked, and who could quietly and accurately translate his scribbled sketches and blotchy instructions into neat constructions. He soon also discovered that Tom Watson was somebody to whom he could confide all his scientific dreams. "His head," recalled Watson, "seemed to be a teeming beehive out of which he would often let loose one of his favorite bees for my inspection. A dozen young and energetic workmen would have been needed to mechanize all his buzzing ideas." One day, while walking along the beach at Swampscott with Watson, Alec noticed a dead gull. He "spread it out on the sand," recalled Watson, "measured its wings, estimated its weight, admired its lines and muscle mechanism, and became so absorbed in his examination that . . . he did not seem to notice that the specimen had been dead some time." Then Alec startled the younger man, who was trying to keep downwind of the rotting corpse, with speculation that someone would soon build a flying machine. Three decades later, Alec himself would try to do just that. In 1875, however, the Scotsman's most stunning revelation to his assistant was his Brantford brainwave: the possibility of an instrument that would allow people to speak directly to each other over the telegraph wires.

In 1875, the capital of the United States was a shabby, unfinished city, with its half-built Washington Monument.

But the "talking telegraph" was not what Alec's backers were paying him to produce, and Gardiner Hubbard was impatient with Alec's tendency to be distracted by his dreams. Gardiner wanted Alec to travel to Washington to apply at the U.S. Government Patent Office for patent protection for his ideas for the harmonic telegraph. This would establish that he was the first inventor to have a workable idea for his invention and would allow him a specified period (usually seventeen years) in which to develop a working prototype. The patent application would be kept secret from other inventors, unless one of them filed a competing application. Alec was reluctant to make the journey: he worried that his ideas were too sketchy and his experimental models too amateurish to submit. The trip would also cost money that he did not have, and would mean missing a week of teaching. But finally, under pressure from Hubbard, he traveled to Washington in late February 1875.

The capital of the United States did not impress its first-time visitor, particularly as compared to Edinburgh or Boston. Ten years after the end of the Civil War, Washington was still just a provincial southern city with pretensions—long unfinished avenues, sparsely settled streets in swampy lowlands. Cows grazed within sight of the Capitol, and cheap stores and hotels lined the muddy expanse of Pennsylvania Avenue. The handful of imposing government buildings made the rest of the city seem even shabbier. The unfinished Washington Monument, on which work had begun in 1848, stood, in Mark Twain's words, like "a factory chimney broken off at the top." Alec completed his patent applications as quickly as possible, but then decided to stay an extra day and meet the illustrious Dr. Joseph Henry, who had become first secretary of the Smithsonian Institution in 1846.

Dr. Henry had been playing around with electricity for half a century. In the late 1820s he had helped to demonstrate that if an electric current is passed through a metal coil, it creates a magnetic field around it. His work with electromagnets had led to the construction of an early type of electric motor, which used an electromagnet to convert electrical energy into mechanical energy. Alec was eager to show his telegraph equipment and explain his ideas to this venerable old warrior. An imposing figure with an unnervingly direct gaze and a clipped manner of speech, the seventy-eight-year-old physicist listened silently as the young inventor explained his approach to telegraphy. He nodded thoughtfully, and suggested Alec set up his "harmonic telegraph" equipment on a table in his office. Then he pulled up a chair himself and bent his ear to the coil so he could listen to the different tones that emanated from the receiver. He seemed impressed, so Alec decided not to stop there. "I felt so much encouraged by his interest," Alec later wrote to his parents, "that I determined to ask his advice about the apparatus I have designed for the transmission of human voice by telegraph."

The elderly scientist was even more impressed with these ideas. He told the eager young inventor that his idea was "the germ of a great invention," and he gruffly advised him to keep working at it. A huge grin spread over Alec's face, but then he remembered his own limitations. Since he lacked the necessary electrical knowledge, he asked Dr. Henry, should he allow others to work out the commercial application of the idea? Memories of his own career must have sprung to the elderly scientist's mind. Although Henry had been the first person to convert magnetism into electricity, it was British scientist Michael Faraday who had earned the credit for it because he was the first to publish his results, in 1831. Forty-four years later, Dr. Henry didn't pause for a minute. If this young Scotsman was going to get the commercial pay-off from his invention, he simply had to acquire an understanding of electricity. "Get it!" he barked at the twenty-eight-year-old.

"I cannot tell you how much these two words have encouraged me," Bell wrote to his parents. But his backers balked. They wanted results on the telegraph experiments that would allow two or more messages to be sent simultaneously in Morse code. The wild notion of transmitting the human voice along the wires still seemed impossible to achieve.

Alec also demonstrated his apparatus to the fearsome President Orton of Western Union. Rumors that Elisha Gray was close to perfecting his telegraph equipment had intensified, and Orton made it plain that he preferred dealing with Gray rather than anyone sponsored by Gardiner Hubbard. Under pressure to win the telegraphy race, Alec once again put aside his theories about transmission of the human voice and devoted all his spare time to the experiments he and Watson were conducting. But the stress of increasing debt and growing discouragement soon resulted in headaches and sleepless nights. It was coming up with breakthrough ideas that he enjoyed, not the endless fiddling and fine-tuning they required before they could be taken to market. "I am now beginning to realize the cares and anxieties of being an inventor," he wrote home gloomily. He felt himself becoming almost unhinged by the pressure. Watson also began to flag as the multiple telegraph refused to perform consistently. He watched Alec lose confidence and work in silence instead of spurring them both on with his accustomed battle cry, "Watson, we are on the verge of a great discovery!"

There was something else unsettling Alec during these months. He was dimly conscious that his feelings toward Gardiner Hubbard's daughter were undergoing a sea change. He rarely gave her voice lessons these days, but his business relationship with her father meant frequent visits to the Hubbards' generous home in Cambridge. He began to see Mabel, now seventeen, in a role other than that of an eager pupil. Perhaps he was suddenly vulnerable to gusts of sentiment because the excitement of the multiple-telegraph race had brought him to an emotional boil. Or perhaps it was because Mabel was no longer a child: she was a confident, good-looking young woman with her hair in an elegant chignon and her shapely figure molded by very adult corsets. Often when Alec spent the evenings in Brattle Street he could not take his eyes off Mabel as she sat on the floor, leaning her head back against her mother's knees so that Gertrude could play with her daughter's thick hair and stroke her arm.

In late spring, Mabel hosted a small soirée at 146 Brattle Street. "I wish you could have seen her," Gertrude wrote to Gardiner, "so fresh, so full of enjoyment and so very pretty. She wore her peach silk and looked her loveliest." She received her twenty guests "with the greatest ease and self-possession," and when Alec, who had been invited to provide the music, sat at the piano stool, Mabel "led off in a waltz with Harcourt Amory. He is a nice pleasant looking fellow and . . . one look at her face told how happy she was." As Alec dutifully stayed at the keyboard, switching from Strauss to Chopin, Mabel twirled through waltzes, gallops, and lancers. When oysters and ices were served at ten, Alec remained quietly in the background, watching men several years younger than he was pay court to Mabel. At the end of the evening, he "finished his kind offices by seeing our friends safely into the cars," reported Gertrude to her husband. Alec may have assumed the part of a stuffy older brother, but his heart was reaching toward another role.

As Alec's visits to Brattle Street continued, fourteen-year-old Berta and twelve-year-old Grace Hubbard realized, with the wicked insight of younger sisters, that Mabel's elocution teacher had fallen for her. They regarded this tall, earnest man, with his hollow cheeks, uncombed hair, and dark eyes, as far too old and serious for their sister (he must be at least *forty*). His manners were too formal, his speech too punctilious, and besides, he wasn't a Yankee. Various McCurdy cousins joined in the teasing of Mabel's secret suitor. "We were all anxious to teach the young Scotchman the latest American customs and slang," recalled Mabel's cousin Augusta. One day, Alec mentioned that he was going to stay at the Massasat House hotel in Springfield, Massachusetts. When he was advised that he had to try the waffles served at this famous New England establishment, he looked blank: he had never heard of waffles. "Try them with oil and vinegar," a giggling fifteen-year-old suggested.

Mabel, meanwhile, saw Alec as nothing more than her father's business associate and her own teacher (although her lessons were usually with Miss Locke). She had been surprised to discover that he was only twenty-eight, and when she sighted the tall figure striding purposefully down Brattle Street, it was hard to imagine that such a serious man nurtured romantic hopes. He was wrapped up in science; he seemed to relax only when Mabel's mother urged him to sit down at the piano and play some Scottish airs. He was censorious of the way that Mabel's sisters and cousins spoke—Yankee voices, he insisted, were "harsh and nasal," and he was sure they could improve their diction if they would only let him help them. And he was obsessed, despite endless setbacks, with getting his complicated sound machines to work.

Alec himself, consciously or unconsciously, was trying to suppress his own emotions. An offhand remark in one of his mother's letters, about the possibility that he had a mystery girlfriend, prompted a rare burst of self-analysis: "I do not think I am one of the marrying sort. When there is anything of the kind on the tapis [carpet] I shall consult you *first*. I am too much given over to scientific experimenting and to my profession to think of anything of that kind yet and I *doubt very much if I ever shall*. I say this not in joke but in earnest. If I ever marry at all it will not be for years yet."

Through April and May, there was no time for romantic daydreams. Alec toiled away, both in the Sanders home in Salem and in the Court Street attic in Boston, trying to put his multiple-telegraph equipment into commercially acceptable working order. He was frustrated by his lack of technical education in electricity, although he tried to reassure himself with the knowledge that "Morse conquered his electrical difficulties although he was only a painter, and I don't intend to give in either till all is completed." Each day, before striding off to fulfill his Boston University duties, he gave to the loyal Watson notebooks filled with sketches. Shortage of money and the problems of interference of frequencies and faulty transmitters bedeviled him, but he did not dare ask Gardiner Hubbard for a further investment. He was forced to borrow from Watson, whose weekly wage of $13.25 was just over half what Alec earned from his university duties. And though he knew that his backers wanted him to focus on the invention that might challenge the Western Union monopoly, his thoughts kept straying back to the electrical transmission of human speech. "In spite of my efforts to concentrate my thoughts upon multiple telegraphy," he recalled later, "my mind was full of it."

As summer approached, the temperature rose in the attic workshop, and blueprints, electromagnetic coils, tuning forks, and steel rods accumulated on the wooden workbench. Next to the attic's grime-encrusted arched window, oblivious to the quiet hiss of an overhead gas lamp, Alec crouched over his latest prototype for the telegraph transmitter. He was trying to tune the receiver reeds. One minute he would summon Watson to help him fix a detail, the next he would send him to the next room to see if a receiver there was working. The breakthrough came on June 2, when the two men, sleeves rolled up and foreheads glistening in the oppressive heat, tested out a trio of transmitters and receivers. As Watson adjusted the spring of one of the receivers in the adjoining room, Alec realized that the corresponding transmitter in his laboratory had spontaneously begun to vibrate. He put his ear next to it, and heard a distinct "twang." He dashed next door to see what Watson was doing to cause the vibration. Watson explained that he had plucked the spring. "Keep plucking it!" Alec snapped, as he returned to the laboratory.

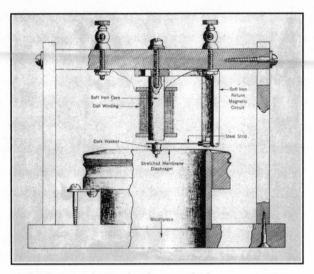

The first membrane diaphragm telephone, June 1875.

He had immediately grasped the import of that "twang" for the "speaking telegraph" that had continued to haunt his imagination. The plucked spring was inducing the wave-shaped current he had postulated the previous summer, in Brantford. During the intervening months he had been stymied in developing this intuition because he had believed that only a substantial continuous electric current, generated by an external battery, could convey the complex harmonics that constitute human speech. Now he realized that his assumption was wrong: a weak alternating current, induced by the vibrating spring, had transmitted a precise sound from one room to the next. As the spring vibrated, it changed the width of the air gap between the spring and the end of iron core of the electromagnet. Varying this air gap changed the magnetic field in the coil, which in turn induced an electric current that corresponded to the motion of the vibrating spring. Alec and Tom Watson worked far into the night, repeating the discovery with all the steel reeds and tuning forks that they could find. "Before we parted," recalled Watson, "Bell sketched for me the first electric speaking telephone, beseeching me to do my utmost to have it ready to try the next evening." With several refinements, these sketches were the designs for the instrument that would eventually become the first telephone.

That night, Bell wrote a quick note to Hubbard: "I have succeeded today in transmitting signals without any battery whatever!" and added that he thought this was a more important direction in which to proceed than the multiple telegraph. He wrote to his parents in a similar state of euphoria: "At last a means has been found which will render possible the transmission . . . of the human voice." He told them he hoped to have an instrument "modelled after the human ear, by means of which I hope tomorrow (but I must confess with fear and partial distrust) to transmit a vocal sound. . . . I am like a man in a fog who is sure of his latitude and longitude. I know that I am close to the land for which I am bound and when the fog lifts I shall see it right before me."

When Alec arrived at the Court Street workshop the following evening, Watson triumphantly produced the apparatus he had sweated over all day. The first telephone was a simple mechanism, consisting of a wooden frame on which was mounted one of Alec's harmonic receivers—a tightly stretched parchment drumhead to the center of which was fastened the free end of the receiver reed. The new element was a mouthpiece arranged to direct the voice against the other side of the drumhead.

Alec connected this apparatus by wire to another harmonic receiver in the metalworking shop downstairs, now empty of the men who kept it buzzing all day. An early experiment seemed to fulfill his hopes: when Alec shouted into the first prototype transmitter, Watson came dashing upstairs to report he had heard his boss's voice over the wire and could almost make out the words. But when the two men switched places and Watson spoke into the transmitter, Alec heard only a babble of unintelligible sounds. His heart sank. He leaned back in his chair, brooding on this frustration and realizing that he might have to spend hundreds more hours at his workbench doing what he least enjoyed—making laborious and minute adjustments to delicate membranes and lengths of wire.

Further discouragement came from Gardiner Hubbard, who arrived at Court Street to inspect the diaphragm device. As he climbed the grubby, narrow staircase to the attic in which Alec and Watson worked, he grunted with effort and disapproval. He was deeply in debt and badly needed a commercial success. Orton, of Western Union, had announced his conviction that Alec's rival, Elisha Gray, would win the multiple-telegraph race. And, the irrepressible Thomas Edison was telling everybody within earshot that his breakthrough in multiple telegraphy was imminent. Hubbard was painfully aware of Orton's personal vendetta against him since he had tried to challenge the Western Union monopoly in Washington the previous year, but he was also confident that Orton was too good a businessman not to grab the first commercially viable telegraph invention produced, whoever its inventor. Preoccupied with these concerns, Hubbard was visibly skeptical as he examined his protégé's vibrating diaphragm. "I am very much afraid," he announced, "that Mr. Gray has anticipated you in your membrane attachment." Alec could not convince him that it was anything more than a distraction from the more important pursuit of the multiple telegraph. Gardiner urged Alec to expend his energies on the multiplex telegraph. *That,* he declared, was the "ne plus ultra" of communication technology.

Alec was almost paralyzed with despair. For the previous year, he had been working night and day to fulfill his teaching obligations and pursue his research interests. His financial worries, the dripping heat of a Boston summer, and his frustrations with the telegraph experiments had already taken their toll. But now he suffered two further blows. First, Watson fell ill with typhoid, further impeding progress on either the harmonic telegraph or the telephone. And then Gardiner Hubbard casually mentioned that Mabel was leaving Cambridge in a couple of weeks and would be out of town for the entire summer.

The news of Mabel's departure forced Alexander Graham Ball to confront the feelings he had repressed for months. The idea of Mabel spending weeks so far removed from Cambridge horrified him: he realized how much he had come to depend on their casual conversations, either in the Hubbards' mansion or in his rooms on Beacon Street. He admitted to himself that, these days, he daydreamed for hours about a time when he might take that "fair young girl" into his arms, a time "when I should feel those soft white arms around my neck or dare to touch that glorious hair." His emotions boiled over. He penned an anguished letter to his parents: "It is now more than a year ago since I first began to discover that my dear pupil, *Mabel Hubbard,* was making her way into my heart. . . . It has been my object up to the present time to conceal from her—from you—and from myself how deep has been the interest I have been taking in her welfare. I say this that you may understand that the sentiments so recently expressed to you on the subject of marriage were *true* when they were written." The news that she was going on an extended visit to Nantucket, the idyllic island thirty miles off the south coast of Cape Cod, destroyed his self-imposed charade. "It was useless striving longer against Fate—I must either declare myself or lose her. I did not know what to do."

So he declared himself, in a letter to Mabel's mother. On June 24, 1875, three weeks after the coiled spring of the transmitter had uttered its fateful twang in his attic workshop, Alec sat at a desk in 292 Essex Street, the Sanderses' Salem house, and stared at a blank sheet of paper. He dipped his pen's metal nib into the ink and watched the liquid drip back into the inkwell. He scratched his beard nervously. His hand shook and his mind raced as he searched desperately for the right words. He had more than enough trouble expressing himself when he wrote accounts of scientific experiments; describing his own emotions was almost beyond him. Finally he began to write. But there would be many false starts and discarded drafts before he perfected his opening paragraphs:

Dear Mrs. Hubbard,

Pardon me for the liberty I take in addressing you at this time. I am in deep trouble, and can only go to you for advice.

I feel sure that whatever answer you may vouchsafe to this letter, your sympathies at all events will be with me.

I have discovered that my interest in my dear pupil, Mabel, has ripened into a far deeper feeling than that of mere friendship. In fact I know that I have learned to love her very sincerely. Could you suspect how desolate the announcement of her early departure from Cambridge has made me, you would indeed pity me.

Chapter 7

Nantucket Passion
1875

Gertrude Hubbard sat down and reread Alec's untidy scrawl, her eyes wide with astonishment. Then she stared out of the library window at the foxgloves, larkspurs, and bleeding hearts in her Brattle Street garden and considered her response. She had suspected that Alec was sweet on Mabel, but she had no idea that he harbored feelings of such intensity, such passion. She was suffused by a wave of protective feelings toward her seventeen-year-old daughter: Mabel might have the demeanor of a sophisticated Brahmin debutante, but Gertrude knew she was still young for her age and very dependent on her family. Although she had never been allowed to feel marginalized by her hearing loss, she was rarely alone among strangers. Away from Cambridge's familiar turf, she was at risk—of runaway street vehicles, because she could not hear shouts of warning; of house fires, because she could not hear alarms.

The day before Alec's letter arrived at Brattle Street, Gertrude Hubbard had invited him to listen to the speeches and watch the fireworks with her and her daughters at Harvard University's Graduation

Day festivities. Alec would be arriving any minute, anxious to hear Gertrude's reaction to his declaration of love and to see Mabel before her departure. "It is my desire," his letter had continued, "to let her know now how dear she has become to me, and to ascertain from her own lips what her feeling towards me may be. Of course I cannot tell what favour I may meet with in her eyes. But this I do know, that if devotion on my part can make her life any the happier, I am ready and willing to give my whole heart to her."

Alec ended his letter of entreaty to Gertrude Hubbard, "I am willing to be guided entirely by your advice, for I know that a mother's love will surely decide for the best interests of her child." But Gertrude was too astute a woman to believe that someone as highstrung as Alec would always do as he was advised. Much as she liked this impulsive young Scotsman and admired his talents, she deplored his tendency to work himself into terrible states over his experiments and his teaching, and, now, over Mabel. He was so gauche. In New England, gentlemen didn't behave like that. And there was so much she didn't know about him. What was his religion, for example? The Hubbards were well-respected members of the local Presbyterian church; only a suitor of sturdy Protestant faith would be acceptable for any of her daughters.

Alec, his hair slicked down with brilliantine and his hands trembling, mounted the steps on that sticky summer day, and once again rang the doorbell. The maid showed him to the sitting room where Mrs. Hubbard, beautifully groomed as always, with her fine white hair neatly pushed under a cap, awaited him. She could not have been friendlier—or firmer. After Alec had perched himself on the velvet-upholstered ottoman, she gently told him that, at seventeen, Mabel was too young to know her own mind, and should be given the opportunity to enter into "Society" and meet other young men. She asked Alec to wait a year before he spoke of his feelings to her. "She promised me," Alec wrote in his regular letter home to Brantford, "that I should have opportunities of seeing Mabel and said that if my mind was still the same at the end of a year I might speak to Mabel herself."

At twenty-seven, Alec Bell was intense, shy, and overwhelmed by passion.

By now, there were sounds of giggles and chatter outside the parlor's closed oak door. Gertrude's four daughters were eager to walk over to Harvard Yard and watch the sons of America's elite take their degrees. There would be so much to see on the university's leafy grounds! Some of the most distinguished professors on the continent, including Charles Eliot Norton and William James, would circulate through the crowd. Local celebrities like poet Henry Wadsworth Longfellow and author Henry James might make appearances. The cream of New York society would be there, twirling their parasols as their offspring swaggered through the quadrangles for the last time. The parade of Fifth Avenue fashions, provided gratis by Social Register families, would provide conversational fodder for weeks.

As Alec walked down Brattle Street with the Hubbard ladies, he managed to subdue his nerves sufficiently to start a conversation with Mabel. In Harvard Yard the afternoon sun cast a rosy glow on the old red brick of the dormitories, and the two young people chatted between the various graduation orations. Alec was captivated, as usual, by Mabel's sweet smile and serious manner. But this was the first time

Mabel's father, Gardiner Greene Hubbard, was horrified by Alec's impulsiveness.

he had conversed with her as a suitor rather than a teacher. His feelings overflowed: he was almost incoherent with self-consciousness. As she kept her eyes trained on his lips, paying earnest attention to what he was saying, she seemed to him so much more mature and intellectual than her lively younger sisters. He felt, he told his parents, that "she evidently took some pleasure in my society." He could see "that she had not been informed of what had passed between her mother and myself [and] was perfectly unconscious of what was passing in my mind."

Gardiner Hubbard had been in New York when Alec had declared his feelings about Mabel to Gertrude. When Mabel's father came home two days later and heard what was going on, he was much less sympathetic toward his protégé than his wife had been. Alec noted tersely, in a diary, that Hubbard "thought Mabel much too young. Did not want thoughts of love and marriage put into her head. If Mrs. Hubbard had not said *one* year, he would have said two. . . . Did not think I was ready to be engaged. No objections personally."

The Hubbards were far from hostile to Alec. Gardiner Hubbard knew he could not afford to alienate his daughter's suitor. Alec represented his best hope of beating Orton to a viable multiple telegraph, and thus to repairing his own fortunes. Mabel's father was torn between his view of his daughter's best interests and his support for Alec's scientific endeavors. "He said he had felt quite an affection for me from the very first time he saw me," Alec told his parents. "That this had led him to offer his assistance in regard to the 'Telegraph'—that he believed I had great talents—and that his object in aiding me in the Telegraphic Scheme was not alone a speculation on his part, but in the hope of encouraging me to devote myself to science."

According to Alec's diary for this anguished period, Mabel's deafness was never mentioned in the many intense conversations between Alec and her parents. Instead, like his wife, Hubbard wondered about Alec's emotional stability. And despite his ostentatious belief in the principle of meritocracy, he knew that the Bell family was, by Boston standards, rather shabby. How could this impoverished young man, who had no family money or status, support Mabel? A little caution was the least that Gardiner Hubbard asked.

Mr. Hubbard's haughty remarks dampening Alec's hopes were delivered in the Brattle Street parlor. To soften the blow, Hubbard suggested Alec join the ladies in the garden. Alec stepped out of the French windows into the warm evening air and saw Mrs. Hubbard's canna lilies glowing in the dusk. There he found Mabel and suggested they take a quiet stroll down one of the gravel walks. He meant to keep silent, as the twilight was too dim for Mabel to see his lips clearly, but with Mabel's hand resting on his arm he could not prevent a few awkward words escaping.

Then Mabel's sister Roberta and cousin Mary Blatchford appeared. Under Cousin Mary's caustic gaze, mischievous Roberta began to play "he loves me, he loves me not" with some of Mrs. Hubbard's asters. The flower she selected for Alec came out, to her delight, as "Love." In his private diary, Alec recorded how Mabel innocently asked him of whom he was thinking. Alec stared helplessly at her, unable to speak. Berta and Mary burst out laughing and said *they* knew who was in their guest's thoughts. Alec was left suffused with embarrassment, and wondering who would spring to Mabel's mind in similar circumstances. They all retreated to the brightly lit veranda, and he heard himself blurt out, "If you could choose a husband, what would you wish him to be like?" Mabel turned her large gray eyes on him in confusion: her teacher's inquiry seemed indelicate and impertinent. She feared he was mocking her. Alec winced at his own faux pas. In his diary he wrote, "Betrayed my feelings—fear Mabel may be laughed at on account of my foolish proceedings."

But at least he had spoken out. He no longer felt like an emotional pressure cooker about to burst.

On July 1, Mabel left as planned with her McCurdy cousins, to spend the summer under the watchful eye of Mary Blatchford on Nantucket. With its clapboard architecture, rambling roses, and history as a whaling center, Nantucket was a favorite vacation spot for wealthy Bostonians. Cousin Mary, now in her late twenties, was a woman in the mold of Olive Chancellor, the character described by Henry James in *The Bostonians* as "a spinster as Shelley was a lyric poet, or as the month of August is sultry. There are women who are unmarried by accident, and others who are unmarried by option; but Olive Chancellor was unmarried by every implication of her being." Mary Blatchford could be relied upon to see that the young people behaved themselves. Like Olive Chancellor, she "knew her place in the Boston hierarchy" and had carefully honed instincts about who would—or wouldn't—"do." No disreputable beaux would cross her threshold.

Alec planned to remain in Boston to pursue his telegraph experiments. It was an awful month. He remained isolated and fretful because he felt too unsettled to pay a call on the Hubbard household in Cambridge. His assistant, Watson, was sick. He continued alone with his telegraph experiments, but he could not concentrate and he was too clumsy to make the minute adjustments at which Watson excelled. He felt himself becoming almost unhinged by the tension and apprehension. Only the elderly Mrs. Sanders was at hand in

Salem to comfort the distracted young Lothario. Her lodger's growing agitation convinced her that he was suffering from "brain fever brought on by telegraphy."

In late July, Alec could bear the tension no longer. He summoned his courage to return to Cambridge. As he noted in his diary, "Called on Mr. and Mrs. Hubbard. Told them that I could not help myself. Should tell Mabel what was in my mind if they gave me the opportunity. Thought she *should* know it. I could only avoid telling her by avoiding *her.*"

The Hubbards were horrified as Alec, quivering with emotion, blurted out his feelings. They had hoped he would calm down, dutifully devote himself to science, as Gardiner had suggested, and suppress his feelings for Mabel for at least another year. Now here he was, striding about in their parlor, running his long fingers through his lank black hair, even more frenzied than he had been four weeks earlier. The intensity of his passion shook them, especially when Alec announced he would go to Nantucket to speak to Mabel himself. Such a step, they insisted, would startle and distress their daughter. Alec must take control of himself and wait until Mabel returned to Cambridge in early August. Once she was reunited with her mother, Alec might speak to her. A chastened Alec dutifully returned to Salem and tried to abide by the Hubbards' advice.

In Nantucket, Mabel had now been under the wing of bossy and overprotective Mary Blatchford for four weeks. Cousin Mary had divined the awkward Scotsman's feelings, and she disapproved. She didn't like Alec's shiny black suit and lack of pedigree. She also realized that Mabel was unaware of his inclinations. So as she and Mabel, clad in baggy knee-length bathing dresses and rubber bonnets, plunged into the icy Atlantic, she decided to meddle. She casually asked Mabel how she felt about Alec. She encouraged her impressionable cousin to denigrate her teacher. She reported to Mabel's parents, "I . . . more or less told her she did not care for him, nor did she seem to care." In particular, Mary made fun of Alec's conduct in the garden. Like any vulnerable young woman, Mabel felt that Mary was ridiculing her as well as Alec, and she instinctively shrank from the incident. Coached by Mary, she confided to her mother in her next letter home that she really *disliked* her former tutor, particularly his long greasy dark hair and black eyes. Blue eyes, she insisted, were more to her taste.

In an effort to dampen Alec's hopes, Mrs. Hubbard summoned him to Cambridge and read him passages from Mabel's letter. Although Gertrude meant well, it was a cruel thing to do, and it sent Alec into a fury of despair. "Don't know what is best to do—quite distressed about it—as such dislikes have a tendency to be permanent," he noted in his diary. His head ached, and he tossed and turned through the sultry August nights. As he guilelessly explained to Gertrude Hubbard in yet another emotional letter, "I do not know how or why it is that Mabel has so won my heart. It is as great a mystery to me as it must be to you. . . . I should probably have sought one more mature than she is—one who could share with me those scientific pursuits that have always been my delight. However, my *heart* has chosen." He desperately wanted to explain his feelings to Mabel himself, to tell her "all that is in my heart, and let her deal with me as she chooses."

But then Alec had to undergo a further turn of the screw. Another letter from Mabel arrived at the Hubbards' home, and Mrs. Hubbard felt obliged to summon him back to Brattle Street again so she could read yet more extracts to him. Mabel herself was now in a spin. Manipulative Mary had continued her campaign: she had told Mabel that Alexander Graham Bell wanted to marry her and had already spoken to her parents. Mabel was shocked, and her letter reflects her own youth, her confused feelings, and her dependence on her mother for advice. Yet there is also an undercurrent of independence in her response, and a belief in her right to be informed about matters concerning her. Her letter reveals a much more stable temperament than Alec's.

> My darling Mamma,
>
> . . . I think I am old enough now to have a right to know if he spoke about [marriage] to you or Papa. I know I am not

much of a woman yet, but I feel very very much what this is to have as it were, my whole future life in my hands. . . . Oh Mamma, it comes to me more and more that I am a woman such as I did not know before I was. I felt and feel so much of a child still so very young I had no idea of marrying or being sought in marriage yet a while . . . I . . . feel more and more how unfitted and unworthy I am of such a position as a woman and wife. Of course it cannot be, however clever and smart Mr. Bell may be, and however much honored I should be by being his wife I never never could love him or even like him thoroughly. But this does not lessen the feeling of responsibility it gives me—even awe. Oh, it is such a grand thing to be a woman, a thinking, feeling, and acting woman, to be thought fit to be the life's companion and mother of the children of a man. But is it strange I don't feel at all as if he did it through love. . . .

I wish I had you here that you might speak to me sensibly and clearly and tell me all how I ought to feel. . . . You need not write about me accepting or declining this offer if it should be made. I feel almost as if I'd die first. . . . I must hear from you to clear me up, I feel so misty and befogged I don't want to think or feel things I ought not. . . . Help me please,

Your own, Mabel.

Mabel's letter convinced Alec he *had* to talk to Mabel himself. He strode back and forth on the Hubbards' parquet floor as he overruled Mrs. Hubbard's objections to such an impulsive gesture. "I said I was *ill*," Alec noted. "Further delay and anxiety would entirely unfit me for anything." Gertrude tried to persuade him to return to Canada and calm down, but Alec was now fixated on rushing off to Nantucket and pressing his suit personally. "I feel it is my duty to *go to her,* instead of waiting till she returns home. It would be treating her as a *child*

Nantucket's famous whaling fleet.

to think it was *necessary* for her to be near her mother in order to receive my explanations. The letter which was read to me . . . was not the production of a girl, but of a true noble-hearted woman, and she should be treated as such."

The Hubbards were not impressed by Alec's frenzied behavior. "You go against the wishes of Mrs. Hubbard and myself," Gardiner wrote sternly, "being carried away by the same unmanly feeling which marked your course on a former occasion. You will regret this new burst of passion."

But the burst of passion overwhelmed Alec, who was now way, way beyond regret. On Saturday, August 7, he stood on the platform at Boston station, waiting for the 11:15 train to Woods Hole, on the southwestern corner of Cape Cod. The tall, unkempt Scotsman, alone and carrying no baggage, must have attracted curious glances from the noisy clans that had emerged from Boston's tenements to set off with sunhats, spades, and swimwear for a weekend on the cape's sandy beaches. But he was oblivious to their stares as he tapped his foot on the ground and repeatedly looked at his watch. His mission consumed him: he was going to make the laborious journey toward his beloved by train and two steamers.

Alighting from the train at Woods Hole, he took a steamer to Oak Bluffs on Martha's Vineyard. But when he arrived there in mid-afternoon, he discovered to his chagrin that the next steamer to Nantucket did not leave until evening. For four hours, tense with determination, he paced the wharf, impatiently watching yachts, dinghies, dories, fishing smacks, and pleasure craft mill around in the sheltered waters of Vineyard Sound. By the time he reached Nantucket, it was almost midnight. He checked into the Ocean House Hotel, lay on the bed, and did not sleep a wink all night. "Unfitted for anything," he scrawled in his diary.

The following day, his spirits sank still further. He discovered that he was trapped. Siasconset, where Mary Blatchford had her cottage, was eight miles away on the other side of the island and there was no public transport on a Sunday to take him there. Moreover, outside his hotel window a fierce summer storm raged, and steamer service to the mainland was canceled. Feeling utterly defeated, Alec sensed the approach of a severe headache. He gloomily imagined what might happen if he *did* manage to see Mabel: "Her eyes would be full of tears and she would not understand what I should say." His life, he felt, had been full of bad judgments and clumsy blunders. His head throbbed. He risked losing the young woman he loved before he had even talked to her, and at the same time losing the patron, her father, on whom he was desperately dependent for funds.

Alec sat at the end of the bed in the cramped old hotel and stared out of the window at driving rain and lashing branches. There was only one solution, he decided. He must put his feelings on paper.

> Dear Miss Mabel,
>
> I have come to Nantucket in hopes that you will see me, and let me tell you all that I long to say. . . . You [do] not know, Mabel, you [are] utterly unconscious, that I [have] long before learned to respect and to love you. I have loved you with a passionate attachment that you cannot understand and that is

> to myself new and incomprehensible. . . . I . . . wish to make
> you my wife, if you would let me try to win your love. . . .

As dusk fell and the storm abated, Alec's pen scratched on and on. He explained that he had spoken to her parents, that they felt she was too young to be confronted with his feelings. He wrote of how he believed that she would be comforted to know there was someone who wanted to devote his life to her happiness, and how he did not think she was too young to handle the situation. He confessed that a year's wait was impossible because he could no longer suppress his emotions. He feared that Mabel did not altogether trust him, because she did not know what was going on. "You do not know, you cannot guess, how much I love you—how much I desire to have the *right* to shield and protect you. Had I *your* love I feel that you would mould my life into any form you will." He made no mention in the letter of her deafness, which was utterly immaterial to him.

By now, darkness had fallen, the rain had stopped, and the lights of the boats in the harbor were twinkling. Still Alec continued to write, with blots and crossings-out littering the pages, explaining how he respected Mabel's parents but felt they were wrong to try and keep him away from her. "I know you are surrounded [here] by young companions who would ridicule and annoy you did they know I was here. Believe me, I have not come to Nantucket to wound and pain you. I respect and honour you too much for that."

Finally, as the first hint of dawn touched the sky, Alec signed off: "Whatever may be the result of this visit, believe me, now and ever, Yours affectionately, A. Graham Bell."

Early the following morning, Alec set off to deliver his letter. The gentle Nantucket landscape had no charm for him as he bounced along the sandy track to Siasconset in an uncomfortable pony cart driven by a local boy. He scarcely noted the occasional glimpses of gleaming white sand, or a summer sea now slumbering in the hazy August light. "Long dreary drive. Flat country. No trees," he noted. When he arrived at Siasconset, he found a hotel and wrote a note to Mary Blatchford announcing his arrival. He told her he would remain at the hotel all day in the hope that he might call on Mabel but would leave if she preferred not to reply. He did not have to wait long for a summons to arrive.

What did Mary Blatchford make of Alec, who had the wild-eyed look of someone who had not slept for several nights? She certainly gave him little cause for hope, as she coldly informed him that Mabel was extremely distressed at the mere idea of seeing him. Alec meekly acknowledged that he could scarcely expect to see Mabel, given the circumstances. He simply handed Mary an envelope thick with his literary outburst of the previous night to give Mabel, then turned to take his leave.

Even Mary was taken aback at Alec's whipped-dog manner. It is possible that, being the busybody she was, she had not mentioned to Mabel that her suitor was here. Now she relented a little and told Alec that he could spend the afternoon on the beach with Mabel and her cousins. But this sounded worse than the moonlit evening five weeks earlier in Cambridge—mischievous young women smirking at the sight of "poor Mabel" with her awkward old Scottish teacher. Alec declined the invitation: "Miss Blatchford looked very much surprised at my being willing to go without seeing Mabel." Now that he had fulfilled his mission, however, he couldn't get out of there fast enough. He was back at the steamer dock to catch the 1:15 p.m. boat direct to Woods Hole, and he reached Boston at 7:45 p.m. That night, he confided to his diary, he "[s]lept for the first time for *ten days*. Not rested since July 29th when I first heard of Mabel's dislike."

Alec's assessment of his sweetheart's character was accurate. A few days later, he received a graceful note from her:

> Dear Mr. Bell:
>
> Thank you very much for the honourable and generous
> way in which you have treated me. Indeed you have both my
> respect and esteem. I shall be glad to see you in Cambridge
> and become better acquainted with you.

I am sure that even if my feelings remain unchanged by
another year it will always be a pleasure to look back upon
your visits and friendship.

I thank you for your promise not to approach this subject
again until I have had more time to know you and myself,
and for the delicate regard for me that prompted it, as well
as for your assurance that you are no longer suffering on my
account. . . .

> Believe me,
> Gratefully your friend,
> Mabel G. Hubbard.

Despite her inexperience, Mabel was already a woman of dignity
and judgment—and with a steadier temperament than Alec.

At the end of August, Mabel returned to Cambridge, ready to receive
her suitor. When he arrived one evening at Brattle Street, she quietly
invited him to accompany her into the greenhouse, where there was
no danger of being overheard by her parents and sisters. There the
clear-headed seventeen-year-old told her teacher that she liked him
but as yet she did not love him. She did not, however, extinguish his
hopes that such a love might develop. She watched Alec's lips quietly
and sympathetically as he poured out his own anguish of the previous
two months. That evening, as Alec took the streetcar back into Boston
and then the last train to Salem, he felt his heart almost burst with joy.
As soon as he reached home, he sat down and reached for a pen. "This
evening is the happiest of my life," he scribbled to "Miss Mabel." "The
clouds that have long been hanging round my heart are all *dispelled.* I
feel that I may speak to you as freely as I *think*—and you will hear me."
He was almost certain that Mabel would learn to love him.

There was, though, one loose end that Alec knew he had to tie
up. He must tell his parents about the new state of his relationship
with Mabel.

He had unburdened himself of his passion for Mabel in a letter to his parents on June 30. He had described her as beautiful, accomplished, and affectionate, and as belonging "to one of the best families in the States." But he acknowledged that his parents might have reservations about her: "Her deafness I felt to be a great bar. . . . I felt that such an attachment would grieve you very much when you should discover how sadly she had been afflicted." Recalling the sarcasm with which Melville dismissed earlier confidences (his affection for Marie Eccleston in London, his decision to learn sign language), he begged his father to treat his emotions gently.

Each day, Alec looked for a letter postmarked Canada; each day, as nothing arrived, his emotional turmoil increased. The silence convinced him that his parents disapproved of the match. Then he received a brief note from his mother that shook him to the core: "You are of course the best judge of the sort of person calculated to make you happy, but if she is a congenital deaf mute, I should have great fears for your children."

Why did Eliza, who was deaf herself, react like this? In particular, why did she use the derogatory label "deaf-mute"? Was she jealous of the new woman in Alec's life? Perhaps she was only trying to sound a note of caution because she knew Alec's inclination to be impulsive. As her own hearing had deteriorated, she was increasingly entombed within her disability; she had never learned to read lips or to use sign language. Alec had not mentioned that Mabel's hearing loss was the result of a childhood infection, and Eliza knew that women who were congenitally deaf were likely to have deaf children. Her son's suggestion the previous June that Mabel's "affliction" might be considered a "great bar" must have set off alarm bells. But Eliza was more than keen for Alec to find a woman who would make him happy and give some stability to his life. "Remember," she had written in April, "we had much rather you would marry a woman than science."

Eliza's comments, particularly the phrase "deaf-mute," wounded her son deeply, and he exploded with wrath. "You would be justly hurt and indignant were any one to allude to *you* in that way," he replied to Eliza, "and certainly so would she." His father's lack of response had enraged him even more: "He has treated it with *entire silence* as if the matter was not worth alluding to at all. . . . You and he *should* know that such a subject as that is, to a young man, of the deepest and most vital importance. The step I proposed taking was to affect my *whole after life,* and I feel that my communication deserved to be treated with more consideration than has been shown it."

Melville and Eliza's silence had so upset Alec that in late August he confided his dismay to Mabel's mother. He announced that he had no intention of writing again to his parents, or of visiting them, until they had properly acknowledged his June letter. By now, Gertrude Hubbard knew that Alec was a highly emotional young man rather than the staid middle-aged Scot the Hubbards had first taken him for. She also may have suspected that Alec's doting mother feared she would lose her one remaining son should he marry. She urged Alec to write again to his parents. He did, but it was a grudging letter: "Should this also be received with the same shameful neglect as my last I feel that there is danger of a complete alienation of my affections from home."

In fact, there was a simpler explanation for the Bells' apparent neglect. Melville's brother David and his family had immigrated to Canada and joined them in Brantford, so their social life had become even more frenetic. David's arrival had meant that Eliza and Melville could entrust the farm to his care, and in early July they had departed on a trip to the Adirondacks. "We took the steamboat for a cruise on Lake George and spent two nights on a rugged Scotland-like hillside in a tiny hotel," Eliza wrote in late July, blithely unaware of her son's feelings. When Alec's grudging letter finally arrived in Brantford, the misunderstanding and his tone of resentment horrified his parents.

"Poor fellow, how can you for a moment doubt our sympathy?" replied Eliza, who went on to explain that their silence was due to his request for discretion, particularly as he had never said whether the

Hubbards were prepared to accept him. "You being now a full man, your own master, and the best judge of what is fitted for your happiness, we should never have thought of biasing you. . . . We had no wish in the matter adverse to your own, for your happiness would make ours." They certainly had no objection to Mabel's deafness: "[I]t would be ridiculous for us to object to your following your Father's example. . . . There is no reason why you and a wife who is deaf should not be as happy as Papa and I have always been."

Eliza's letter was loving and gentle, but she was hurt. She knew her son all too well. "We are the victims," she suggested, "of your own excited imagination." She urged Alec to remember that "there are not so many of us left that we can afford to take up unreasonable offense against each other, or imagine affronts where none were thought of or intended. We should . . . cling the more closely together for mutual strength and support."

In the privacy of Tutelo Heights, Eliza had had a hard time persuading her husband not to take "unreasonable offense." Melville Bell had been outraged by his son's letter: as impulsive as his son (and equally sensitive to unintended slights), he had immediately sat down and written a furious and aggrieved letter to Alec. After all he had done for his son, how could Alec treat him like this? Given the sacrifices he had made for his son's health, how could Alec accuse him of neglect? Melville boiled with wrath, but once his rage was vented, he reluctantly allowed Eliza to tear up the note and to end her own letter in a gentler tone: "We both feel thoroughly bewildered, and fear something must have upset your mind."

His mother's explanation soothed Alec. His equanimity restored, he gave his parents the news they most wanted to hear: he would arrive at Tutelo Heights in early September for a prolonged visit. And when he stepped through the parlor's double windows on a warm, late summer day, he found an eager audience for both his consuming interests: the multiple telegraph and Mabel Hubbard. Melville was particularly anxious to heal the rupture. "If he has not killed his fatted calf for me,"

Alec confided to "Miss Mabel" in a letter, "he has done everything else to make me happy—and to show his affection for me."

It had been a nerve-racking few months for a naive twenty-eight-year-old who had, up until now, buried his emotions deep. Those close to him already knew that too much work and too little sleep could tip him into an overwrought state. The severe headaches to which he was increasingly prone were almost certainly migraines, as he suffered an abnormal sensitivity to light. He put tremendous demands on himself, not least because, with the deaths of his brothers, he felt the weight of family expectations resting on his shoulders. His tendency to work round the clock, and to alternate between states of fierce focus on one goal and an inability to concentrate on anything, suggest a lack of balance in his temperament.

Yet for all Alexander Graham Bell's neurotic intensity, there was nothing wrong with his brain. He was erratic in his habits and intellectually obsessive, but it was his unconventional mind that made him a genius. He refused to be hemmed in by rules; he allowed his intuition to flourish; he relied on leaps of imagination, backed up by a fascination with the physical sciences, to solve the challenges he set himself. It was perhaps inevitable that when such a brilliant young man fell in love, his emotions would spiral out of control.

It was Alec Bell's good fortune that he had fallen in love with a woman who would prove able to handle this obsessive personality. In 1875, Mabel was still barely more than an adolescent, tucked under the protective wing of her family and ten years younger than her suitor. But the way she had met the challenge of her hearing loss proved both her intelligence and her strength of character. She knew her deafness set her apart from others, and she had learned to handle herself with grace and confidence. In the years ahead, she would learn to use the same qualities to shield Alexander Graham Bell from stress, emotion, and self-doubt. Thanks to Mabel, the genius of his mind was allowed to flower and the potential for instability in his temperament was never allowed to explode. Their union would provide him with a rich, well-rounded family life, a safe haven in which he could reach for his dreams. In the years ahead, Mabel would have to make many compromises and sacrifices. But in return, his disregard for her deafness meant she would live, as her parents had always intended, completely in the hearing world.

Perhaps Mary Blatchford and Mabel's other relatives found the partnership of their quiet, rather awkward cousin and the gawky Scotsman slightly ridiculous. But Alec was too consumed by his fascination with sound to notice, and Mabel was used to being regarded (and sometimes dismissed) as "different." Each emerged from the summer of 1875 with a simple trust in the other's good faith.

THE ATLANTIC COAST OF NORTH AMERICA

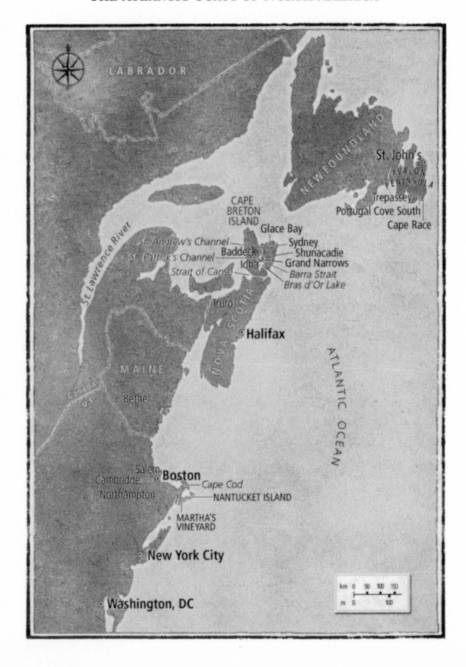

Chapter 8

PATENT NO. 174,465
1875–1876

From now on, Mabel Hubbard was a steadying influence in Alec's life—a source of good sense, insight, and reassurance. "You know my dear," she teased him at one point, "you are one of those men who can't live without a little wholesome tyranny from their womenkind!" She admired his protean imagination, but she deplored his volatility. "You are a man of brilliant talents . . . but your mind is so fertile it is always drawn off by every new idea that comes up; you like to fly around like a butterfly sipping honey, more or less from a flower here or another flower there." When Alec was hell-bent on a project, Mabel was the only person who could divert him. When he was distraught, she was the only person who could calm him down.

The first test of their relationship came when Alec returned to Boston from Canada in October 1875. Mabel's father expected him to buckle down with Tom Watson and produce the multiple-telegraph apparatus immediately. Gardiner Hubbard wanted a commercial application of Alec's theories as fast as possible. But Alec was still preoccupied with his discovery of the "magneto-electric current" generated by a vibrating spring. He wanted to write the specifications for the apparatus that would transmit "vocal utterance telegraphically." And he still had to earn his living, so he committed himself to a heavy teaching load, including a teacher-training course for Visible Speech teachers and lectures at several local schools.

His patron was furious. "Your whole course since you returned," Gardiner Hubbard wrote to his protégé, "has been a great disappointment to me and a sore trial. . . . I have been sorry to see how little interest you seem to take in telegraph matters." At this rate, he would never be able to afford to marry Mabel. So Hubbard told Alec that he had to choose between Visible Speech and telegraphy—in other words, between his father and his potential father-in-law. "He even went so far as to say," Mabel confided to Mary True, her former teacher, that "Alec should not have me unless he gave it [Visible Speech] up." The threat was plain: Hubbard would dissolve their business arrangement and shut the door of 146 Brattle Street in Alec's face if he didn't buckle down.

Hubbard had learned nothing from the events of the previous summer. Then, his warning to Alec against a "burst of passion" had exactly the opposite effect: Alec had rushed off to Nantucket. Now, his equally heavy-handed attempt to lock Alec into telegraphy research goaded the younger man into angry resistance. Once again, Alec was suffused with emotions he could not control, and there was an ugly showdown in the Hubbards' parlor. Hubbard was at his most Boston patrician, stroking the long gray beard that flowed over his starched white collar and speaking with ice-cold authority about "duty." Alec, his knuckles white and his voice shaking, reacted with all the fervor of a free-thinking Scotsman: he refused to be browbeaten into abandoning either his father's lifework or Hubbard's daughter. The following day, he wrote to Hubbard apologizing for "anything in my conduct that might have seemed disrespectful, or that might have given you offense." But he stuck to his principles. He would not abandon Visible Speech or his commitment to his deaf pupils. "Should Mabel learn to love me as devotedly and truly as I love her, she will not object to [my] work. If she does not come to love me well enough to accept me *whatever my profession or business may be,* I do not want her at all. I do not want a half-love." He was not going to let the fact that his major investor was also the father of the woman he loved steer him in any other direction.

At this crucial juncture, Mabel stepped in. She already knew that she cared for Alec and that, as she confided in a letter to Mary True, "I would sooner or later marry him, but I did not know if I loved him sufficiently to tell him so yet." She also resented being a pawn in her father's business plans. So she closeted herself with her closest adviser: her mother. Gertrude Hubbard told her, Mabel later recalled, that "I could not go on as I was doing, I must be engaged or give him up." But Mabel's mother had a soft spot for the passionate pianist and teacher who had stood up to her husband. She made sure that Alec was invited to 146 Brattle Street on November 25, to join the celebrations for Mabel's eighteenth birthday, which had fallen on Thanksgiving Day. When he appeared on the doorstep, obviously wretched, Mabel quietly piloted him into a small second-floor room that she used as a study, for another tête-à-tête. As she revealed in her letter to Mary True, "I told him that I loved him better than anybody but Mamma, and if he was satisfied with so much love I would be engaged to him that very day!"

Alec was completely taken by surprise. He felt duty bound to suggest that perhaps she was too young—"He almost refused to let me bind myself to him, he reminded me how young I was and how I had not seen other men." But Mabel insisted that her love was real. Alec could barely contain himself for the rest of the evening, and by the time he reached home in the early hours of the following day he was almost delirious. He scribbled a note to Mabel: "I am afraid to go to sleep lest I should find it all a dream, so I shall lie awake and think of you. . . . It is so cold and selfish living all for one's self. A man is only half a man who has no-one to love and cherish." When he returned to Cambridge two days later, he and Mabel affirmed their engagement. Mabel's parents apparently accepted the fait accompli with grace, although they did not allow the couple to set a date for the wedding yet.

Once the betrothal was official, the relationship settled into a pattern that would last a lifetime. Mabel wrote to her newly minted fiancé, "My darling, I warned you before we were engaged that though I might love you very much, I could not do so in the passionate hot way you did." Mabel recognized that not only was Alec's temperament far more impulsive than hers, his emotional needs were also greater than hers, and perhaps beyond her capacity to fulfill. "I love you all I can and my powers of loving seem to increase daily. You must be satisfied with this. I would give you more if I could but I cannot help my nature."

While Alec provided the passion, Mabel supplied stability and structure. This was as true of the practical details of his life as it was of his sprawling emotional hinterland. She took one look at his desk, which was almost submerged in unanswered letters and half-written articles, and got to work—sorting papers into piles and replying to his correspondents. "We are so much obliged to dear Mabel for acting as your Amanuensis," wrote Eliza Bell. Insisting that his scientific pursuits should not swallow up his health and other interests, Mabel tried to cure him of all-night working sessions. One day a parcel arrived at Alec's laboratory with a note from Mabel explaining that this was a portrait of him that she had painted. He eagerly tore the brown paper wrapping off and found himself staring at a picture of a night owl. She also told him that she was going to buy him a piano, to lure him away from his experiments. Alec wrote to his parents that she wanted him "to play for her *every evening* when we are married!" Eliza replied, "If the fever in which she lost her hearing did not utterly destroy the nerve, perhaps she might hear the piano by resting a piece of solid stick on the sounding board and holding it there with her teeth. [I found that] by such means the noise would reach me."

Most important, Mabel gently steered her fiancé toward fulfilling his obligations to her father. Gardiner Hubbard was finally convinced of the value of Alec's mechanism to transmit the human voice, and wanted to apply for a patent for the idea as well as continuing with the multiple telegraph transmitter. But he was frustrated on both counts. In Washington, Alec's patent application for the multiple telegraph from the previous February was blocked by similar applications from three other inventors: Elisha Gray, Thomas Edison, and a Danish scientist named Paul La Cour. It would take the patent examiner some time to sort out whether the applications overlapped, and who had first conceived of the idea. And Alec, in a typically muddle-headed, unbusiness-like move, had already taken the initiative to patent his "speaking telegraph" in Britain. While staying in Brantford the previous September, and feeling very short of money, he had sold a share of the rights to his invention to his parents' neighbor, the Canadian publisher-politician George Brown, for $500. The deal was that Brown would file for a British patent while he was in London the following January.

This meant that Alec could not file a patent application in Washington until Brown had been granted the British patent. But, unknown to Alec, Brown had lost interest in the invention after being told by a London lawyer that it was worthless. While Alec eagerly awaited news of his British patent, the application lay forgotten at the bottom of Brown's suitcase as he dined with old friends at the Travellers Club on Pall Mall. Back in Boston, Gardiner Hubbard struggled to control his irritation.

This time, however, Hubbard realized it was smarter to avoid another showdown with his excitable protégé. In February 1876, without consulting Alec, he went ahead and filed on Alec's behalf for a patent on the "speaking telegraph." He dropped off at the U.S. Patent Office in Washington the application with its descriptions of Alec's electrical theories and of the crude apparatus that Alec and Watson had built.

At first, Alec was furious with Hubbard for taking matters into his own hands. But Hubbard's action was crucial. Two hours after Hubbard's call on the U.S. Patent Office, Elisha Gray filed a caveat there. The caveat, or warning, served notice to other inventors that Gray was working on the transmission of speech by electricity. Since

Alec, had not yet succeeded in transforming his theories
by constructing the apparatus that would demonstrate
...es, he could not apply for a full patent. On March 3, Alec's
twenty-ninth birthday, the United States Patent Office announced that
it had granted Patent No. 174,465 to Alexander Graham Bell for a
"speaking telegraph." His patent gave his invention seventeen years of
protection in the market. If anybody else was working on a "speaking
telegraph," they now had to prove that their idea was completely different from Alec's. Alec Bell had won the first stage of the race. He
acknowledged that Hubbard had done the right thing, and he proudly
presented the patent certificate to Mabel, who hung it in her second-
floor study in Brattle Street.

But there was no time to lose in self-congratulation. Both Alec and
Tom Watson knew that their invention was far from fully developed:
it had yet to transmit a clear message. They moved their workshop
to the privacy of Alec's new Boston lodgings at 5 Exeter Place, and
spent hours fiddling with the equipment in order to make the vocal
transmission distinctly audible. Alec thought a different kind of battery,
which would strengthen the electric current carrying the sound, would
improve clarity. "I had made for Bell a new transmitter," described
Watson later, "in which a wire, attached to a diaphragm, touched acidulated water contained in a metal cup, both included in a circuit through
the battery and the receiving telephone. The depth of the wire in the
acid and consequently the resistance of the circuit was varied as the
voice made the diaphragm vibrate, which made the galvanic current
undulate in speech form." On the evening of March 10, 1876, Alec and
Watson set up the receiver on the bureau in Alec's bedroom, then took
the transmitter with its battery connected to the cup of sulfuric acid
into the room next door. Watson barely had time to return to the bedroom before he heard Alec's voice emanating from the receiver, saying,
quite clearly, "Mr. Watson, come here, I want you!"

In Alec's account of this historic moment, Watson rushed out of the
bedroom declaring he had heard clearly the message. Tom Watson's

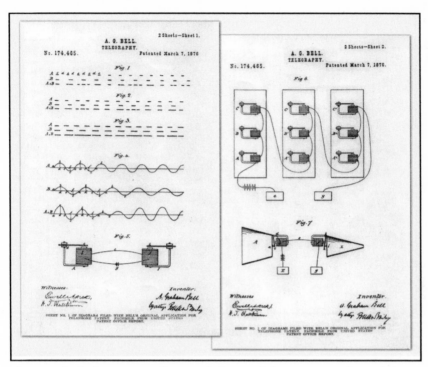

The crucial telephone patent illustrated Alec's undulating current theory (Figs. 1-4), reed receivers (Fig. 5), and circuitry for the harmonic telegraph (Fig. 6) and his telephone (Fig. 7).

account is far livelier. He recalled not only the message, but also a note of urgency in his boss's voice: "I rushed down the hall into his room and found he had upset the acid of a battery over his clothes. He forgot the accident in his joy over the success of the new transmitter when I told him how plainly I had heard his words. . . . His shout for help on that night . . . doesn't make as pretty a story as did the first sentence, 'What hath God wrought!' which Morse sent over his new telegraph . . . thirty years before, but it was an emergency call—therefore typical of the great service the telephone was to render to mankind."

Whatever the truth of the matter, there is no doubt that inventor and technician exploded with excitement at the breakthrough they had wrought. Then they changed places, so Alec could hear Tom read from a book into the mouthpiece. The two young men spent the rest

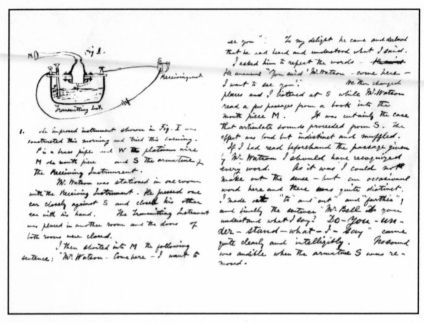

Alec's notes on the first successful transmission of speech.

of the night laughing and testing the transmitter over and over again, until Alec ended the world's first telephone conversation with the words "God save the Queen!" The solemnity of this sign-off lasted only a moment. Loyalty to Queen Victoria was supplanted by the exuberance of the Mohawk war dance that George Johnson had taught him several summers earlier in Brantford. Now, as the sun rose beyond the dusty windows of Exeter Place, Alec gleefully entertained Tom Watson with war whoops and stamping.

Success was sweet. Alec told his father, "This is a great day with me." Alec might have exasperated his patrons with his scatter-brained enthusiasms, but the reach of this young, self-educated man's vision is illustrated in his next remark: "The day is coming when telegraph wires will be laid on to houses just like water or gas—and friends converse with each other without leaving home." Today we take such communication totally for granted. In 1876, when three-quarters of the population of North America spent their entire lives in the isolated rural communities in which they were born, when electric lighting was still a decade away, and when a reliable mail service was regarded as a miracle, most people dismissed such a notion as fantasy.

Alec's fiancée quickly recognized the world-shaking significance both of the technical breakthrough and of the patent. Alec, Gardiner Hubbard, and Thomas Sanders now had a monopoly on an extraordinary new technology—as long as they moved quickly into production and the patent wasn't contested. Mabel wrote to Alec, "How can I tell you how very glad I am for you, I am so glad you have at last obtained this patent for which you worked so hard and have now as you say a fair prospect of success. I am just as happy as can be." She herself had worked hard to understand the science behind Alec's instruments. ("I want to know all about yourself and your ideas," she instructed him, although, tongue in cheek, she added, "so far as you consider possible for a poor ignorant, feeble-minded *woman* to understand.") Her father's faith in her fiancé had paid off, and the brilliance she knew he possessed had been recognized in Washington. "You don't know how much my heart is in your work or how anxious I am that you should succeed," she wrote from Cambridge a few months later. "I want so much that you should take your place among scientific men."

But Alec continued to be torn three ways—between demands from his own father to keep working on Visible Speech, pressure from Mabel's father to concentrate on telegraphy, and his own commitment to the deaf. Despite the excitement of the breakthrough, he clung to his self-selected mission. "You must settle down to the conviction," he replied to Mabel, "that whatever success I may meet with in life— pecuniary or otherwise—your husband will be known as a teacher of the deaf. . . . I see so much to be done—and so few to do it."

Mabel was more hard-headed than Alec. Her own deafness did not convince her that Alec should devote his life to her fellow sufferers— indeed, she never thought of herself as a sufferer, and she rarely mingled with other deaf people. In later life, she would admit that "when I was young and struggling for a foot-hold in the society of my natural equals, I *could not* be nice to other deaf people. It was a case of self-preservation." Right now, she shared her father's fear that Alec would let slip his immediate opportunity to make his fortune with the telephone because of this yearning to help the afflicted. Although she herself had a good working knowledge of Visible Speech, she didn't really see much future in it and thought it a waste of Alec's talents—although she never dared admit this to Alec. Visible Speech "will be of greatest value to learners, deaf or hearing, but I think it will be a long time before it will come into general use," she gently informed him. "There may be money in it but you are not the one to get that money because you love your work and cannot bear to ask pay for that labor of love." Success with the telephone, she pointed out, would give him the financial freedom to devote himself to the interests of the deaf, "[s]o for that reason I think it would be better if you made Visible Speech secondary to the telephone." In letters and speech, Mabel had a certain tone of voice to which, over the years, Alec would learn to listen carefully.

The Hubbard household in Cambridge was now Alec's second home. Several evenings a week, he would lope along Brattle Street to be welcomed at the door of Number 146 by a radiant Mabel. After dinner, Gertrude Hubbard would ask him to play the piano, or Mabel's sisters would beg him to tell them about Scotland. Alec, once so somber and awkward in this wealthy, youthful company, now reverted to the boy who had entertained his own cousins in Edinburgh. Surrounded by Gertrude, Grace, and Berta Hubbard and by Augusta, and Caroline (Lina) McCurdy, clad in shimmery silks and taffetas or well-starched lace and linens, Alec recited the poetry of Robbie Burns or recounted the plots of Walter Scott's novels. Augusta remembered how "[w]e would all gather round the piano while he played and sang, or . . . he would hold us spellbound, with his wonderful tales of old Scotland, for he was then as always a most charming story-teller." After the loss of his own brothers and his years of intense and lonely labors over his lab bench, Alec enjoyed being, in Augusta's words, "one young man among a lot of pretty young girls." And Mabel, perched close to her fiancé and watching his lips carefully, reveled in Alec's popularity with her family.

That spring, Alec had to submit to the prenuptial obligation of meeting Mabel's relatives, as was expected of every well-bred young couple in the 1870s. Knowing Alec's eagerness to focus on his invention, Mabel kept visits to the minimum. However, she worried about the impression that Alec, who rarely paid much attention to his appearance and often forgot to get his hair cut, might make. "Be sure you keep looking very nice and clean, and do me credit my dear," she advised her future husband. Her cousins were all amazed that Mabel, aged only eighteen and unable to hear, should be the first of their number to become betrothed. And they were intrigued that she was marrying outside their circle. "International marriages were not as common then as now," recalled Augusta McCurdy half a century later, "and even Canada seemed far away."

In New York, Mabel's McCurdy grandparents were most impressed with her clever young Scotsman. Robert McCurdy, a self-made man who had built up his own dry goods business, didn't share Gardiner Hubbard's Brahmin distaste for uncombed hair, shiny suits, and poverty. The McCurdys were also more realistic than Mabel's doting parents about her marital opportunities. They knew that there would be prejudice against Mabel's deafness and strange speech within Boston's upper crust. Robert McCurdy thought that Mabel had done well to find a man so sympathetic to the challenges of hearing loss.

Alec had refused to continue as Mabel's speech instructor once they were engaged. Clasping her hand and looking deep into her wide gray eyes, he explained that he wanted to be her lover, not her teacher. But despite the improvements he had already made to her speech, even her family sometimes found the contorted vowel sounds of her speech hard to understand. "Mamma says I try so manfully to read your letters [aloud] aright that no one can understand a word I say," she told Alec on one occasion. Strangers often assumed that this well-dressed young woman was stupid when she just stared at them; they rarely realized she was struggling to read their lips (a task made more difficult by the thick beards then in fashion for men). With Alec, however, she was totally uninhibited. When she accompanied Alec on a walk, no one remarked on the lively chatter between the tall, thin professorial-looking figure and his high-spirited, long-haired companion, whose eyes were fixed so lovingly on his face. Passersby smiled indulgently at the obvious mutual affection, not realizing that Alec and Mabel were always holding hands or linking arms because touch was such an important form of communication for them. The only unusual aspect of their conversations emerged on dark nights. On evening strolls, Mabel would do all the talking between street lamps, and then the two would pause under the popping gaslight while Alec replied. If the couple took an evening drive, they often carried a candle in the carriage so they could converse without interruption.

Alec's mother, Eliza, was the person most concerned about Mabel's deafness. She had no way to communicate with her future daughter-in-law. Neither of them knew the Abbé de L'Epée's sign language, and Eliza could not read lips. She depended increasingly on the double-handed finger language with which Alec had communicated with her since his Edinburgh childhood, and which her husband now used regularly with her. Mabel depended entirely on lip-reading, so she would be able to "hear" what Eliza said to her but Eliza would not hear her replies. Still, Eliza longed to meet this young American woman with whom her son was so besotted: she could see that Mabel was a wonderful influence on her only son. She sent Alec some moonstones for Mabel's engagement ring. How, though, would she be able to understand Mabel, particularly if her speech was indistinct? "Ask her to learn the double-handed finger alphabet for me," Eliza pleaded with Alec. "She need not fear yielding to the temptation of using it too frequently, for few besides ourselves understand it." But Mabel never learned the double-handed alphabet, relying instead on Alec to be the intermediary between her and her future mother-in-law.

Alec was now far more cheerful than he had been the previous year.

Still, when he reflected on his future, he felt shackled by poverty. The Hubbards would not entertain any discussion of a date for their daughter's wedding until Alec was in a position to support Mabel. Mabel urged him to focus on the commercial potential of the telephone because she knew it could secure their future together. But Alec's business partners, Hubbard and Sanders, were not eager to sink funds into developing and manufacturing the telephone until they were confident that a market existed for it. Too many people regarded it as a toy that could never replace the telegraph because it didn't leave a printed record. At Western Union, William Orton was pumping money into Thomas Edison's version of the quadruple telegraph, by which two messages could be sent in one direction and two in the other. A Western Union electrician had sent the company president a memo suggesting that "[t]his 'telephone' has too many shortcomings to be seriously considered as a means of communication. The device is inherently of no value to us."

In May 1876, Alec began giving a few public demonstrations of his prototype telephone with the help of Mabel's twenty-six-year-old cousin William Hubbard, a Harvard student. This very first telephone certainly had shortcomings. It could carry sounds only one way, over a short distance, and the quality of transmission was unreliable, so demonstrations always included some well-known musical passages as well as the spoken word. The first was to "a dignified assemblage of greyheads" at the American Academy of Arts and Sciences in the Boston Athenaeum, and the second, a few days later, to the Society of Arts at MIT. Alec and Willie would rig up their primitive transmitter in a nearby room, then connect it by wire to the receiver in the lecture hall. At the end of Alec's lecture, Willie would transmit some music (the hymn "Oh God Our Help in Ages Past" was a popular choice) and then a familiar passage of poetry. Few of the words in Willie's rendition of a soliloquy from *Hamlet* were clear, but according to the *Boston Transcript*, the audience was impressed. "I feel myself borne up on a rising tide," Alec wrote to his parents.

A poster entitled "Stride of a Century" caught the atmosphere of the Philadelphia Exhibition.

Ever the entrepreneur, Gardiner Hubbard knew that a few lame demonstrations in front of a bunch of university professors were not going to raise any capital. He was looking for major investors. He was eager for Alec to use a much more ambitious launch pad for a scientific breakthrough: that summer's much ballyhooed World Exhibition. Held in Philadelphia, the city in which the Declaration of Independence had been signed a century earlier, the exhibition was going to celebrate the hundredth anniversary of the American Revolution. It had already involved ten years of planning; it would cost more than $11 million, and Gardiner himself was one of three prominent Bostonians charged with organizing the Massachusetts contribution to the education and science section.

The Philadelphia Exhibition put America on exuberant display to itself, as the nation emerged from the economic depression and wrenching

Bartholdi's Electric Light gave Americans a glimpse of the future.

dislocation that had followed the Civil War. The 285 acres of fair-ground in Fairmount Park teemed with invention, appetite, avarice, and optimism as Americans showed off their industry and creativity. There were seven major exhibition buildings plus nine foreign pavil-ions and seventeen state buildings (only two, however, from southern states, which had been devastated by the Civil War). The buildings were linked by a narrow-gauge railroad. The exhibits reflected the tastes of the times. The Horticultural Hall represented the Victorian love of exotic gardens under glass, while anti-liquor reformers made their point with their monumental Total Abstinence Fountain. Among the 22,742 exhibits, familiar artifacts (machine-made shoes, bone cor-sets, sewing machines) jostled with exciting innovations (canned foods, dry yeast, mass-produced furniture, linoleum, ready-made clothing). In the Women's Pavilion, Mary Potts's patented flat-iron and Margaret Calvin's Triumph Rotary Washer illustrated the application of new technology to traditional women's work.

There was a heady, forward-looking momentum, as large num-bers of visitors sampled for the first time such novelties as bananas (separately packaged in tinfoil and costing ten cents), popcorn, Hires Root Beer, and ice cream sodas. The most thrilling outdoor feature was Bartholdi's Electric Light, or the Colossal Arm of Independence. This monumental arm, holding aloft a brilliantly lit torch and created by French sculptor Frédéric Bartholdi, was a stunning spectacle. Ten years later, Americans would be even more dazzled by it when the complete Statue of Liberty of which the colossal arm was a component was erected in New York harbor.

The exhibition opened on May 10, with a march composed by Richard Wagner and an address by President Ulysses S. Grant. President Grant's most distinguished guest for the ceremony was an engaging, progressive-minded eccentric: Pedro II, emperor of Brazil. Dom Pedro, as he insisted on being called in the American Republic, had inherited the throne of Brazil in 1831, when he was only five years old. At home, the forty-one-year-old emperor was usually decked out in a splendid Gilbert-and-Sullivan-style uniform festooned with ribbons, medals, outsized epaulettes, and gold braid, but on this occasion he sported a frock coat with a heavy gold watch chain strung across his well-filled waistcoat. An energetic autodidact, as a youth he had mastered French, Hebrew, Arabic, and Sanskrit, as well as many of the Tupi-Guarani languages indigenous to South America. He was now a passionate opponent of slavery within his realm.

Dom Pedro could barely take his eyes off the exhibition's central attraction: the Corliss steam engine—the largest steam engine in the world. He puffed up the ladder onto the engine's platform after President Grant, and stared with amazement at the mighty cylinders. He watched the president hit the levers that allowed steam into the cylinders, heard the engine hiss, and felt the floor of the thirteen-acre Machinery Hall tremble. He stared in wonder as the huge walking beams slowly started to move up and down, feeding the giant flywheels that spun around, gaining momentum and storing energy. Seventy-five miles of belts jerked into action, causing the shafts and pulleys of the steam engine to drive lathes and saws, drills and looms,

The 70-foot-tall Corliss steam engine, weighing 56 tons and producing 1,400 horsepower, drove all the exhibits in the Centennial Exhibition's Machinery Hall.

presses and pumps. Thanks to the Corliss, the *New York Tribune,* the *Boston Herald,* and the *Philadelphia Times* printed their daily editions at the fair.

Dom Pedro was not the only observer left slack-jawed by the Corliss steam engine. Over the next four months, almost ten million people came to watch the mechanical behemoth that symbolized the raw power now driving America's industrial revolution. But it was another aspect of the Philadelphia Exhibition that captivated Gardiner Hubbard, who made repeated visits to this huge national birthday celebration. He had secured space for Alec's prototypes of both the multiple telegraph and the speaking telegraph, along with a display of Visible Speech charts and books. Now he watched scientists and wealthy investors pacing down the Machinery Hall's aisles, comparing different innovations. He saw Elisha Gray's and Thomas Edison's diagrams and proposed inventions displayed prominently, and Alec's rivals often present in person, drumming up interest in their work. Gardiner sent an urgent telegraph home. Alec, he insisted, must drop everything and appear in Philadelphia for Sunday, June 25, when the Committee on Electrical Awards was scheduled to decide which piece of electrical apparatus should receive a medal. Sir William Thomson, the famous British physicist from Glasgow University who had helped design the first transatlantic cable, was chairman of the committee. Alec *had* to be there to demonstrate that his prototype of the telephone, with the makeshift mouthpiece and tin cup of acid at the transmitter end, was the communication device of the future.

But Alec had other ideas. He was preoccupied with preparing his students and his teachers of the deaf for their final examinations. He had already suffered the interruption of a visit to Boston's School for the Deaf by Dom Pedro, during which he had explained Visible Speech to the genial ruler, and arranged for two of his father's books to be sent to Brazil. Why should he now desert his professorial duties for the hoopla and frenzy of Philadelphia, which he knew would be hot, sticky, and crowded? He ignored his future father-in-law's telegrams.

So Hubbard tried a different tack. He sent a telegram to Mabel and her mother, explaining the vital importance of Alec's presence in Philadelphia. In Cambridge, Mabel recognized the validity of her father's arguments. She also knew that Alec was a born tinkerer. Unlike Thomas Edison, the ruthless self-promoter who saw science as Darwinian competition and who always announced his inventions before he had even got them working, Alec hated revealing anything until he was confident of its success. Edison was an ambitious self-made American; Alec was a cautious Scot more interested in scientific progress than commercial success. And now he was putting loyalty to his students and his father (who had recently reproached him with "poking over experiments" instead of promoting Visible Speech) ahead of capitalizing on his breakthrough and his future father-in-law's investment.

Mabel took matters into her own hands. She tucked her hair into a bonnet and asked for the Hubbard carriage to be brought to the door of 146 Brattle Street. Then she told the coachman to drive her to Alec's rooms, and insisted to Alec that he stop drafting exam papers

Replicas of the liquid variable transmitter (left) and reed receiver (right) displayed at the Philadelphia Exhibition.

and accompany her on a drive. Before Alec realized what was happening, he found himself at the railroad station, where his fiancée handed him both a ticket to Philadelphia, via New York, and his valise, which she had secretly packed. A bewildered Alec stared helplessly at Mabel, then refused to get on the special Saturday excursion train, which was already getting up steam for the journey.

The exchange between Mabel and Alec as they stood on the station platform that hot June day encapsulates the nature of their relationship, and would become part of Bell family lore. According to an account that their daughter Elsie wrote years later, Mabel "burst into tears and said, 'I think you might do this just to please me. If you won't do a little thing like this now, I won't marry you.'" Then she turned her eyes away so she could not "hear" Alec's objections. A frowning and extremely reluctant Scotsman climbed onto the train. He wrote to his mother from New York that, seeing "how pale and anxious [Mabel] was about it, I could not resist her and here I am. . . . What I am going to do in Philadelphia . . . I cannot tell."

Alec's black mood did not lift when he arrived in Philadelphia. He found he had to share a hot little room in the pretentiously named Grand Villa Hotel ("six or seven private dwellings united together") with Gardiner Hubbard's brother James. He wrote a miserable letter to Mabel:

> Oh! My poor classes! What shall I do about them! You don't know what a horrid mean thing it is for me to leave them at this time. . . . Then there is the expense which I *cannot afford*. Travelling expenses, hotel, rent of room for classes, and in addition I have to pay Miss Locke *five* dollars for every day I am gone. You really don't know what you do when you make me come here! . . . I must confess I don't see what good I can accomplish in Philadelphia unless I stay for a long time and as far as telegraphy is concerned, I shall be far happier and more honoured if I can send out a band of competent teachers of the deaf and dumb who will accomplish a good work, than I should be to receive all the telegraphic honours in the world.

Mabel had already learned that Alec often had to let off steam when he was unsettled before he could focus on the task in hand. By the time this grumpy letter reached her in Cambridge, he had begun to enjoy himself in Philadelphia. He was pleased to meet Sir William Thomson (it was a pleasure, he reported, to hear "a good broad Scotch accent") and other notable scientists. He still found the heat and the crowds insufferable, but he sent for Willie Hubbard, to help him demonstrate his telegraphic and telephonic apparatus on the crucial award day.

The atmosphere in the exhibition's Machinery Hall was steamy on June 25. Because it was a Sunday, Fairmount Park was closed to the public. The sun blazed mercilessly through the glass roof of the echoing, cavernous building as a small, solemn party of judges slowly inspected the exhibits. Alec Bell and Willie Hubbard had been up since dawn, arranging the reed receiver in a gallery behind a huge Hood and Hastings Organ, to ensure the best reception. As usual, the delicate instruments required endless fiddling to get the desired effect. "In the narrow gallery," recalled Willie later, "the heat was intense . . . and there were no ventilators. We were obliged to work in the scantiest possible attire, and even then were dripping with perspiration." Alec became even more sweaty and nervous when he heard the unsettling news that the judges would have examined both Edison's and Gray's devices before they reached him. As the tension mounted in lockstep with the temperature, Alec's head began to throb. One of his debilitating headaches seemed inevitable. Then, with the sun overhead and the heat almost unbearable, came the really bad news. The judges announced that they were going to call it a day as soon as they had finished with the exhibit before his.

Had the judges quit then, Alec would have given up the fight and caught the next train to Boston. But nobody had reckoned with Dom Pedro, for whom tropical heat was no problem and who had been asked to help judge the electrical devices. Dom Pedro had taken a liking to the earnest young teacher on his visit to Boston's School for the Deaf two weeks earlier. Catching sight of the lanky, wilting Scotsman in the Machinery Hall, he bustled over to find out what he was doing in Philadelphia. Once Alec began to explain his invention, Dom Pedro started bouncing up and down on his toes and repeating, "Ce n'est pas possible!" The jovial emperor's excitement attracted Sir William Thomson; within minutes all of the judges were crowded around Alec. They were intrigued to hear that Alec, unlike his rivals, proposed to give an actual demonstration of the speaking telephone. Willie climbed up to the gallery behind the organ, where Alec had positioned the receiver, with Dom Pedro in perspiring pursuit. In the stifling heat, hidden by the organ casing, Willie and the emperor stripped to their undershirts as they prepared to listen to Alec. Willie checked the reception, then invited Dom Pedro to put his ear to the receiver. Out of sight below them, Alec cleared his throat, then began to talk into the mouthpiece that was connected to the cup of acidulated water. A storm of emotions crossed the Brazilian emperor's face—uncertainty, amazement, elation. Lifting his head from the receiver (a small cylindrical device with a wooden base that rested on a table), he gave Willie a huge grin and said, "This thing speaks!"

Soon a steady stream of portly, middle-aged men were clambering into the gallery, stripping off their jackets, and bending their ears to the receiver. "For an hour or more," Willie remembered, "all took turns in talking and listening, testing the line in every possible way, evidently looking for some trickery, or thinking that the sound was carried through the air. . . . It seemed to be nearly all too wonderful for belief." Dom Pedro asked if this new machine could speak Portuguese, and was delighted to hear his own language over the wires.

Dr. Joseph Henry, who had been so impressed by Alec's ideas when he had met him at the Smithsonian the previous year, was among the first to be convinced. In the *General Report of the Judges,* written for the Centennial Organizing Committee, Dr. Henry declared that Alec's telephone was "the greatest marvel hitherto achieved by the telegraph." His British colleague Sir William Thomson noted, "I was astonished and delighted." As Alec's was the only "speaking telegraph" that actually worked, the panel of judges agreed that young Mr. Bell had to receive the gold medal for electrical equipment, and asked him to do another demonstration the following day. At the same time, Alec heard he had also been awarded a gold medal for his Visible Speech display.

Alec's chief reaction to the excitement was relief that it was all over. He and Willie Hubbard wearily made their way back to their room in the Grand Villa Hotel, stripped off their sweat-stained suits, donned their pajamas, and slumped on their beds. They were surprised when a bellboy appeared at their door, and even more surprised when they read the name on the business card he handed them: Elisha Gray. Alec asked the bellboy to show up the rival inventor, but he was assailed by doubts. He had examined Gray's telephone in the Machinery Hall and realized that the main difference between his and Gray's prototypes was that Gray had not found a way to strengthen the undulatory current. But Gray was a qualified electrician. Had Gray tracked him down now to pick his brains, so that he could rush back to his own equipment and make the adjustments required for it to function?

Gray, a middle-aged man with a serious but friendly manner, soon swept Alec's doubts aside. As soon as he entered the room, he congratulated Alec on the glorious achievement of a vocal telegraph. He appeared to accept that Alec had won the telephone race: their conversation focused on the multiple telegraph and the advantage to each if they worked together. A couple of days later, Alec would write to his parents that he and Gray "explained away all matters in dispute, and have decided that it may be advantageous to both of us to unite our [multiple-telegraph] interests so as to control the Western Union Telegraph Company, if those associated with us can be brought to a mutual understanding." In the end, the two inventors did not join forces, but the conversation convinced Alec that Gray was "an honourable man." This rosy opinion would come under heavy pressure in the years ahead.

After Gray had left, Alec picked up a letter from Mabel, mailed a couple of days earlier. She had written, "It was very hard to send you off so unwillingly but I am sure it was for the best and you will be glad of it by and by. Don't get discouraged now. If you persevere, success must come. . . . How I miss you!" Alec barely drew breath before he was stuffing his sweaty clothes into a valise and heading toward the railroad station. So what if Sir William and Lady Thomson had requested a special demonstration the following day? Willie Hubbard could handle that. Alec had a train to catch, exam papers to mark, and a fiancée he could hardly wait to see.

Part 2

THE STRUGGLE FOR BALANCE
1876–1889

I have my periods of restlessness when my brain is crowded with ideas tingling to my fingertips when I am excited and cannot stop for anybody. Let me alone, let me work as I like even if I have to sit up all night or even for two nights. When you see me flagging, getting tired, discouraged . . . put your hands over my eyes so that I go to sleep and let me sleep as long as I like until I wake. Then I may hang around, read novels and be stupid without an idea in my head until I get rested and ready for another period of work. But oh, do not do as you often do, stop me in the midst of my work, my excitement with "Alec, Alec, aren't you coming to bed? It's one o'clock, do come." Then I have to come feeling cross and ugly. Then you put your hands on my eyes and after a while I go to sleep, but the ideas are gone, the work is never done.

Alexander Graham Bell to Mabel Bell, March 1879

Chapter 9

RING FOR THE FUTURE
1876–1877

It was now seven months since Alec and Mabel had become officially engaged, and as the couple went through the Boston betrothal rituals—the at-homes, the polite calls on distant relatives, the interrogations about future plans—Hubbard friends and relatives slowly came to terms with Mabel's intended. Maybe Alec was a little stiff and his sideburns often untrimmed, but everybody could see that he made his former pupil happy. There was a new glow in Mabel's cheeks and a new spring in her step as she glided from one social event to another in the slender, ruched skirts that were so fashionable in this period. And Alec? Well, Alec protested at being displayed like a trophy, but Mabel's sisters pointed out that he always acceded to her wishes. He bowed politely to the cousins and the aunts, and offered simplified explanations of what he was trying to do with his "talking telegraph."

There was, however, one person who had *not* warmed to this shabby Scottish stranger. Soon after Alec returned to Boston from Philadelphia, Mabel accompanied Cousin Mary Blatchford to her

Nantucket cottage for a few days. Mabel urged her fiancé to visit her there, but Mary stuck her nose in the air and did not look Alec in the eye when she issued a vague summons to "light suppers." Alec had little stomach for Mary's haughty manners and lingering disapproval. Mabel vented her frustration in a letter to her cousin Lina McCurdy: "I feel so cross, ill-tempered and out-of-sorts, I think I'll inflict my crossness on you. . . . Alec hasn't, unheard-of thing, been here since Thursday, and I don't believe he'll come today! Oh dear, I do hate the man's ceremonious politeness!" Mabel chose to see Alec's reluctance as cultural rather than social. "If he were an American he would be content with Cousin Mary's general invitation, but being a Scotchman he will take care to come late tonight to avoid coming to supper!"

Mabel's mother, Gertrude Hubbard, had none of Mary's Brahmin snobbery. She welcomed her future son-in-law to 146 Brattle Street while Mabel was away, because she knew Alec's heart was in the right place. At the same time, she was appalled at his volatility. "We have all of us our hands full with Alec," she wrote to Gardiner. "He has not yet sent in his report to the [Philadelphia] Bureau of Awards [required before the Philadelphia Exhibition medals could be sent to him] & I can't make him *do* it. He says that his brain won't work & he cannot make it. Then he has had applications from two Lecture Bureaus . . . for lectures on Acoustics & Electricity. He wants to lecture because he enjoys it, & as a matter of dollars & cents." Gertrude, the mother of four demure daughters, couldn't resist giving her husband a staccato account of Alec's frantic style. "He is crazy at the idea of Mabel's going away . . . & wants to be married. Then he would give up V. S. [Visible Speech] or the Telegraph, he says he cannot & I believe he ought not to try and carry them on together. Which shall it be? . . . Then he ought to go to Portland to see . . . about starting a [public school for the deaf]. . . . Then he wants to stop at Toronto & see George Brown, and he must be in Brantford on Tuesday—& more than all he wants to talk with you. . . . Poor May will have a busy life if she attempts to keep him up to present duty."

A few days after she mailed this letter, Gertrude gave a sigh of relief as Alec packed into his battered leather bags all his notebooks, electrical supplies, Visible Speech papers, and telephone equipment and left Boston for Brantford. Maybe he would calm down in Canada's cooler climate.

Alec was distracted because his intellect was so deeply focused on his experiments. "There is a sort of telephonic undercurrent going on [in my mind] all the while," he wrote to Mabel. He was certainly too preoccupied to pay much attention to travel arrangements. In Rochester he got into the wrong carriage of the northbound train, only to discover that his mistake meant he was headed for Buffalo while his baggage was en route to Canada via the Niagara Suspension Bridge. And when, at one o'clock in the morning, the train finally reached Paris, eight miles from Brantford, he charged off into the night rather than staying at a hotel until dawn. Arriving at his parents' house in the small hours, he first had a snooze in the barn, then discovered that the window of his father's study was unfastened and crawled through it. The following morning, a maid who had never met Alec walked into the study and was horrified to discover a large, disheveled young man, with bits of straw sticking to his coat, snoring on the couch. She ran out, calling for Mrs. Bell. "I was awakened by a kiss from my mother," he explained to Mabel, "and found nearly the whole family congregated around me."

That summer, with six hundred miles between them, Mabel and Alec wrote to each other two or three times a week. In Brantford, Alec labored over his letters, often making two or three drafts before carefully penning a final copy. ("The midnight hours I spent over letters that were copied *in a hurry* to impress you with my ability," he admitted a few years later. "Oh! Cupid! Cupid! Cupid had succeeded in hitting me very hard.") Letter-writing was such agony for Alec because he was now exploring his emotions for the first time in his life, and he struggled to express himself truthfully. For all the anguish of composition, however, they were suffused with spontaneity and ardor. "Do you know, Mabel," he wrote on July 27, separation from you renders me as nervous and miserable as can well be imagined. I fear—I know not what. . . . Distance somehow seems to throw a veil between us and hides you from my sight. I long to be near you—to have you all to myself again—if only for a moment, to feel your dear little head nestling upon my shoulder and know that your affection for me is not all a dream, a myth, a creation of my own fancy, but that there is something tangible and real about it all. Oh! Mabel dear, I love you far more than you can ever know or than I can ever tell. Don't let anybody or anything take you from me now. I am jealous of the distance that separates us and shall not be happy till I have you in my arms once more.

There! You see what a goose I am—but I can't help it! Read my letter through, laugh at it as much as you like, and then—as you value the reputation of your staid [?] and learned [?] "professor"—tear it up and destroy it, so that no prying eye may put together the scattered fragments and find out what a foolish lovesick individual I am.

The days in Brantford dragged by for lovesick Alec. He wrote to "My darling May" every two or three days. "My thoughts are flying Cambridgewards this morning," he told his fiancée on August 13,

[a]nd I long to be with you. I only wish the intervening space between us would vanish "like the baseless fabric of a dream and leave not a wrack behind[.]" But facts are stubborn things, and space persists. Railroads and steamboats may cut it down by half, telegraphers may ignore it altogether, but alas to flesh and blood it still remains a hard impenetrable fact—dark and opaque. . . . Imagination must bridge the gap . . . and I can only say that I shall be very glad to get back to you once more and tell you the old old thing that you know so well—that I love you far more than I can possibly tell you.

Mabel's letters are equally loving, but they reveal a rapidly maturing young woman anxious to balance passion and pragmatism. Ardor alone would not get them to the altar: Mabel recognized that until her fiancé started making some money from the telephone, her parents would not permit them to start making wedding plans. She wrote, "I long so much to have you back again. . . . All I want you to do is to work away at electricity steadily at present, and do your utmost to induce some one to take up your foreign patents. . . . Having to support me will give you an object, and will help to give you that stability and perseverance that you lack."

Meanwhile, there was plenty to keep Alec busy in Brantford. For the first time in his life, he met his mother's brother, Edward Symonds, and Edward's three daughters, who had all arrived from Australia. He attended the wedding of one of his Bell cousins. ("Grace Church was filled with people. . . . I can't say that I like so much display about a wedding. I think it so much nicer to have the ceremony performed at home in the midst of one's own family and friends.") He visited his friends on the Six Nations Reserve. Chief George Johnson allowed him to be photographed wearing the young chief's ceremonial buck-skin coat, covered with silver trade brooches, and a hat of eagle and ostrich feathers that Alec had always admired. Alec could not resist a bit of Wild West theatrics: "Dressed in full costume and with a toma-hawk in my hand I stalked majestically into the room, frightening my mother nearly out of her senses."

But most important, Alec followed Mabel's advice and "worked away at electricity steadily." He began by testing his telephone equip-ment between the house and the barn at Tutelo Heights. To the amusement of neighbors, Alec was seen walking backward through the grass away from the house, unwinding coils of stovepipe wire as he went. (He exasperated his mother by winding the extra wire around the newel posts of the house's staircase, gouging spirals into the wood in the process.) Once he was satisfied with these experiments, he took his telephone apparatus and yet more stovepipe wire to the little village of Mount Pleasant, five miles from Brantford, and set it up in Wallis Ellis's general store, which was also the local office of the Dominion Telegraph Company. At an agreed time, his uncle David Bell declaimed Shakespeare into the membrane transmitter that was attached to the Dominion Telegraph Company's line in Brantford. Alec, perched on sacks of dried beans and with his ear pressed to the receiver he had exhibited at Philadelphia, was able to make out a few words. Triumphantly, he noted, "My undulatory current can be used on telegraph lines."

On August 4, demonstrations of Alec's telephone equipment constituted the main entertainment at a party Melville gave for his brother-in-law Edward Symonds at Tutelo Heights. The guests included the frock-coated elite of Brantford: two members of parliament, the principal of the Ontario Institute for the Blind, two bank managers, and three doctors. After a hearty dinner and many toasts, the guests assembled on the front porch, where they smoked their cigars and listened to Alec explain his theories. Once again, at an agreed time, Uncle David Bell, in Brantford, began declaiming "To be or not to be" into the metal mouthpiece of the battery-powered transmitter containing acidulated water. Two miles away, on the porch of Tutelo Heights, the Bells' guests took turns with the receiver, a small cylindrical device containing the diaphragm that converted the signal back into sound, so that they could listen to Hamlet's soliloquy. This was followed by a greeting in the Mohawk tongue from Chief George Johnson, then by a selection of hymns sung by Uncle David's daughter, Lily. The guests marveled at Alec's equipment.

Melville was pleased with his son's success, noting in his diary, "Gentlemen's supper. 23 guests. Telephone to Brantford." But the *Brantford Expositor* was more interested in the presence of Brantford's haute bourgeoisie than in the telephone test. Toronto's *Globe* took a week to catch up with events, perhaps because its owner, George Brown, was not present. Halfway down page 3, on August 11, 1876, there was a brief item: "At a party at the residence of Professor

A. Melville Bell, on Friday evening, a rare treat was afforded to the guests in the experimental explanations made by Prof. A. Graham Bell, of the new system of telephone invented lately by that gentleman."

Lack of publicity did not deter Alec. He was much more interested in science than in salesmanship. Within days of the party, he was busy planning his most ambitious Canadian telephone test between Brantford and Paris. This time, the apparatus would be tried in one-way communication over a much longer distance, powered by much stronger batteries, and it would use regulation telegraph wires, rather than stovepipe wire.

On August 10, Alec climbed into the Bells' buggy and drove to Paris with the cylindrical metal receiving device. Uncle David was left in the Dominion Telegraph Office in Brantford with an assortment of young relatives eager to sing into the transmitter. Alec set up his equipment in the Paris telegraph office, which occupied a corner of Robert White's boot and shoe shop on Grand River Street. The Bells' old friend Reverend Thomas Henderson, with whom they had stayed when they first arrived in Canada, arrived to watch the fun, and was soon followed by Paris's mayor, the Grand Trunk Railway agent, and several local factory owners. By 8 p.m., most of Paris was hovering around the shoe shop, smoking and gossiping as darkness fell, and speculating on whether a telegraph could really be expected to talk. At the agreed time, Alec put the receiver to his ear and was horrified to hear "perfectly deafening noises proceeding from the instrument." In the notes he wrote next day, he described how "explosive sounds like the discharge of distant artillery were mixed with a continuous crackling noise of an indescribable character." The crackling almost drowned the sounds of his cousins' patriotic chorus of "The Maple Leaf Forever."

Alec realized right away that the telegraph signal was too weak: it needed a stronger battery. Since communication on the telephone was still only one-way, he had to telegraph new instructions to Brantford, eight miles away, asking the operator to switch from low-resistance to high-resistance coils on the transmitter electromagnets. As he noted the following day, "The vocal sounds then came out clearly and strongly, and the crackling sounds were not nearly so annoying." Alec recognized various voices, including that of his Uncle David as the latter did a recitation, this time from *Macbeth*. But then there was a pause, and a different voice began, "To be or not to be. . . ." Alec was taken aback: he knew that his father had had an appointment in Hamilton, twenty miles away. He telegraphed to Brantford, "Change has improved transmission greatly. Whose voice did we hear speaking Hamlet's soliloquy? Was it my father's? Important." Pressing his ear to the receiver, he heard his father admit that he had found it impossible to tear himself away from the telephony experiment and keep his Hamilton appointment. Yes, it was he who had launched himself into Hamlet's speech.

By the time various excited citizens of Paris had taken their turn at the receiver, Uncle David was almost hoarse. The Brantford singers had exhausted their repertoire of popular songs such as "The Little Round Hat," "Maggie May," and "Oh, Wouldn't You Like to Know," and the clock had struck eleven. But as Alec wearily drove home in his buggy through the warm summer night, he could congratulate himself on having obeyed Mabel's instruction to "work away at electricity."

In late August, Mabel and Alec were at last reunited in Boston. In one of her summer letters Mabel had told her fiancé, "I am in rather a hurry to be trying to do something to help you." Soon she was spending each afternoon in his boardinghouse, replying to letters from strangers asking about the telephone apparatus. Meanwhile, Alec and Tom Watson were refining the equipment, and Alec was lecturing regularly on his invention. On October 13, at Boston University, he delivered "the best lecture I ever gave." But he still hadn't completed the report to the Philadelphia Bureau of Awards that would allow his name to appear in the list of Exhibition medalists, and a reproach from Mabel triggered an uncharacteristic flare-up. "The real reward of labour such as mine is *success,* and medals and certificates of merit only lower it to a vulgar level," Alec shot back. "However disappointed you may be in me, don't think my failure to answer those questions and to secure a medal arose from lack of love for you or from slight disrespect for your wishes."

During the long months of their engagement, these two individuals from very different backgrounds learned to accommodate not only each other's characters but also each other's beliefs. Mabel Hubbard was a conventional Boston Presbyterian who naively assumed that her fiancé shared her religious beliefs. When she was staying with Hartford friends where the father was Protestant and the mother Roman Catholic, she confided to Alec, "I am so glad we are of the same religion. I cannot understand how there can be perfect confidence and oneness between two people holding such different opinions on such deep and important matters." Alec, a skeptical Scot whose family never attended church, gently informed her that he believed "[m]en should be judged not by their religious beliefs but by their lives." He respected Mabel's beliefs, but he himself couldn't accept the notion of life after death: "Concerning Death and Immortality, Salvation, Faith and all the other points of theoretical religion, I know absolutely nothing and can frame no beliefs whatsoever." Mabel quietly accepted Alec's agnosticism, although she firmly informed him, "It is so glorious and comforting to know there is something after this—that everything does not end with this world."

Alec's views on another issue, Mabel discovered, were equally unorthodox. When she raised the issues of women's rights (a topic frequently discussed in the best Beacon Hill drawing rooms), Alec proved to be unusually progressive on property issues. He agreed with her that, contrary to the laws of the period, married women were *not* their husbands' property. They should be allowed to keep their own capital, he believed, and a couple should share their income equally. However, the extremes of late-nineteenth-century feminism appealed to neither of them. In Philadelphia the previous July, Susan B. Anthony, president of the National Woman Suffrage Association, had led a counter-Centennial demonstration outside Independence Hall to demand the right to vote. "We ask justice," the well-dressed and compelling Mrs. Anthony declared from her soapbox to a crowd of enthusiastic followers. "We ask equality, we ask that all the civil and political rights that belong to citizens of the United States be guaranteed to us and our daughters forever." Such a demand was a little too strong for both Alec and, especially, Mabel, who wrote to her fiancé, "I . . . would not have the public sentiment that forbids women to appear in public life, or to assume duties hitherto belonging exclusively to men, outraged if it could be helped."

Long engagements were not unusual in this period, but by November 1876, Mabel and Alec had been engaged for a year, and there was no marriage ceremony in sight. The American prototype of the telephone still needed further work. With the American patent secured, the way was clear for Alec to apply for a British patent, from which, if a manufacturer took it up, the inventor might earn a respectable income. But the application involved writing up specifications, and Alec hated paperwork. "Even your love," he confessed to Mabel, "fails to overcome the procrastinating spirit that leads me to put off to the future the accomplishments of every difficult task." Mabel could not resist comparing the work habits of her father and her fiancé. "Why cannot you work like Papa," she asked Alec, "who goes on steadily day by day doing as much as he can and only *so* much, so never getting tired and having to give up?" In November, Gardiner Hubbard tried a different tack in his efforts to chivy Alec into action. Recently appointed chairman of a special federal commission to look into railroad mail transportation, he swept Mabel off on a trip across the American continent. Each day, Mabel sat by the train window, recording the passing scene in letters to Alec. She described Chicago's "disagreeable, unfinished look," the "dreary yellow Prairie stretching way to the barren bluffs," and the number of houses in Salt Lake City occupied by the wives of Mormon leader Brigham Young. Gardiner's tactic worked: a letter from Alec reached Mabel in Colorado, reporting that the specifications for the British patent had finally been sent to London. At the same time, Alec finally completed the report for the Philadelphia Exhibition Bureau of Awards, so that his two gold medals could be sent to him.

Alec continued to remodel both the speaking and the hearing equipment of his telephone and to manipulate different kinds of materials to improve clarity. In October, he decided to try out a two-way conversation, and sent Tom Watson plus both transmission and receiving devices off to a factory in Cambridge. He remained in Boston, in the Exeter Place boardinghouse where both he and Tom were now living, waiting for Tom to communicate. Within hours, the two men were enjoying the first two-way long-distance conversation. When Watson got back to their boardinghouse in the early hours of the morning, he and Alec were so excited that they took off their jackets, started whooping, and performed the Mohawk war dance together. The following morning, an indignant landlady told Watson, "I don't know what you fellows are doing up in that attic but if you don't stop making so much noise nights and keeping my lodgers awake, you'll have to quit them rooms."

The local press soon started to take note of Alec Bell's activities. When he and Tom Watson managed an easy two-way conversation between Boston and Salem, eighteen miles away, on November 27, the *Boston Post* carried an item about it. The reporter noted incredulously, "Professor Bell doubts not that he will ultimately be able to chat pleasantly with friends in Europe while sitting comfortably in his Boston home." A more substantial article appeared in *Scientific American*, giving Alec the credibility he needed with manufacturers and fellow inventors. Soon he and Tom Watson had done sufficient work on their prototype telephone to justify applying for a new patent in Washington. With uncharacteristic focus, Alec managed to complete the application in early January.

His new application covered refinements to the transmitter and receiver, which were housed in a wooden box, and used a steel horseshoe-shaped permanent magnet with shorter coils of a finer wire than the earlier designs had. Issued on January 30, 1877, Patent No. 186,787 was the second of the two fundamental telephone patents that secured Alec's ownership of his invention. At this stage, the Bell telephone looked like a box camera, with a single lenslike opening that served as both mouthpiece and earpiece. Callers spoke into the wooden box, then pressed their ears to it to listen to the response. But Alec and Tom were already working on the next generation of their prototype: the wall-mounted "butterstamp telephone." This featured a hand-held receiver-transmitter that the caller could move between ear and mouth. Its diaphragm was enclosed in a flat disk atop a wooden handle, resembling the kitchen implement that was used to stamp butter pats, through which the wire from the battery was threaded.

What with research, teaching commitments, and an ever-increasing volume of correspondence, Alec was overwhelmed by demands on his time. It seemed as though *everybody* wanted to purchase a telephone or to invite Alec to speak or to offer their own theories on how telephones worked. One man turned up at Bell's door insisting that the telephone was obsolete because two prominent New Yorkers were already beaming their voices directly into his head. If Dr. Bell wanted to find out how this worked, the stranger insisted, he should take off the top of his skull and study the mechanism. "I knew," Tom Watson recorded, "I was dealing with a crazy man."

Each time Mabel visited Alec's rooms in Exeter Place, she was more appalled at the clutter: ramparts of books barricading his bed; paperwork piled so high on his desk that it was impossible to find a blank sheet on which to write, let alone a pen. Eighty-three letters arrived within a four-day period in early March. In April, despite Mabel's efforts to keep on top of the correspondence, there was a stack of sixty letters awaiting replies. And Alec was still dissatisfied with his telephone prototype. Applications for telephones were arriving from entrepreneurs in cities such as Detroit, Akron, and Syracuse, but Gardiner Hubbard (who handled them) was reluctant to manufacture and supply copies of the prototype until it was completely reliable.

As the pressure mounted on Alec, his headaches returned. He cleared a space on his desk and scribbled a miserable note to his beloved.

> When will this thing be finished? I am sick and tired of the multiple nature of my work, and the little profit that arises from it. Other men work their five or six hours a day, and have their thousands a year, while I slave from morning to night and night to morning and accomplish nothing but to wear myself out. I expect that the money will come in just in time for me to leave it to you in my will! Oh! How I long for a nice little home of my own—and a nice little wife in it, and some time to rest. Don't scold me dear . . . I am sad at heart, and keep my feelings bottled up like wine in a wine cellar—only they don't grow any better by keeping!

But Alec never lost faith in his invention. Two days later, in another note to Mabel, he described to her how, very soon, people would be able to "order everything they want from the stores without leaving home, and chat comfortably with each other . . . over some bit of gossip." Then, he assured her, "Every person will desire to put money in our pockets by having telephones." Then, he and Mabel could afford to get married.

Alec, like his father before him, understood that scientific demonstrations still provided some of the most popular entertainments available. In the pre-movie era, members of the expanding, increasingly well-educated middle class were happy to spend a few dollars to improve their knowledge with a magic lantern show on the "Wonders of the World" or a talk on Darwin's theory of evolution. Often the "science" was pretty suspect: for years, for example, the famous Fox sisters, Maggie, Kate, and Leah, from Rochester, New York, had been holding audiences in thrall with demonstrations of table-tapping and spirit-rapping. So Alec wasn't surprised when requests for demonstrations started arriving at Exeter Place. Mabel's mother was horrified

Alec's theatrical talents drew large crowds to demonstrations of his invention.

at the notion of her future son-in-law, a university professor, joining a throng of clairvoyants, hypnotists, and levitationists on the stages of small towns across America and Canada. "I heartily disapprove of Alec's giving telephonic concerts. I think it undignified, unscientific, mercenary—claptrap, humbug, Barnum, etc." But Alec, who only months earlier was decrying medals as "vulgar," now realized that public lectures would at least replenish his wallet.

On February 11, Mabel traveled to Salem and joined Mrs. Sanders in the Lyceum Hall there to see Alec give a telephone lecture and demonstration. This was the first time she had seen Alec perform in public; knowing how high-strung he was, and that he had been stricken with a migraine a few days earlier, she was almost paralyzed with nerves herself. The lecture was scheduled to begin at half past seven, but by ten to eight Alec had not appeared and the overflow crowd was getting restless. The stuffy wooden hall was filled with the fug of damp woolen overcoats, oil lamps, and the wood-burning stove. Mabel sat in the second row, smiling graciously but twisting her gloves between hands clammy with nerves. She was convinced something dreadful had happened. Mrs. Sanders nudged her, and mouthed a dismissive comment that she had heard from someone in the crowd: "Oh! Telephony and blue glass [for which there was then a fashion] will both go the same way!" A rising cacophony of stamping and catcalls began.

Finally, to Mabel's intense relief, Alec appeared on the stage and the audience broke into thunderous applause. Always at his best when teaching, he spoke fluently and easily, without notes. He described to his audience not only how his invention worked but also its purpose. It was not just an electric toy that could link two speakers, he insisted. He envisioned a central office system that could connect telephones in various locations with a "switch." His listeners shook their heads skeptically as he went on to suggest that, within a few years, people in different cities would be able to call each other up and have a chat.

Then it was time for Tom Watson, sitting in the laboratory eighteen miles away in Boston's Exeter Place, to speak down the line. Mabel described the excitement to her mother: "Alec requested the people to keep perfect silence and the room became so still that I felt it and hardly dared to breathe. The voice was heard to the farthest end of the hall, and you ought to have seen how they laughed and cheered." The sound of Watson's greeting ("Hoy! Hoy!": the salutation he and Alec always used), renditions of "Auld Lang Syne," "Yankee Doodle," and "The Last Rose of Summer," and remarks about an engineers' strike on a local railroad were clearly audible. "My singing was always a hit," Tom later wrote. "The telephone obscured its defects and gave it a mystic quality. I was always encored!" Various prominent Salem citizens then spoke to Tom. Mabel knew by the crowd's reaction that the demonstration had been a greater success than Alec had even dared hope; she herself, of course, had been unable to hear Tom's disembodied voice.

A *Boston Globe* reporter pronounced the evening "an unqualified success." The *Providence Star's* reporter gushed over the inventor, "a tall, well-formed gentleman in graceful evening suit, with jet-black hair, side-whiskers and mustache." The reporter was not the only person smitten by the star. Mabel reported that "some ladies even spoke into the telephone. Alec could not get rid of them."

Newspapers far beyond Salem picked up the story of Alec's lecture. The *New York Daily Graphic* carried an illustrated account, along with a full explanation of how the telephone worked. A month later, the *Athenaeum* of London picked up the story, and in April, Paris's *La Nature* published an article entitled "Le télégraphe parlant: téléphone de M. A. Graham Bell." After seven lean years, Alec, now thirty, could finally dare to imagine a prosperous future. He decided to charge $200 to any organization that wanted to sponsor one of his lectures. As soon as he had some money in his pocket, he commissioned a special gift for Mabel: a tiny silver model of a telephone, which cost him $85. Tom Watson was shocked at this "bit of extravagance."

Alec's lectures up and down the eastern seaboard, including dates in Boston and Springfield, were usually a hit. Alec had inherited his father's and grandfather's stage presence, and he would stride confidently onto the platform, cutting an impressive figure in (thanks to Mabel's insistence) a well-brushed coat, his hair neatly trimmed. He captivated his audiences with his ability to translate scientific language into simple English before enthralling them with the sound of a voice transmitted from several miles away. In March 1877, more than two thousand people braved a snowstorm in Providence to see a demonstration of the telephone by its inventor. "After the lecture," Alec wrote to Mabel, "I was besieged by 'Autograph Hunters'!!!"

Alongside the widespread enthusiasm, however, there was a sense that the telephone had sinister implications. Alec's act was certainly good entertainment. As the *Providence Star* editorialized, "the sensation felt in talking through eighteen miles . . . leaves the spiritual séance away back in primeval darkness." But hearing voices when no one was present was historically considered evidence of insanity, and few people could understand how electricity, whatever its other miraculous qualities, could produce or convey a human voice. That, after all, was a gift from God. The *Providence Press* suggested that "[i]t is difficult to really resist the notion that the powers of darkness are in league with it." The *Boston Advertiser* stated, "The weirdness and novelty were something never before felt in Boston." The *New York Herald* found the telephone "almost supernatural."

Back in Boston, as Mabel carefully pasted the newspaper reports into Alec's already bulging scrapbook, she must have smiled at the thought of her dearest Alec successfully jolting smug New Englanders out of their skepticism. But she was her mother's daughter. Although she was happy for Alec to give public lectures, she didn't want his dignity diminished. "Why do you let them speak of you as *A.* Graham?" she asked him in a letter. "I personally hate it when I think how handsome the full name is."

The stress of public performance soon took its toll on Alec, and his mood swung between elation and depression. "I have a little of the volcano in my composition and I often feel as if I shall go mad with the feverish anxiety of my unsettled life," he confessed to his fiancée in April. "Thank heaven you do not know what it is to be drifting about the world by yourself, without a place you can call your own." But marriage to Mabel continued to seem beyond his reach. He had no income from the telephone, because Gardiner Hubbard was still developing his market strategy for selling or leasing equipment. Hubbard was also trying—and failing—to find backers for the venture. In Chicago, the inventor Elisha

Gray was now contesting Alec's patent, despite his friendly overtures in Philadelphia the previous year. Alec was so convinced that in the end the Hubbards would not allow Mabel to marry him while his finances remained insecure that he decided to launch himself into another money-raising lecture tour. But the initial novelty of the "talking tele-graph" was wearing off. When Alec spoke in May at Chickering Hall in New York City, a *New York Times* reporter gave a lukewarm account of "Prof. Alexander Graham Bell" and his "exhaustive discussion of the transmission of sound . . . illustrated by a number of complex and not very intelligible figures cast upon a prepared background by means of a stereopticon." The reporter was mollified only when he managed to hear on Prof. Bell's telephone receiver the popular tune "The Sweet By and By" being played on an organ in New Brunswick, New Jersey, thirty-two miles away.

It was the business world that would seize the potential of the speak-ing telegraph. An enterprising engineer set up the first experimental telephone exchange in Boston on May 17, 1877; it connected Brewster, Bassett and Company, bankers, the Shoe and Leather Bank, and the Charles Williams Company. At last, Gardiner Hubbard decided that Alec's invention was sufficiently reliable, and commercial production at the Charles Williams workshop moved into high gear. Gardiner clev-erly opted to lease telephones rather than sell them so that he and his partners could maintain control. In June he drew up the legal docu-ments for the formation of the Bell Telephone Company, with himself, Alec, Tom Watson, and Thomas Sanders as shareholders. "We shall soon begin to reap the rewards of Alec's invention," Gardiner wrote triumphantly to their lawyer. The following month, Alec sold a part interest in his English patent for $5,000 to a Rhode Island cotton bro-ker named William H. Reynolds. In early July, Watson reckoned there were 200 telephones in service. Within a month that figure had risen to 778, and by the end of August to 1,300, although these were almost entirely for dedicated single lines rather than for systems radiating out of manually operated switchboards. In early 1878, the world's first commercial telephone exchange would be installed by eager entrepreneurs at New Haven, Connecticut. Six months after that, Canada got its first telephone exchange, in Hamilton, Ontario.

One of the first customers was Sam Clemens (better known as Mark Twain), in Hartford, who persuaded his boss at the *Hartford Courant* to connect his home with the newspaper's offices. Twain would also be one of the first writers to incorporate the telephone into a short story. In his 1880 sketch "A Telephone Conversation," the narrator eavesdrops on two women speaking about such topics of universal female concern as children, scandal, and church. Within only a couple of years, now that a few privileged users were listening to the familiar voices of friends and family rather than to strange voices in public demonstrations, the telephone had lost its "weirdness."

This explosion of commercial activity finally persuaded Gertrude Hubbard—and, more particularly, her husband, Gardiner—to give their permission to Alexander Graham Bell to marry their daughter.

Once Gardiner Hubbard had announced to Alec that the wedding could go ahead, Mabel and her mother turned their attention to the bridal trousseau. But there were severe restrictions on Mabel's choice of gowns and colors. A few months earlier, Gertrude Hubbard's mother had died, so the rigorous demands of North American mourning protocol had enveloped the Hubbard family. All the women were shrouded in black crepe from head to foot. Mrs. Hubbard and her eldest daughter, Sister, were expected to wear only black and white for two years after Grandma McCurdy's death. Mabel, Roberta, and Grace might wear lavender after they had observed full mourning for twelve months. In a tight little group like the Hubbard women, it was inevitable that Mabel would insist on conforming to family tradition.

"You English, for you are English, wear mourning for a very short time," Mabel explained to her fiancé. "Here it is the custom to put it on seldom but when you do, you wear it for a long time. [Mama] said she felt very strongly about it; if I put on colors, it would seem as if you had taken me away from my family, that I cared no longer about them and their sorrow." Mabel assured Alec that she wished their wedding to be a happy one, but "Berta and Grace wear black until winter, and the Gertrudes until next winter, and I should feel badly if my dress separated me from my own family."

Alec knew that this was a crucial moment in his future marital politics: it was a test of how far Mabel would be prepared to leave her family's circle. But he also knew that it was a test that he, rather than Mabel, must not fail, as her family was so important to her. For once he was moderate. He told her that she should do whatever she felt best—but he did persuade her that her wedding gown need not be black.

Bride and groom agreed that the ceremony should take place in the parlor of the Hubbards' Brattle Street home—the same parlor in which Alec had first demonstrated to Gardiner Hubbard the properties of piano strings, and in which he had poured out his love for Mabel to her mother. Only Mabel's family and a handful of friends were invited to attend the wedding. Alec's parents were not present—probably to Alec's relief, as he knew the social gulf that yawned between the two families. On July 10, 1877, a bower of Madonna lilies saved for the occasion by the gardener was constructed in the bay window at the front of the Brattle Street drawing room, and the dining-room chairs were arranged around the perimeter of the room. The following day, Wednesday, July 11, Tom Watson and Alec's new junior assistant, Eddie Wilson, took the horsecar from Boston to Cambridge, self-consciously clutching the first pairs of white gloves either of them had ever owned. Before knocking on the Hubbards' front door, they surreptitiously pulled the soft kid gloves onto their calloused workmen's hands. Inside, they found Mabel standing like a queen, surrounded by family. Her thick, wavy hair neatly dressed in an elaborate coil, she wore a simple white gown with Alec's wedding gift to her, a cross of eleven pearls, at her throat. Her large gray eyes barely left Alec's face, and she looked astonishingly young. But Alec, now thirty, knew that his nineteen-year-old wife already had far more business sense than he would ever have. In addition to the pearl cross, he had also presented her with all but 10 of his 1,507 shares in the Bell Telephone Company, which constituted about 30 percent of the total. This gesture horrified Mabel's parents, but by now they had learned to be surprised by nothing that Alec Bell did.

From Cambridge, the newlyweds traveled to Niagara Falls for a week's honeymoon. They stayed in Clifton House, the grandest of the resort's hotels, in which well-heeled visitors could enjoy gardens, billiard rooms, baths, and nightly balls, as well as spectacular views of the falls from balconies and verandas. (Clifton House even kept a pack of hounds for any English visitors who wanted a spot of hunting.) Niagara Falls was a popular venue for honeymooners, who thrilled to the raw power of the waterfalls and the feats of human daring they inspired. Only a few years before the Bells' visit, the Italian tightrope-walker Maria Spelterini, sporting a risqué costume of scarlet tunic, green bodice, and flesh-colored tights, had dazzled the crowds by performing her high-wire act with her feet encased in peach baskets. Mabel wrote an enthusiastic letter to her mother about the excitements of the place, including a description of her trip on the *Maid of the Mist,* the pleasure boat that took them immediately below the falls. Although she could not hear the water's thunderous roar, she could feel the boat shuddering and vibrating with its extraordinary force. The untamed power of so much water made Mabel feel "oppressed and frightened," and she was "glad to get away from [the falls] and admire them from a distance." Mabel kept no private journal of her first weeks as a married woman, although she did secrete in a Clifton House envelope a sentimental memento—an unruly tangle of her thick, golden brown hair. Nor did she confide to her mother any details of married life with Alec, except for her exasperation at his slowness in the morning: "I have at last succeeded in waking my Scotchman, a feat which took more than ten minutes to accomplish."

One week after the wedding, Alec and Mabel arrived in Brantford. Melville Bell had made a brief visit to Boston earlier in the year, but this was Mabel's first encounter with her mother-in-law. As Alec helped her down from the cutter that had brought them from the station, Mabel smiled shyly at the plain little woman with a prim bonnet and starched lace collar who awaited her on the porch of Tutelo Heights. Eliza smiled back and then, to Mabel's total surprise, walked down the steps toward her, lifted her hands up, and broke an oatcake over her head. Alec roared with laughter at the startled expression on his bride's face and explained that this was the traditional Scottish welcome to a new bride, symbolizing the promise that she would never go hungry. Mabel brushed the crumbs off her hat and stepped gingerly into the house, wondering what other Scottish customs she would encounter.

Once in the parlor, Mabel admired Eliza's paintings, mostly copies of Old Masters, which covered the walls. But she also noticed the modesty of the Bell establishment—and the distance Alec had traveled when he was accepted by the Hubbard family. "It is so funny to see him here, he is so different from them all, and actually seems a great dandy," she told her mother. A couple of days later, the new bride donned her "brown silk and white grenadine" gown for a large party, at which Eliza served ham sandwiches and a trifle cake ("a queer mixture of sponge cake, strawberry preserves, [homemade] wine and custards"). Champagne was poured for a motley crowd of farmers, lawyers, local politicians, and drivers.

The young Bostonian was quite taken aback, and struggled not to allow her lip to curl like Cousin Mary's. "Such a democratic assemblage I never saw," she wrote home. But she couldn't resist noting that the Mohawk chief George Johnson was present, and was "as intelligent and civilized looking as most of the people there." The downhome Ontarians, for their part, must have found Mabel quite an exotic bird. Tightly corseted, and with ruffles, beaded fringes, and a rosette embellishing her silk skirt, she was far better dressed than any of the female guests. So was Alec, who sported the grosgrain-edged topcoat, creaseless striped trousers, stiff collar, and gray silk bowtie that he had worn for his wedding. But before the evening was out, he had removed the coat, loosened his tie, and sat down at the piano to play a few Scottish favorites, including "Bonnets of Bonny Dundee," and some popular American tunes such as "Comin' through the Rye" and "Green Grow the Rushes, Oh!"

Alec and Mabel returned to Boston by train in early August so that Alec could attend the first shareholders meeting of the Bell Telephone Company. The company, which had been created the previous month, was a voluntary association, unincorporated and without any declared capitalization; it consisted of a declaration of trust in which Alec assigned all his telegraphic patents, past and future, to Gardiner Hubbard as trustee with the responsibility for administering the company. Voting rights and five thousand shares were divided among the board of managers: Alec, Gardiner Hubbard, Gardiner's brother Charles Eustis Hubbard (who had ten shares), Thomas Sanders, and Thomas Watson. Sanders was elected treasurer; Alec, electrician (in charge of research and development); and Tom Watson, superintendent (in charge of manufacturing). By now, Charles Williams's machine shop in Boston, where Alec and Tom had conducted some of their first experiments, was manufacturing telephones at the rate of twenty-five a day. Thanks to Gardiner Hubbard's efforts, there were buyers for Bell telephones all over the eastern seaboard and into the Midwest. Before the month was out, an enterprising Melville Bell (to whom Alec had given the Canadian patents to his invention) had connected the Ottawa office of Prime Minister Alexander Mackenzie to the residence of the Canadian governor general, the Earl of Dufferin, and rented them two telephones, for $42.50 each per year.

By then, Mr. and Mrs. Alexander Graham Bell had left New York City bound for Scotland aboard the 2,713-ton steamship *Anchoria*. Mabel discovered to her delight that the *Anchoria*, a three-masted, one-funnel vessel built in 1874 at a shipyard in Britain's Barrow-in-Furness, was a thoroughly modern ship: "All the bells in our rooms are electric bells. I think this such an appropriate steamer for Alec to sail in, it is so full of electricity." Unfortunately, the bells didn't work, which made them, as Mabel noted in the joint journal that the newlyweds briefly kept during the voyage, "a tantalizing mockery." Before they were halfway across the Atlantic, however, Alec had fixed them. He had also set up telephone wires between the wheelhouse, the captain's bridge, the smoking room, and the drawing room, to the delight of both crew and passengers.

Mabel flourished on board ship—she was immune to seasickness, thanks to her damaged middle ear. Alec temporarily developed a form of sciatica in his leg that Mabel fussed over. "Please don't let Cousin Mary know," she wrote to Cambridge, "or she would think she was right in thinking I have married a 'broken-down invalid' which is by no means the case!" Alec was soon well enough to sit for Mabel while she sketched his portrait. Alec recorded his indignation in their journal: "She represented me as a bald-headed man with very round shoulders and most uninteresting physique."

Chapter 10

LONDON LIFE
1877–1878

Mabel Bell clung to the rail as the *Anchoria* steamed up the Clyde River to Greenock, Glasgow's port, in the early hours of an August morning. The pre-dawn breeze chilled her face and tugged at the tendrils of hair she had hastily swept up into an untidy chignon. She smiled with excitement as she made out the shapes of buildings and hills on the shore and the flocks of seagulls swooping overhead. As the sun slowly rose, she could see the Clydeside shipyards, where one-fifth of the world's ships were being built, extending for miles along each riverbank.

She turned when she felt a gentle tap on her arm. Her smile widened as she realized that the prospect of landfall had triggered a small miracle: her husband had risen before noon and was already shaved and dressed. As first-class travelers, the Bells would be the first passengers permitted ashore. Alec could barely wait to set foot in what he called, in an assumed Scottish accent that made Mabel laugh as she struggled to read his lips, his "ain countri." He listened to the fierce screaming of the gulls, but he was oblivious to the oily stench drifting across the water, which made Mabel wrinkle her nose.

Once the ship docked, the Bells quickly passed through customs and found a hansom cab to take them through the cobbled streets to the Queen's Hotel. Mabel was at first appalled by her surroundings. "A blacker, smokier, foggier place I never wish to see," she confided in a letter to her mother. Smokestacks, brick factories, and fiery, glowing iron and steel foundries ringed the city, which during these years was the industrial workshop of the British Empire. The elegant remnants of eighteenth-century Glasgow, an intellectual and mercantile center, were hidden behind squalid tenements, home to the poorly paid Irish workers who had flooded into the city over the previous seventy years. The air was thick with pollution, and the once-creamy stone of Glasgow's churches, City Chambers, and railroad station was encrusted with soot and grime. Through the cab windows, Mabel saw shabbily dressed women and barefoot children hurrying through narrow streets toward textile mills.

But Alec was jubilant. Just seven years earlier he had reluctantly left Britain, bound for "the backwoods of Canada" in the hope of regaining his health. Now he was a robust and happy man, returning with a new wife and a growing reputation as an inventor. As soon as he and Mabel had eaten breakfast, he disappeared to find some childhood friends and visit the local telegraph office. Mabel unpacked their bags and found her pen so she could write home. As she settled herself at the writing table, she realized she could either spoil her husband's homecoming with grumbles or find a silver lining to Glasgow's clouds of soot. She took a deep breath, then dipped her pen in the ink. "The buildings are all solid, handsome edifices," she informed her mother, who, alongside her husband and three remaining daughters, had recently moved to Washington, D.C. "Their very blackness is a rest and delight after the shiny, showy newness of America."

Mabel had more reason to be positive later the same day, after a trip up one of the hills overlooking the city for a tour of the University of

Glasgow. She admired the "huge and heavy, ponderously gothic" buildings, assuming they predated anything she knew in the New World. She couldn't have been more wrong. In fact, the ancient university had left its original cramped facilities and relocated only seven years earlier, onto a splendid campus designed by the architect Sir George Gilbert Scott. But the stonework was already encrusted with the filth that coated the city.

Alec was too busy being "wild over the wonders of science shown to him" to notice his surroundings. Having struggled and failed to complete a university education himself, he was entranced with the university's state-of-the-art laboratories, which made his little Boston workshop look like a child's science set. He had barely examined the apparatus before he was dragged away by the postmaster and superintendent of telegraphs to give demonstrations of the telephone. The following day, he gave a further demonstration to a larger audience, which included Glasgow's business and civic leaders and reporters from the local newspaper. "He said before the experiment the gentlemen were very polite to him, but after it they treated him with veneration, one gentleman said after this he would believe anything he heard from America no matter how wonderful or incredible it might seem," Mabel told her mother.

Alexander Graham Bell and his "talking telegraph" were soon all the rage, and the Bells were swept along on the wave of celebrity. Alec planned to spend three or four months in Britain, working with his partner William Reynolds, the Rhode Island cotton broker who had purchased a part-interest in the English patent, to set up a British Bell Telephone Company, so the publicity was invaluable: it made finding investors and agents much easier. Nevertheless, inundated by invitations to speak, Alec was torn between his desire to be with his new wife and his obligation to his invention. Mabel was dazzled by the attention from titled grandees ("Alec says I will be Lady Bell yet") but frequently found herself alone in the Glasgow hotel, watching the rain fall on the statues of Queen Victoria and Sir Walter Scott in the square outside her window.

When Sir William Thomson, who had been a judge at the Philadelphia Exhibition the previous year, asked Alec to attend the annual meeting of the British Association for the Advancement of Science, in the English coastal town of Plymouth, Lady Thomson invited Mabel to sit with her. The normally self-possessed nineteen-year-old suffered a momentary loss of confidence: "I am so frightened to think how ill I can act my part as 'Graham Bell's' wife. . . . O I wish I were safe at home. I'd give up all hopes of being Lady Bell and all the glories of being the wife of a successful inventor, everything, but being Alec's wife to come home to quiet humbleness."

Alec himself was beginning to show signs of nervous strain: the sciatic pain in his right leg returned, and he complained of headaches. But his appearance at Plymouth was a triumph. The butterstamp telephones performed beautifully, and Sir William gave his opinion that this was one of the most important inventions he had ever seen. Up until now, most British scientists had regarded the telephone as "a scientific toy of no commercial value," Alec later recalled, "but when Sir William Thomson spoke, the world believed." The *Exeter and Plymouth Gazette* reported that "[t]he telephone is beyond all measure the lion of the Association meeting."

Alec had written to his mother from the *Anchoria*, "I cannot tell you what a longing I have to see again the places I remember so well, London, Bath, Edinburgh and Elgin." At the end of September, he disentangled himself from telephone business and headed north with his wife. They visited the house on Edinburgh's Charlotte Square in which Alec had grown up, then traveled to northeast Scotland so that Mabel could see Elgin, the little market town where a sixteen-year-old Alec had spent a happy eighteen months teaching music and elocution at Weston House.

Alec had also planned a second honeymoon on the Moray coast, with Mabel and himself "roughing it" in a seaside cottage and living off fish they caught themselves. Mabel had some reservations about her husband's romantic fantasy. "Neither one of us ever saw a fish cooked much less ever did it ourselves," she told her mother. "Alec appears to think it a very simple operation but I have an idea that the fish has to be opened and cleaned, and a part of its inside taken out first—Is it so?" But Alec was unstoppable. Undeterred by the sharp autumn wind, he found a pony cart to take them to the tiny village of Covesea, six miles north of Elgin. When they arrived at the straggle of stone cottages perched on the cliff top, he persuaded a surprised local woman, Mrs. Cameron, to rent them a room in her house. Mrs. Cameron watched indulgently as her newlywed American guests proved hopeless house-keepers. Neither of them could get a kettle to boil on the open fire, and the smoke from the fireplace induced a migraine in Alec. Alec's idea of a week's provisions had consisted of only one loaf of bread, half a pound of sugar, and a pound of butter. Their landlady took pity on these American ingenues. "To my great joy," Mabel wrote home, "Mrs. Cameron now boils our eggs, tea and potatoes for us." Mrs. Cameron's "lassie" walked to the next village of Lossiemouth to restock.

The fish-and-oatmeal diet was monotonous, and the bed was too short for Alec, but the young couple were determined to enjoy them-selves. Each day they walked along the wonderful, firm sands of the Moray coast, or over the red sandstone cliffs, until they found a spot where Mabel could sit, sketch, and read while Alec wandered off. Sometimes he collected rocks: other times he took a pistol, to shoot rabbits. One day, back in the cottage, he used up the whole week's supply of sugar lumps by investigating whether it was the sugar itself or the air in the sugar lump that caused air bubbles to rise to the cup's surface as hot tea caused the lump to melt. "What a man my husband is," Mabel wrote fondly. "I am perfectly bewildered at the number and size of the ideas with which his head is crammed. I give you a list: Flying machines to which telephones and torpedoes are to be attached occupy the first place just now from observations of sea gulls. Practicability of attaching telephones to the wire fence. His mind is full of these two. . . . Then he goes climbing about the rocks and form-ing theories on the origins of cliffs and caves. . . ."

In the evenings, Alec sprawled in the only comfortable chair, absorbed in Sir Walter Scott's novel *Fair Maid of Perth* and oblivious to drafts, discomforts, or his wife's gaze. Mabel used Scott's *The Marquis of Lossie* as a writing desk on which to write long letters home. She would occasionally look up to smile at Alec or stare at the forbidding picture of Queen Victoria that hung over the fireplace. They spent the last day on the beach, Alec "wild and full of fun, though rather ashamed that the inventor of the telephone should go wading." Mabel persuaded him that "he should not be the slave of his own position." By week's end, Mabel's skirts were torn and even Alec had had enough smoked herring. But happy memories of Covesea's stiff breezes, panoramic views, and isolation remained with both Bells, shaping their ideas of the ideal retreat from the world.

The second honeymoon was a welcome interlude before the Bells' next plunge into the universe of telephones. During their first four months in Britain, Alec gave at least twelve telephone lectures and demonstrations in England and Scotland, drawing crowds as large, on a couple of occasions, as two thousand people. So many people turned up to hear his talk to the London Society of Arts and Manufactures that it had to be repeated a few days later. He also gave many private demonstrations, including one from the depths of a Newcastle coal mine to the surface, and one in London between divers in the Thames.

Alec was, however, finding it far harder to set up a British Bell Telephone Company than he had anticipated. He faced a thicket of regulations, plus challenges to his patent. The telephones that Tom Watson had made for him to take across the Atlantic were subject to breakages, and there was no one with Tom's skills available to fix them or to manufacture new ones. There were rumors that the telephone might carry diseases as well as messages, and that others on the line could eavesdrop on private conversations. Alec's partner, William Reynolds, was not proving to be much good at getting the business on its feet. Various British instrument makers were infringing Alec's

British patent, and now an emissary from Thomas Edison, Alec's nemesis, had arrived from the United States and was both promoting an Edison telephone and trying to lure Reynolds away from the Bell model. Mabel told her parents that her husband yearned to return to America "but he thinks he can be of much more use here."

There was an additional reason to keep the Bells in Britain longer than expected. "My darling, darling, sweet, little, wee Mother," Mabel wrote to Gertrude Hubbard, in a letter marked *Private*, "O Mamma darling, if I only could put my arms tight around you and ask you if it may really, really be true what Alec and I are just beginning to hope may be." A month before the Covesea adventure, Mabel had started complaining of feeling ill in the mornings, and had then gone on a shopping spree for "little flannel wrappers." In mid-October, Alec insisted that Mabel consult a doctor as to whether a transatlantic voyage was appropriate for a woman in what everybody euphemistically described as "a delicate condition." After the medical consultation, Mabel told Alec that he himself must break the sad news to his own parents that it would be several months before they would see their beloved only son. Alec dutifully wrote to "My own dear Papa and Mama": "I long for one look at your dear faces and for the sound of your voices again. But . . . I write specifically to explain our sudden change of plan." The doctor had said the voyage did involve some risk to Mabel, so "we shall take a furnished house for six months with the privilege of remaining for a year if advisable."

For all her matter-of-fact tone, Mabel was apprehensive. The prospect of giving birth in an unfamiliar country would have intimidated any young woman in the late nineteenth century, when women often died in childbirth from septicemia and when babies often didn't see their first birthdays. It was even more challenging for Mabel, far from her family and still struggling to lip-read a wild assortment of different accents ("bath" was pronounced differently everywhere she stayed). Since husbands never attended doctor's appointments or delivery rooms, who would accompany Mabel through pregnancy, childbirth,

Mabel, aged twenty, in the brown silk dress she wore for the Brantford party.

and nursing, or even help her outfit herself with a new wardrobe during her pregnancy and prepare a layette for a baby? In addition, the Bells now had to find a home and establish their own household in London. A middle-class couple like them would be expected to live in one of London's better neighborhoods (Kensington or Regent's Park were Mabel's preferences) and to have at least two servants. Alec believed that it was important, in status-conscious England, "to make as good a show as possible." Her own helplessness in the face of all these challenges, particularly when Alec repeatedly disappeared to spread the telephone gospel, allowed a note of despair to creep into Mabel's letters home. "Alec is at Birmingham," reads one note, "and I feel so stupid with nothing to do but write of which I am dead tired having done nothing else. If only I had some sewing. . . ."

Gertrude Hubbard's immediate response was to announce that she would leave her husband in Washington and join her daughter in London as soon as possible. However, Alec solved the servant problem by recruiting Mary Home, his grandfather's former housekeeper at Harrington Square, to come and live with them. Mabel explained that "[w]e propose to have one servant to do general work and Mary Home will manage her and relieve me of all the trouble while, Alec says, she will let me have my own way in all things. She knows all about the stores and marketing and would see that I did not get cheated. . . . She can also do a little sewing for me, and go out shopping with me thus saving the expense of a woman like the one I am employing now." The only problem was that Mary's front teeth were missing, which made it hard for Mabel to follow what she was saying. Emma, the little maid hired to help Mary, was no easier, because she had such a thick Cockney accent and was "wonderfully stupid about everything outside her own duties."

By mid-November, Mabel was over morning sickness and back to her own sturdy self. She urged her mother not to come until the spring: "Alec . . . can take care of me." Alec had found a house in South Kensington, and toothless Mary Home introduced Mabel to the challenges of housekeeping in the world's largest and dirtiest city. Mabel was astonished when Mary told her she would need two dozen dusters at least, including "round cloths, kitchen cloths, house flannels and I know not what else." She still had to come to terms with London's "blacks," as Londoners called the flakes of soot that floated in the air, marking everything they touched, inside and outside. Mary Home took Mabel to Hitchcock and Williams's wholesale shop, "where all the best Dresden table damask comes from," to purchase tablecloths, sheets, pillowcases, bolster covers, napkins, and half a dozen damask towels. Alec presented his wife with a housekeeping allowance of £10 a month to cover rent, coal, gas, wages, and food (in today's currency about £700, or US$1,400), and with a book called *Common Sense Housekeeping*. Mabel pored over the esoterica of housekeeping, marveling that "a splendid fire is made with ashes sifted and mixed in a little water with coal dust mixed to a paste," adding, for her mother's edification, "It must be baked in the fire and the poker not used."

The Bells moved into 57 West Cromwell Road on December 1. It was a typical mid-Victorian four-story house, part of a block-long terrace of similar houses constructed of gray brick with creamy stucco facings, each with two chimneys on the roof and a well below sidewalk level with a basement entrance. In those days, West Cromwell Road was an unremarkable street on the edge of London, lined with elm trees and stretching from Earl's Court in the east to Hammersmith Cemetery in the west. It was, however, safely middle-class, within walking district of the more rarefied (and expensive) neighborhood of Kensington High Street, where Mabel would be able to enjoy well-stocked department stores like Barker's. Closer to home were the busy little shops along Fenelon Road, where Mary Home could patronize the butcher, greengrocer, bootmaker, and confectioner. West Cromwell Road also had the advantage of being near Earl's Court Station, on the newly opened District Railway line, which connected the area with the city and all the main-line train stations.

The West Cromwell Road house had seventeen rooms and many modern features, including gas lighting throughout and a bathroom with hot running water on the second floor. To an American eye, the rooms were small and cramped, and Victorian clutter—pots of ferns, china ornaments, knickknacks made of ivory or ebony, occasional tables—added to the claustrophobia. Alec appropriated as his workspaces three rooms, one on each of the upper floors, which he would connect with telephone wires. Only one room was big enough to hold a double bed, and Mabel started fretting about where her mother and sister would sleep when they arrived the following spring. But by the third evening, as she sat in the dining room with the gas jets lit, the heavy curtains drawn, and the brass fender reflecting the flames in the tiled fireplace, Mabel wrote to her mother that life in the new house was "fun."

Mid-pregnancy, Mabel enjoyed a burst of energy. She turned her attention to organizing domestic routines—key to every Victorian homemaker's self-image. She established a rigid cleaning routine, with Emma the Cockney maid dusting the study, drawing room, and dining room every day and the halls and bedrooms every other day. Mabel kept Boston standards firmly in mind: "I cannot see that even Cousin Mary would think the house dirty," she boasted. Next, she drew up a weekly menu of unappetizing monotony: "I had stewed beef for dinner Saturday; Sunday . . . sirloin of beef, potatoes, brussel sprouts, bread pudding and fruit; Monday the beef warmed, potatoes ditto because I was away; yesterday mutton, potatoes, cauliflower; today we are to have the mutton cold with potatoes and macaroni." She asked her mother for domestic tips: "I should like your recipe for . . . floating islands. . . . How do you make fish balls, and what can I do with the remains of cold chicken?" Two months before her wedding, Mabel had written to Alec's mother that "like a true Briton [Alec's] spirit depends on his having a good dinner. I am beginning to learn that my happiness in life will depend on how well I can feed him." Regular, nutritious meals were one way of providing some stability in both their lives.

Mabel even established some control over Alec's unconventional working hours. She forced him to get out of bed for breakfast at 8:30 each morning ("It is hard work and tears are spent over it sometimes"), and she persuaded him that, after they had dined together at seven, he would not return to work in his study until 10:00. His wife's entreaties may have irked Alec, but the daily routine agreed with him. Now that he was married, he was no longer living on his nerves—and it showed. "Why Mamma dear," Mabel reported, "he had his wedding trousers made larger sometime ago, and who would have thought they would so soon be tight again." On Christmas Day, he had stooped to extinguish the candles on the Christmas tree, and his trousers had burst. The man who had been skinny since childhood and had weighed only 165 pounds on his wedding day now topped 200 pounds. From now on, he was invariably described, in that wonderful Victorian euphemism, as "majestic."

After Christmas, Mabel made her first attempt at entertaining. Alec had invited for dinner William Preece, who, as the post office engineer-in-chief, was one of the most learned electrical engineers in the country and therefore an extremely useful business contact. Mabel cast aside cold mutton and macaroni in favor of a much more ambitious menu: jugged hare followed by her mother's famous dessert, floating islands. Her mother had not yet supplied the recipe for floating islands, an exotic concoction of poached meringues on an egg custard, but Mabel fearlessly soldiered on. "Miss Home undertook to stir the custard while I industriously beat the egg whites, of course flurried as I was about Alec, I let her go on until it boiled. That meant ruin. We took four more eggs, and I stirred while she beat. I forgot to stir, and the first thing I knew the milk was all over the floor, that meant ruin No. 2. We tried some more milk and after Herculean efforts the custard got safe through and was sent to cool."

By now, it was five thirty. Mabel ran upstairs to change, thinking that Mr. Preece would arrive at seven. But just as she started to braid her hair, she remembered that Mr. Preece had been invited for 6 p.m. She pulled on her gown and rushed downstairs again, only to meet Mr. Preece coming up. It was an awkward moment: Alec, who had a bad cold, had stumped off to bed with instructions to be woken only after Mr. Preece had arrived; Mary Home hissed that the hare would not be satisfactorily jugged before seven; the drawing room, Mabel realized with horror, was slowly filling with smoke—Emma, the maid, had forgotten to open the damper before she lit the fire. Mr. Preece beat a hasty retreat down West Cromwell Road, assuring his hostess that he had to make a call but would return in an hour.

Poor Mabel. Her thick hair was falling out of its pins, her dress was disheveled, she was five months' pregnant, and her first dinner party was heading toward disaster. Her face often wore a look of anxiety as she tried to follow conversations, but she was now in a panic. Nevertheless, she took a deep breath and rose to the occasion. She finished dressing, then set the table herself to ensure it was done correctly (this was a skill young ladies in Boston were taught) and told Emma to stay in the dining room throughout the dinner so there were no periods of *longueur* between courses. Mr. Preece returned, Alec woke up, the hare was cooked. But just as they were about to sit down, there was a ring at the doorbell. It was the doctor, to check on Alec's cold. Once again, events spun out of control. There was no gravy for the hare ("Mr. Preece was surprised and I was mortified.") The handle fell off a pitcher of water, causing the contents to soak the table. Alec had barely returned to the table after seeing the doctor when he disappeared again halfway through the meal to see a friend. Emma left the room, so Mabel had no one to help her serve. Half the floating islands custard slopped over the rim of the dish in which it was served.

It was a good thing that all present shared a sense of humor. Mabel wrote a cheerfully self-deprecating account to her mother of her first dinner party as "the worst failure I ever saw." Mr. Preece forgave the lack of gravy, and was one of the sponsors, a few weeks later, of the special general meeting of the Society of Telegraph Engineers held in London "for the purpose of welcoming Professor Graham Bell to London." And Alec roared with laughter and hugged his adored wife. Then he described to Gertrude Hubbard his pride in her daughter's accomplishments, and their closeness as a couple: "Mabel has hitherto been so much part of you that it seemed at first like tearing her life to pieces to remove her from your sheltering care. If there is anything that can console me for this cruelty, it is the feeling that the temporary separation has brought her nearer to me—that it has made us more truly man and wife than we could ever have hoped to become in America." He added that marriage had transformed Mabel from "the helpless clinging girl into a self-reliant woman"—an assertion that reflects more his romantic nature than reality. Mabel had never been either helpless or clinging. However, she too recognized the deepening ties of affection between them: "Instead of finding more faults in him, as they say married people always find in each other, I only find more to love and admire. It seems to me I did not half know him when I married him."

But even as the marriage flourished, telephone business wilted. Alec and Reynolds could not find a British manufacturer who could produce models to their specifications ("they say the British take a month to do what the Americans do in a week"). By now Reynolds had paid him in full for British patent rights, and Alec finally had a comfortable income from the capital that his father-in-law had invested for him. Alec continued giving public performances, since they yielded both fees and fame. Thousands of people attended his demonstration at the soaring glass Crystal Palace, designed by Joseph Paxton for the 1861 Great Exhibition in Hyde Park. Yet there seemed to be one hurdle after another before the British Bell Telephone Company could be established.

When the bell rang at 57 West Cromwell Road in early January, the elderly Mary Home, knowing her employer would not hear it, hobbled up the basement stairs and along the passage to open the front door. A boy in Post Office uniform handed over a telegram, addressed to Mr. Alexander Graham Bell. It was from Sir Thomas Biddulph, private secretary to Queen Victoria, asking when it would be convenient for Alec to show the telephone to the widowed queen at Osborne House, her residence off England's south coast on the Isle of Wight. The whole household was galvanized with excitement, as news of this unexpected invitation from the most powerful person in the British Empire filtered up and down the stairs.

Mabel was disappointed to discover that she was *not* invited to attend the demonstration, but Alec was soon preoccupied by the challenge of ensuring crystal-clear transmission between the council room in Osborne House (the Italianate villa designed by the late Prince Albert) and two other sites—a cottage on the Osborne House grounds and the little town of Cowes a mile to the north. On January 14, 1877, singers stationed in Cowes were poised to perform over the wires in the same way that Alec's Uncle David had performed in Ontario. A local electrician had made the queen her own polished walnut receiver, with little ivory nameplates and gold switches. Alec and Reynolds, splendid in white tie and tails, drove up to Osborne House at 8:45 p.m. and were shown into the Council Room. Meanwhile, Sir Thomas and Lady Biddulph sat by the telephone that had been installed in Osborne Cottage. Alec was frustrated to discover that the telegraph line to Cowes was broken, so the four-part harmonies from the singers stationed there would go unheard. But there was no time to fix this: a footman appeared at the door and announced that the queen was approaching. Minutes later, the swish of voluminous, crinolined skirts was heard, and a short figure appeared in the doorway. The courtiers bowed low as Her Majesty Queen Victoria glided into the room, accompanied by her youngest daughter, Princess Beatrice, and her son Prince Arthur, the Duke of Connaught and a future governor general of Canada.

Outwardly, Alec was every inch the respectful subject. But he did not allow the monarch's almost tangible regality to color his impression of her. "The Queen was humpy, stumpy, dumpy," he later told Mabel, who sent on the report to her mother. "Her ungloved hand looked like a washerwoman's so red and coarse and fat, and her face was fat and a bit reddish." Alec was quietly amused at the way that Her Majesty asked her gentleman-in-waiting to ask Professor Bell various questions, which the courtier then repeated to him "as though he were an Indian and couldn't understand." Yet, despite the problem in Cowes and the general hoity-toityness, the demonstration went well. Soon the royal party was chatting into the transmitters to the Biddulphs in Osborne Cottage and nodding enthusiastically as they heard the replies. At one point, Alec crashed through royal protocol by touching the queen's hand in order to get her attention. This was the way he always caught his wife's attentions, but the courtiers were appalled. However, Her Majesty was prepared to overlook the faux pas from this clever, albeit uncouth, "American."

After the royal party left the Council Room, Alec received a request from the queen that he demonstrate his invention to her "upper

servants." That night, Victoria wrote in her journal that the telephone
was "extraordinary," but added, doubtfully, "it is rather faint, and one
must hold the tube close to one's ear." The following day, *The Times*
carried a long account of the demonstration, under the heading "The
Telephone at Court," and reported that "Her Majesty . . . and the
entire Royal Household evinced the greatest interest." Sir Thomas
Biddulph asked Alec if the queen could purchase the telephones.
Alec had always intended to present them to her. "The story of Alec's
touching the Queen," Mabel told her mother, "has grown until the
latest version is that he took hold of her *arm and pulled,* and she
sweetly smiled."

The queen's well-reported interest spurred William Reynolds to
issue a prospectus for the proposed telephone company, with an ambitious projected capitalization of £500,000 (several million pounds in
today's currency). It was none too soon, in Mabel's view. "Wherever
you go, on newsstands, at news stores, stationers, photographers, toy
shops, fancy goods shops, you see the eternal little black box with red
face, and the word 'Telephone' in large black letters," she told Alec's
mother. "Advertisements say that 700,000 have been sold in a few
weeks." These "Domestic Telephones" were simply toys. More seriously, unscrupulous British manufacturers were now selling assemble-it-
yourself copies of Alec's invention: each purchaser bought a box containing the magnet, iron plate, coil, wooden handle, and mouthpiece
in it, with instructions on how to put it together in five minutes. Alec
was sick with outrage at this infringement of his British patent. His
partner, Reynolds, threatened legal action.

There were further pitfalls. Reynolds's proposed company ran into
difficulties because various of Alec's rivals contested his British patents via both the courts and letters to *The Times*. Moreover, the scientists at Plymouth may have been convinced of the telephone's value,
but the great British public was resistant. Since the nation boasted
both the best mail service in the world and an extensive and efficient
telegraph system, why did it need a speaking machine? "In America,"

commented *The Times*, "with long lengths of single wire, and a fine dry climate, the telephone may perhaps come into use practically. But in England, with most of the telegraph wires already overweighed, it is hardly likely to become more than an electrical toy, or a drawing-room telegraph, or at most a kind of electrical speaking tube." In the end, Reynolds raised only £20,000 through the sale of shares, and the company limped along, badly managed and undercapitalized.

By now Alec was realizing how little he liked the commercial side of invention. He far preferred developing a new call-bell system, or improving telephone reception. And the less attention he paid to the British Bell Telephone Company, the less confidence his fellow directors placed in him. At the same time, he failed to apply for a German patent, and he nearly lost the French patent. Meanwhile, Thomas Edison had devised a new kind of receiver, using carbon instead of acidulated water, that did not infringe Alec's patent and that worked better than the Bell model. Edison started selling Edison telephones in England. Alec grew steadily more depressed. "Business," he wrote to his father-in-law, "is hateful to me at all times." As the months in England dragged on and his hopes of cashing in on his invention slipped away, business became even more odious.

There was, however, one new role in which Alec delighted. "My dear Mama," he wrote to his mother, in Brantford, on May 10, 1878. "Our little baby has at last come into this world—unlike its Papa and Mama, punctual . . . such a funny black little thing it is! Perfectly formed, with a full crop of dark hair, bluish eyes, and a complexion so swarthy that Mabel declares she has given birth to a *red Indian!* . . . I can't say much about good looks . . . but she is our own baby and that is enough for us." Gertrude Hubbard had arrived in West Cromwell Road a few weeks earlier, along with Sister, to help Mabel with her confinement. Everything had gone smoothly, although "poor Mabel had a fearful time, and even begged to be killed, but . . . the pain and anguish have been quite forgotten in the happiness of the mother in the possession of her child." The baby was named Elsie May—Elsie (a Scottish variant of Eliza) after Alec's mother, and May for both Mabel and her own birth month.

Mabel quickly adapted to motherhood. Gertrude Hubbard was impressed by her daughter's domestic competence. "Mabel superintends her household," she wrote to Alec's mother, "as quickly and with as much ease and self-reliance as though she had been mistress of a house for years." Mabel kept Elsie in a pillow-lined drawer on her bed so the vibrations of the baby's movements at night easily woke her. And Alec privately assuaged his own unspoken fears about his firstborn child. A few days after his daughter's birth, he stole into Mabel's room, and standing behind the canopied bed, he blew a loud blast on a trumpet. He later told a friend that "Mabel never moved, but the little one flung out its arms and legs and shrieked in terror."

Sadly, financial anxiety undermined the domestic bliss. Both Alec and Mabel recognized they were spending money faster than Alec was earning it. Public interest in telephone lectures was dwindling, and the British Bell Telephone Company had still not been launched. In the late summer of 1878, Alec was angered to hear that both Elisha Gray and Thomas Edison were publicly challenging his telephone patent in Washington. Moreover, the unscrupulous William Orton had decided to ignore Bell's patent and manufacture a telephone that incorporated the inventions of Thomas Edison, Elisha Gray, and another Boston-area inventor, Amos E. Dolbear. Western Union was now illegally selling these telephones, which used the company's extensive network of telegraph lines throughout the United States. The American Bell Telephone Company, with Gardiner Hubbard at the controls, launched a suit against Western Union, accusing the company of infringing Bell's patent. It was a wildly uneven battle—an underfinanced, sickly David against a mighty, well-capitalized Goliath.

Mabel was outraged by these developments. "The more I think of it," she wrote of the accusations and the infringements of her husband's patent (he was giving telephone demonstrations in Ireland at the time), "the more it seems like a deliberate attempt not only to rob you of all credit in your own discovery, but also to convey the impression that you are a thief. . . ." At first, Alec tried to brush off the accusations, assuring his wife that "Truth and Justice will triumph in the end." But he gradually realized that his competitors were seriously damaging his reputation. By the spring, his father-in-law was suggesting to him that he should return to the United States to fight for his patent. Later that summer, Gardiner Hubbard's suggestion had become as close to a command as he dared to make it.

Predictably, Gardiner's manner and Alec's own dread of conflict and paperwork triggered a volcanic explosion. "I am sick of the telephone," he announced to Mabel, "and have done with it altogether, excepting as a plaything to amuse my leisure moments." Instead, he told her, he was going to return to his first love: teaching the deaf. Asked by a school in Greenock, Scotland, to find a Visible Speech teacher for the deaf, the celebrated inventor rushed north and installed himself into the dusty, unfurnished classroom as teacher for three young girls. "I have been absolutely rusting from inaction—hoping that my services might be wanted somewhere," he tried to explain to Mabel. Escape from telephone trials made him giddy and melodramatic.

> Now I am needed and needed here. I am not going to forsake
> my little school just when it is struggling for existence—
> though the telephone should go to ruin—and though my wife
> and child should return to America and leave me here to
> work alone.
>
> Let me go back to the work I love if I can find support for
> us both. Don't let me be fettered to an unwelcome task. I shall
> always work at Telephony, but let it be from a love of science,
> and from a wish to help on the advancement of knowledge.
> Don't let me be bound hand and soul to the Telephone, I
> don't want to make it my sole means of support. The position
> of Inventor is a hard and thankless one. The more fame a
> man gets for an invention, the more does he become a target

for the world to shoot at—while no one seems to think the
inventor deserving of pecuniary assistance.

Alec's pen scratched on as he worked himself into a fever of self-
sacrifice. "Of one thing I am determined and that is to waste no more
time and money on the telephone. . . . Let others endure the worry,
the anxiety and expense. I will have none of it. . . . A feverish anxious
life like that I have been leading since our marriage would soon change
my whole nature. Already it has begun injuring me and I feel myself
growing irritable, feverish and disgusted with life." Once a permanent
teacher had arrived at the Greenock school, he loftily informed his
wife, they would return to Brantford and he would find a teaching job
in Ontario. One of his final engagements in England would be a series
of lectures that he agreed to give in Oxford on "Speech."

Mabel must have been momentarily floored by this outburst, but
she knew her husband well by now. A mollified Alec reappeared in
London three weeks later to help her and Mary Home pack up their
household. Mabel bit her tongue when Alec booked a passage for
his family on October 31, 1878, aboard the steamship *Sardinian* to
Quebec instead of Boston or New York, as her father was expecting,
so that he could travel straight to Brantford without calling at the Bell
Telephone Company headquarters in Boston. Once at sea, Alec's temper improved as he watched their fellow passengers fuss over Elsie
May. He noted that Elsie's daily bath in the Ladies' Cabin was "a feature of attraction to all the ladies on board." Their fellow passenger the
Hon. Edward Blake, a prominent Canadian politician, "was so fond of
her that he would take her in his arms and trot up and down the saloon
with her many times every day."

From a business point of view, the Bells' year in Britain had not
been a success. Despite Alec's lectures and the demonstration for
Queen Victoria, the telephone was still regarded by the general public as a toy and by his rivals as an invention to be pirated. Moreover,
the British post office was eager to protect its monopoly on telegraph

traffic, and downplayed the advantages of voice-to-voice communication. Then there was the class issue. Those who, in Britain, could afford a telephone usually had servants to run errands and deliver notes for them, and were reluctant to absorb this newfangled, democratic American invention into the elaborate etiquette of the late Victorian years. In any household with aristocratic pretensions, nobody wanted to deal with a crucial question: If a telephone bell rang, should master or servant answer it?

The following year, Gardiner Hubbard would travel to London to rescue the faltering British Bell Telephone Company. He reorganized and recapitalized it, and found London lawyers to protect its patents. It was soon thriving, although Britain would long lag behind other Western nations in the spread of the telephone. But in the fall of 1878, Alexander Graham Bell's smooth passage across the Atlantic was a mere ten-day period of calm before the storm—a vicious and protracted legal battle in the United States.

Chapter 11

LITIGATION BATTLES
1878–1880

In early November 1878, the *Sardinian* steamed up the mighty St. Lawrence River, docking at Quebec City on November 10. Alec and Mabel, along with Mary Home, Annie (the English nurse-maid who accompanied them to Canada), and her charge, five-month-old Elsie, gathered at the top of the gangplank, staring up at the cliff above them and at the cramped streets, gray stone houses, and slender-steepled churches that clung to it. Alec, his woolen coat flapping in the wind, was happy to be home. All of a sudden, his attention was caught by the sound of a familiar voice yelling his name above the babble of porters, sailors, and reunited families. Searching the crowd, he picked out the boyish face and unruly dark hair of Thomas Watson, his Boston assistant. Sixteen months earlier, Watson had been named superintendent of the Bell Telephone Company, as well as one of its five shareholders. Alec's spontaneous smile of recognition was soon replaced by an angry frown. He knew why Watson was here.

Watson elbowed his way through the crowd toward the Bells as they disembarked. The young machinist was struck by how a year in

Europe had changed the thirty-one-year-old inventor. Alec's thick wavy hair was still inky black, but the once-bushy whiskers were now neatly trimmed, his frock coat had a stylish cut, and his greater girth gave him a gravitas he had previously lacked. There was a new confidence about Mabel too, as she directed Mary to help find their bags and checked that Elsie was still asleep in Annie's arms. But Watson could see, from Alec's furrowed brow and the grim set of his mouth, that this would be a difficult encounter.

With a one-tenth interest in all of Alec's patents, Tom Watson had a considerable stake in the Bell Company's future—as did Alec's two major business partners, Gardiner Hubbard and Thomas Sanders. Desperately needing Alec's testimony for their suit against Western Union, they had agreed that Watson should intercept Alec in Quebec City and bring him to Boston. The Patent Office required a preliminary statement by each claimant, setting forth his claim to the invention: the dates when he had conceived the idea and when he had first constructed a working instrument. Hubbard, Sanders, and Watson had filed supporting statements, but Alec was the inventor, so his evidence was pivotal. The Patent Office had already extended the deadline for Alec's statement several times; if he did not file it by late November, the suit would die and Western Union would be able to ride roughshod over Bell's patent claims. Gardiner Hubbard knew, however, that Alec had told Mabel he was determined "to waste no more time and money on the telephone. . . . Let others endure the worry, the anxiety and expense." This horrified his business partners, who faced bankruptcy unless he got involved. Thomas Sanders had already invested $110,000 (well over $2 million in today's currency) in Alec Bell's invention and had still not seen a cent in return.

"I found Bell even more dissatisfied with the telephone business than his letters indicated," Watson recalled in his autobiography. "He told me he wasn't going to have anything more to do with it." Standing quietly in the chilly November air, Watson listened to Alec rant that he hated the world of commerce and the slurs on his integrity. Why should he waste time with greedy lawyers when he could do far more good helping deaf people? He was an honest inventor, not a business-man; if the other shareholders were so keen on this business, why couldn't they deal with all of this?

When Alec had finally blown off all his steam, Watson assured him that the telephone business was going much better in the United States than in Britain, that it had a great future, and that everybody stood to make a lot of money if only he would come to Boston now. Alec's black eyes flashed with exasperation as he turned his gaze on his wife. Mabel remained silent as she watched his lips, but her husband knew that she agreed wholeheartedly with Watson. He could feel her quietly willing him to join her father in the fight to save their company—the company that bore his name. Alec gave a sigh, then agreed to go to Boston, but only after he had delivered Mabel, Elsie, Mary, and Annie to Brantford. "I went with him," noted Watson, "because I didn't want to run the risk of losing him."

Throughout the voyage on the *Sardinian,* Alec had looked for-ward to seeing his parents. Knowing how much his mother still mourned the loss of two sons and a grandson, he longed to see her face when he presented her with her first granddaughter. He had been imagining her face for weeks, and now Gardiner Hubbard was spoiling his homecoming. Would the chilly Boston lawyer never leave him alone? Would he never acknowledge that Alec had obliga-tions to his own family, as well as to his wife's? Paperphobic to the last, Alec's heart sank as he thought about the laborious task of writ-ing out the whole history of his telephone experiments. There were two other matters on his mind, too. He had virtually no money left: he insisted Watson send a telegram to Hubbard asking, "Will com-pany pay Bell's expenses incurred in its services to Boston and back?" (The reply, "Yes," arrived the same day.) And he was suffering acute stomach pains. The following day, during the long railroad journey to Montreal and on to Toronto, he had two shivering fits and nearly fainted several times.

Mabel anxiously watched her husband, white-faced and preoccupied, as the train steamed through the monotonous forests of dark evergreens punctuated by fields of bare earth or of dried cornstalks. Was his illness brought on by stress, or was it something more serious? When the Bells finally arrived in Brantford, Alec's mother was so perturbed by her son's obvious discomfort that she barely looked at baby Elsie before she sent for the local doctor. He announced that two abscesses were the cause of Alec's pain and fever; he lanced one, but could do nothing for the second. The following day, Mabel reported to her mother, Alec was better, but still "very weak and looks so dreadful I am frightened." Yet Thomas Watson was pacing the porch at Tutelo Heights, and Mabel knew her father might be ruined if her husband didn't get on the train to Boston immediately. After a tearful farewell, she watched Melville Bell drive Alec and Watson off in the pony cart to Brantford Station. She could see Alec wince every time a wheel hit a stone.

A terrible sense of anticlimax descended on Tutelo Heights once the pony cart was out of sight. Both Mabel and Eliza Bell had tears in their eyes, but each was too shy to share her misery. They still barely knew each other, and they were shackled by their difficulties of communication. After staring helplessly into each other's face, they retreated in different directions. Eliza went into the kitchen to look after dinner. Mabel checked that Elsie was safe with Annie, then wearily climbed the stair to her bedroom and poured out her anxieties to her own mother in a penciled scrawl: "O Mama I feel so unhappy to let him go alone," she wrote. But Alec had insisted she remain in Brantford, to allow his mother to enjoy Elsie, "so I can say nothing. . . . O I want so much to be with you, see your little house in Washington."

Mabel and her mother-in-law were right to worry about Alec's health. His condition deteriorated during his journey, and as soon as he arrived in Boston he went straight to Massachusetts General Hospital to have the second abscess lanced. When this news reached Brantford, Mabel insisted on leaving Canada with Elsie and Annie, and rushing to his side. At the hospital in Boston, she found Alec close to panic. He was in great pain, but he had refused to face the risk of surgery, during which ether would be administered, until she was with him. Happily, once the surgery was performed, his condition and spirits improved rapidly. "He likes his nurses extremely and threatens not to come back to me at all," Mabel told her mother-in-law. "He is very glad he came on, it was the most fortunate thing because in Brantford he could not have had the skill and care he had here." He recovered enough to dictate to Watson from his hospital bed the preliminary statement that his business partners had been so impatient for him to file with the Patent Office. It arrived in Washington just before the final extension to the deadline expired.

Alec left hospital on Monday, November 31, and joined his wife and daughter in Mary Blatchford's house in Cambridge. At first, Cousin Mary was snappy with the invalid: despite his fame and success, her lip continued to curl at his pedigree. But her icy snobbery melted as she discovered she and Alec had two things in common: insomnia and a love of literature. And Alec was *awfully* good at recitation: you could almost *see* the characters he was portraying. While Mabel slept on the sofa, Alec read to Mary with all the passion and intensity that had made his father such a successful public performer. Mary always picked up her needlework when he began, but she was soon far too mesmerized to ply her needle. It would be hours before Alec might start to yawn, and Mary would look at the clock and realize it was almost dawn and they *must* go to bed. But Alec's convalescence was brief and the demands of the court case pressing. "Alec has been hard at work," Mabel reported to Eliza Bell in December, "at [rewriting and expanding] his preliminary statement for the Patent Office, looking over old models and letters."

The clash between the Bell Telephone Company and the Western Union Telegraph Company had turned very nasty, because demand for telephones had exploded. When the Bells had sailed off to Glasgow a year earlier, there had been about 1,100 telephones, the majority rented from Bell, in use in North America. A year later, there were several thousand telephones in all parts of the United States. Western Union, with more than $40 million in capitalization and $3 million a year in net profits, had set up the American Speaking Telephone Company. This subsidiary was outselling the Bell Company, with telephones that incorporated Edison's patented carbon transmitter, which made reception much clearer. Western Union crews responsible for new lines had the huge advantage of being able to piggyback onto the telegraph network already in place—they could easily string telephone wires to the existing telegraph poles. Bell crews had to work around the clock, erecting new poles as well as stringing new wires. Western Union employed all the ruthless tactics of a company bent on a monopoly. It cut telephone rates (and, it was rumored, Bell Company telephone wires) and assured its customers that they would be reimbursed in the unlikely event that the Bell Telephone Company would win the patent suit. By the time the Bells sailed home, most major U.S. cities had two competing telephone systems in use.

Meanwhile, the Bell Telephone Company was in a serious cash crunch. Gardiner Hubbard, an entrepreneur and promoter who always operated on a wing and a prayer, had been blithely ordering telephones from Williams's shop before he had customers ready to rent them. Thomas Sanders, who had deep pockets, was getting nervous. While Alec was still in England, Sanders wrote to Gardiner Hubbard, "How on earth do you expect me to meet a draft of two hundred and seventy five dollars without a dollar in the treasury and with a debt of thirty thousand dollars staring at us in the face?" Bell Company employees had not been paid for weeks, and were lending each other carfare and sharing their lunch-buckets. Williams and other suppliers were all pressing for payments owed to them. Before Alec and Mabel had returned to North America, Sanders had managed to persuade Hubbard to reorganize the company and sell some stock, in order to stave off collapse, but it was still touch-and-go. The Bell Telephone Company's future depended on victory in the suit against Western Union.

Soon after Christmas, Alec joined his father-in-law first in Boston and then in Washington for endless meetings with the company lawyers. At first he was facetious about events. "My darling little wife," he wrote to Mabel, who was with her grandfather in New York City, in January 1879. "Telephonically a storm is brewing! Thermometer ever so far below zero—and Bradley [an investor], Sanders, Vail [the company's general manager] etc. shivering over the ashes of the Bell Telephone Company!" But his good humor evaporated as he realized that Gardiner Hubbard's rather slippery reputation was injuring the confidence of some of the other investors. "My sweet darling wife," he wrote a week later. "I am troubled and anxious and don't know what to do."

Mabel rushed to reassure him: "My darling I long for you so much you dear big black fellow (take care that the worry and excitement now doesn't turn your hair gray!)." Gardiner Hubbard, with Alec's support, managed to stare down his challengers, but he was forced to play a less public role in the company. And soon Mabel was finding frequent separations as painful as Alec, and realizing that she simply didn't care if he was the untidiest person she had ever met—she wanted him and his mess back. In March she wrote, "I miss you dreadfully every moment, but manage to get along until I think that you are gone for a long time and not for a few days and then my heart and courage go down into my boots. . . . I am writing in your study now, such a frightfully good order as it is in, swept and varnished as if you were dead and buried. I hate the sight of it and wish I had left it as it was this morning."

Alec's spirits were not improved by Western Union's attacks on the Bell Company. Articles started appearing in newspapers that suggested that Bell had stolen the telephone from Gray, and that he was not a sufficiently skilled electrician to have made the scientific breakthrough. Alec was outraged: "I can't bear to hear that even my friends should think that I stumbled upon an invention and that there is no more good in me." There were even whispers that the U.S. Patent Office had shared information illegally with Bell, making his patents invalid. Bell Telephone stock prices crept downward.

The Bell Company's suit for patent infringement against Western Union's telephone subsidiary opened on January 25, 1879. Western Union had built its case, in the words of Bell biographer Robert V. Bruce, "around the proposition that undulatory currents were not new with Bell and that Dolbear, for example, had preceded him in consciously achieving that effect." But the Bell Company lawyer, a canny old bird named Chauncey Smith, had decided to base his strategy not on Bell's focus on undulatory currents but on the whole principle of using electricity for transmitting speech, as exemplified by the instruments described in the 1876 patents. "Smith's strategy," notes Bruce, "was to be of fundamental significance in the outcome. Through it the Bell interests would win control of the basic principle of telephony, not merely of some particular devices in it."

First, however, each side had to amass and present its evidence. Mabel enlisted the help of Cousin Mary to go through the accumulation of her husband's notebooks and papers in the Hubbards' home on Brattle Street. Thomas Watson spent days preparing models of the original telephones. Alec hunkered down in a Boston library, hunting up references to earlier experiments. The lawsuit was particularly irksome for him because he already had another invention in mind. "Oh! That those lawsuits were ended!" he wrote despondently in March. "I am afraid to make more inventions, for fear of being dragged into an interminable business connection with the Company."

In Boston, Mabel realized that if the Bell patent was successfully defended, it would be thanks to her father's efforts with lawyers and evidence rather than to her husband. A letter from her mother, in Washington, gave her a sense of the hysteria surrounding the case. "We hear Mr. Edison is working indefatigably," Gertrude wrote. "He now has twenty-five different kinds of telephones, the last a water telephone, a marvel of loudness and distinctness. The W. U. [Western Union] have engaged all his time and efforts for five years and pay him a weekly sum beside furnishing him unlimited means for electrical instruments, chemical collections of minerals and whatever can aid him. He has a stenographer at his side taking down every new idea or experiment. The W. U. are bent upon improving Alec's patents, or making a telephone which shall wholly supplant his. Alec cannot afford to be an agent, he needs his mind free to meet and overcome a most powerful and unscrupulous foe." But Gertrude concluded her letter on a reassuring note: "Papa has no anxiety as to the final issue."

Elisha Gray gave his evidence in New York in April. Alec was not impressed. "First day of Elisha Gray's Cross-Examination just concluded," he wrote his father on April 4. "Everything coming out in our favour!" Three days later, the Bell team triumphantly produced the pièce de résistance of their case: a letter that Gray had written to Alec back in March 1877 and that had turned up in the wastepaper basket at Bell's old Boston lodgings in Exeter Place. In the letter, Gray congratulated Bell on his work, and wrote, "I do not, however, claim even the credit of inventing it, as I do not believe in a mere description of an idea that has never been *reduced* to *practice* . . . should be dignified with the name invention." The Western Union lawyers winced as the letter was read in court. Gray confirmed the authenticity of the letter, then apologized to his counsel: "I'll swear to it, and you can swear at it!"

Alec sympathized with a man whom he felt was fundamentally decent, if misguided. "Poor Mr. Gray. I feel sorry for him," he told Mabel. "I feel sure he would never of his own accord have allowed himself to be placed in the painful position in which he is now." But the admission in Gray's letter went to the heart of the defense argument—an argument that the Western Union lawyers were beginning to realize was distinctly wobbly. The issue was whether Alexander Graham Bell's patents and legal proofs were superior to those of the Western Union's stable of inventors: Elisha Gray, Thomas Edison, and Amos Dolbear. None of the Western Union witnesses had stood up well under cross-examination; their indignation often made their stories sound confused and contradictory. When Alec had filed his patent application in February 1876, Gray was still only playing around with possibilities and had filed a caveat announcing his intentions rather than a patent application describing an invention. Thomas Edison did not design his transmitter until after he had seen Alec's at the Philadelphia Exhibition. Amos Dolbear's claim that he had identified the undulatory current as the means to achieve transmission of speech before Alec did was made irrelevant by the fact that he had not patented an invention before Alec was granted Patent No. 174,465.

In contrast, Alec's majestic bearing, steady gaze, and straightforward manner made him a formidable witness. With his almost photographic memory, educated British accent, and clear articulation, he radiated an unrivaled authority as he methodically laid out the sequence of events leading up to his patent submission. His testimony, taken in July 1879, filled nearly a hundred pages of the six-hundred-page printed record, and looked unbeatable.

Even before the hearings wound up in September, the Western Union lawyers had decided that the risks of pursuing the case were too great. They informed their client that the Bell patents were watertight, and they opened negotiations with Chauncey Smith to settle the case. Their first suggestion was that each side should have an equal share in the combined patents of both. Smith rejected the suggestion out of hand. The final out-of-court deal, signed on November 10, 1879, was an expensive blow to the mighty company: Western Union transferred at cost all telephones, lines, switchboards, patent rights in telephony, and any pending claims to the Bell Telephone Company. In return, the Bell Company agreed to stay out of telegraphy and to pay Western Union 20 percent of all telephone receipts until protection guaranteed by Patent No. 174,465 ran out. What a victory for Bell! The shareholders in the Bell Telephone Company held title to a monopoly on a wildly popular invention, and there were at least fourteen more years of patent protection ahead. They also now had access to a network of established agencies and customers.

"I felt," Thomas Watson recorded later, "as if a crushing weight had rolled off of me." If Alec had been in Boston instead of Washington when the deal was reached, Watson would have insisted on his former boss joining him in an exuberant, whooping, stamping Mohawk war dance—the kind that had once infuriated their landlady. Instead, Watson celebrated by catching the train to Marblehead and walking along the beach, declaiming to the skies all the poetry he could remember. "It was an undignified thing for the Chief Engineer of the Telephone Company to do," he later wrote. "But I certainly felt better for it next day."

The settlement of the Western Union suit was far from the end of the litigation battles. The telephone business was now so profitable that Alec Bell had to deal with all manner of people who wanted a share of the action. During the nineteenth century, groundbreaking inventions were repeatedly litigated: the 1834 patent on Cyrus McCormick's harvester was challenged nine times; the 1840 patent on Samuel Morse's telegraph, fifteen times. These numbers pale into insignificance, however, compared to the more than six hundred separate cases involving the telephone. Most owed their origin to Gardiner Hubbard's vigilance, because they were brought by the Bell Company against infringements of Alec's patents. The long list of unsuccessful litigants included serious inventors, such as Gray and Dolbear, who had convinced themselves, despite the evidence and the court judgments, that they had had the idea of a telephone first. Then there were the rascals, such as Daniel Drawbaugh, a Pennsylvanian machinist who claimed he had invented a telephone before Bell, although he told the court, "I don't remember how I came to it." He had not applied for patents, he explained, because he could not afford to pay the fees or build the models.

The clamor of claimants was accompanied by a parade of professors as witnesses, a gallery of drawings proving prior invention, and a medley of exhibits, tin dippers, teacups, and mustard tins with which inventors claimed to have antedated Alec's discovery. In most of the cases, the Bell Company's opponents gave up the fight before they got to the higher courts. Twelve of the suits reached decisions in the circuit courts, and five went as far as the Supreme Court in Washington. "Out of the dispute over Bell's claims," wrote Judge Brower in an 1892 decision, "has come the most important, the most protracted litigation that has arisen under the patent system in this country."

Every single case that went to court was decided in Alec's favor. Alec was usually spared the necessity of appearing as a witness because his lengthy and careful 1879 testimony was simply inserted into the court record on subsequent occasions. On his lawyers' advice, he told a later correspondent, he would "writhe in silence under the unscrupulous attacks which were made upon me."

There were, however, two cases that required his appearance in court.

The first was a suit alleging patent infringement that the Bell Company brought against the People's Telephone Company. A group of unscrupulous New York and Cincinnati businessmen had created this company by paying $20,000 to Daniel Drawbaugh for his claim to have invented the telephone, and then selling first stock in the company and then pirate telephones. Over the course of four weeks in March and April 1883, Alec was cross-examined on the deposition he had given in the original Western Union case. His own lawyers, from the Washington firm Mauro, Cameron, Lewis and Kerkam, marveled at Alexander Graham Bell's "capacity for long-sustained mental effort" when giving a deposition. It began in mid-morning, and by two that afternoon the lawyers mustered the courage to ask Alec if he cared to pause for lunch. "I don't lunch," said Bell, and proceeded with his deposition to the end of the day. The People's Telephone Company claim was finally dismissed in 1891, when the Bell patents had almost expired—by which time Mauro, Cameron, Lewis and Kerkam had earned an estimated $50,000 in fees from their client. Drawbaugh himself was reprimanded by the court for blatant falsehoods, but, undeterred, he popped up a few years later to claim he had invented the radio before Marconi.

The second case that required Alec to appear in court was a much murkier, more complicated one, involving political shenanigans. Two shady Tennessee businessmen had banked on the assumption that as long as a legal challenge to the Bell patents was working its way through the U.S. courts, any alleged infringers might be able to continue to pursue their business. The gamble, for Dr. James W. Rogers and Casey Young, was to keep spinning out the legal process until the Bell patents had expired. The two men set up the Pan-Electric Telephone Company, persuaded various local southern heroes, including General Joseph E. Johnston and Senator Augustus Garland, former governor of Arkansas, to become directors, and began to sell stock. The celebrity names on the letterhead quickly attracted investors. The Pan-Electric Telephone Company appeared well launched when Garland was appointed attorney general in 1884 by newly elected president Grover Cleveland. The company's Tennessee backers promptly asked the attorney general to sue for the annulment of Bell's patents, on the grounds that they were obtained by fraud and that Bell was not the original inventor.

Even Mr. Garland realized that he was a little too close for comfort to Pan-Electric, so he left his deputy to respond to the request while he was absent from the capital. His deputy, the solicitor general (a Virginian), and the secretary of the interior (a Mississippian) agreed that the United States government should sue the American Bell Telephone Company. The main argument was that there had been fraud in the U.S. Patent Office when Alec's original application for the telephone patent had arrived in the office on February 14, 1876, and this allegation must be properly investigated. The case rested on the claim that Alec had not mentioned the 1861 Philipp Reis telephone to the Patent Office examiner, Zenas F. Wilber, and that Wilber had showed the Elisha Gray caveat to Alec's lawyers so they could insert the substance of it in the Bell application. The suit was launched in January 1887, at government expense. Meanwhile, Pan-Electric directors continued to line their pockets, and headlines alleging that the

Bell telephone patents had been obtained under false pretenses were splashed across New York newspapers.

Lawyers for the American Bell Telephone Company moved fast: they successfully obtained injunctions for patent infringement against Pan-Electric and its subcompanies, which effectively stopped Pan-Electric from selling pirated telephones. But Alec himself was hurt by the allegations of fraud and by the slurs on his name. In 1892, he would have to spend a total of nine weeks, off and on, in court, once again reviewing his role in the history of the invention of the telephone. Eventually, owing to lack of evidence and the death of the government's chief counsel in 1896, the whole case quietly expired.

"The real excellence of your deposition and its naturalness," one of Alec's lawyers reassured him, "lie in the fact that in telling your own history you are telling the story of the man who invented and who knew that he had invented, the electric speaking telephone." But the experience deeply wounded the inventor, and taught him a lesson he never forgot. From now on, he kept scrupulous records of every idea, every experiment, every piece of equipment he constructed. There are sixty volumes of "Laboratory Notes" in the Bell archive.

In May 1879, six months after returning from England, the Bells were scrambling for money. "We have not a penny to call our own," Mabel wrote anxiously to her father. Alex was "down with a bad headache," and they were living on "the grocer's confidence . . . as there are many things we need at once we are pretty hard up." But as the law case progressed and victory began to seem likely, Alec and Mabel Bell realized that there was a chance of real wealth ahead. Telephones would never prove as lucrative as railroads, steel, or oil wells: the Bells, along with other major shareholders of the new National Bell Telephone Company, including Gardiner Hubbard, Thomas Sanders, and Thomas Watson, would be millionaires several times over, but they did not join the Carnegies, Morgans, or Vanderbilts within the ranks of the nineteenth-century mega-rich. After the various refinancing maneuvers that the

Bell Telephone Company had gone through, the Bells had ended up with 1,106 shares of the new National Bell Telephone Company's total of 7,250 shares. In March 1879, before Alec had even given his testimony in court, Mabel's father informed her that stock priced at $50,000 three months earlier now had a market value of $71,890. It is always difficult to calculate today's equivalents of such sums, allowing for inflation, but assuming a twenty-fold increase in value of the dollar using the consumer price index, this meant that Mabel's stock in the Bell Telephone Company was worth $1,318,370. After living hand to mouth since their marriage two years earlier, Alec and Mabel might now never have to worry about money again. For Mabel, who was still only twenty-one, the prospect was quite delicious.

That year, Mabel recorded in her journal a "long discussion on riches" with Alec. The conversation took place in the parlor of 1509 Rhode Island Avenue, a Washington house they had rented so they could be together during the long court case, and where Mabel could be close to her family. Alec was sitting at his desk with an untidy pile of old letters and notes in front of him. With much harrumphing and grumbling, he was trying to find material for his deposition. Mabel sat on the floor, oblivious to the dust and disarray surrounding her, unpacking trunks of books and clothes and musing on their future. Looking up at her husband so she could read his reaction, she told him that she would like "fifteen thousand a year, my fine house and carriage." Alec's Scottish parsimony rose to the surface. He told her that five thousand a year seemed more than adequate to him, and surely she would be able to keep a carriage on that? Or did she want him to give up his scientific work and devote himself to making money so that she could "lie in [a] carriage and dress in velvet?"

Alec's dismissal of such dreams as frivolous and self-indulgent brought out a streak of Gardiner Hubbard in Mabel. With a wicked smile, she asked sweetly, "What is there higher than making money?" Alec brushed aside her sarcasm as he waxed lyrical on one of his favorite subjects: "Science, adding to Knowledge, bringing us nearer to God." He put down the papers he was leafing through and, fixing his black eyes on his wife's face, he went on, "Yes, I hold it is one of the highest of all things, the increase of knowledge making us more like God." It was a strange way for a self-confessed atheist to frame his ambition. But Mabel just grinned as he asked dramatically, "Will you bring me down and force me to give up my scientific work?" They both knew that this would never happen. The previous month, he had bought a set of the new *Encyclopaedia Britannica* and had announced he was going to read it from start to finish. However reluctant Alec was to exploit his inventions, nothing would dampen his irrepressible urge to explore, discover, improve. "No," Mabel replied amiably. "Only I want money too if I can get it." Husband and wife exchanged knowing looks. "So you shall my dear," Alec smiled, "and doubtless you will by and by."

It didn't take long. Now that the Bell Telephone Company had a monopoly on telephone service, shares worth $65 each in March rose to $337 in September and to $525 in October. When the deal with Western Union was announced, the share price topped $1,000. "We are beginning to realize that we have wealth," Alec informed his father just before Christmas. By now, he and Mabel were worth, in today's equivalent, about $20 million, and the numbers kept rising.

In 1880 there was a further corporate reorganization, and the National Bell Telephone Company became the American Bell Telephone Company. Mabel (to whom Alec had handed over his original one-third interest in the embryonic Bell Company, at their wedding) had been selling small numbers of shares ever since the price started rising early in 1879, and she continued throughout 1880. Nonetheless, she remained the company's largest single shareholder at the end of 1880, owning 2,975 of the company's 73,500 shares. By Christmas 1880, Alec could write to Melville that their income was now $24,000 a year, when a loaf of bread cost two cents and a housemaid earned twenty-five cents a day plus room and board. In today's terms, their annual income amounted to nearly half a million dollars, in an era when there was no income tax. How standards had risen after exposure to wealth! "We should be able to live on that," wrote the man who had previously thought $5,000 a year would be more than sufficient.

Yet wealth was not a major priority for either Mabel or Alec. Alec was most eager to pursue some of the other ideas he had for the exploitation of electricity. He filled his notebooks with ideas for switchboards, phonographs, and an underwater distress signal. While he was still giving evidence in the Western Union battle, he had started to explore the idea of using light waves rather than wires to transmit speech. He recruited a young man called Sumner Tainter, a manufacturer of optical instruments who had once worked in Boston alongside Thomas Watson in Charles Williams's shop, to help him develop what he had already named a "photophone."

In a laboratory Alec had set up near his Rhode Island Avenue house, Alec and Tainter constructed a device that would use mirrors to reflect sunlight into the photophone's transmitter. When a speaker's voice made another mirror in the transmitter vibrate, a varying beam of light was reflected toward a saucer-shaped receiver. Inside this receiver was a crystal of selenium, a photoconductive element. When light shines on selenium, the current it conducts increases. In the photophone receiver, the selenium crystal, a battery, and a telephone receiver were connected together such that they reproduced the sound that vibrated the mirror in the transmitter.

Alec was soon as obsessed with the photophone as he had once been with the telephone, and he reverted to his unconventional work patterns. He would work through the night, fiddling with the apparatus and making notes. Then he would snore all morning, the curtains in his bedroom drawn tight. When his wife reproached him for his antisocial behavior, he begged for acceptance of his night-owl habits. According to Mabel's journal, he pleaded that "I have my periods of restlessness when my brain is crowded with ideas tingling to my fingertips when I am excited and cannot stop for anybody." He begged to be left alone when he was in this high-strung state: he didn't need to eat or sleep, he assured her. Any attempt to make him do either would bring him crashing back to earth, his train of thought derailed and his ideas vanished. Such behavior today would be regarded as hypermanic.

Mabel had her own concerns—the concerns of a young wife and mother who wanted to settle down. Initially, Alec resisted the idea of making their permanent home in the capital, which he found too busy, too hot, and too pretentious. "I am afraid a quiet life in Washington is too good a thing to be hoped for," he wrote to Mabel in 1879, when he was staying in the Hubbards' Brattle Street house, "so I think we all had better settle down together here in Cambridge." Boston remained much more congenial for an inventor: Alec could mingle there with university professors and discuss his own ideas with the machinists in Charles Williams's workshop. But Gertrude and Gardiner Hubbard now spent far more time in their Washington home, on Connecticut Avenue close to Dupont Circle, than in Cambridge. Mabel wanted to be close to her mother and sisters, and her gentle pressure slowly paid off. "Alec has stopped railing at Washington and is beginning to find there are nice and scientific people here," she told her mother. Early in 1880, Gardiner Hubbard, who was in charge of a trust fund that Alec had set up at the time of his marriage to handle telephone rights outside North America, agreed to invest some of the monies in a permanent home for the Bells in Washington. Residence in Washington meant that Mabel would now have to learn the geography—physical and social—of a city that (greatly to its own satisfaction) was rapidly becoming one of the world's most powerful political capitals.

By the late 1870s, the city on the Potomac River had finally escaped from the dark shadow of the Civil War. As the federal government, under President Rutherford B. Hayes (the first president to use the telephone), flexed its political muscle, people flooded into the city. The Washington Monument was still an incomplete stump, but there was a building boom, and the District of Columbia embarked on an ambitious program to grade streets, install streetlamps, lay sewers and gas mains, and plant trees. "It is a very unfinished city," Gertrude Hubbard commented to Eliza Bell, with all the asperity of one accustomed to the sophistication of Boston and New York City, "but certainly [it] will be a very beautiful one and is quite unlike any other."

Before the Civil War, the majority of members of Congress lived in boardinghouses, or "messes" as they were called, but with the advent of railroads and the end of the war, congressional wives began to accompany their husbands to the capital. Government officials were joined by a surge of western mineral kings, department store millionaires, prominent real estate speculators, and powerful industrialists from the northern and Midwestern states. These nouveaux riche newcomers commissioned architects like Adolph Cluss and Henry Hobson Richardson to design imposing mansions for themselves northwest of the Capitol, along K Street, Connecticut Avenue, 16th Street, Massachusetts Avenue, and New Hampshire Avenue. So many mining millionaires, including Nevada senator William Morris Stewart, bought land around Dupont Circle in the 1870s that it was nicknamed "the Honest Miner's Camp." The elaborate new mansions boasted arched carriage entrances, massive chandeliers, polished mahogany paneling, cavernous stone fireplaces, forests of ferns, and all the technical improvements of the time—gas lighting, central heating, speaking tubes, ventilators, hot and cold running water, and dumbwaiters.

Politics and business still dominated Washington conversations, but there were now enough wives and daughters in the city to constitute "Society," with its own rituals, hierarchy, and rigorous calendar. During the "Season" (November until the start of Lent in the spring), the Washington elite attended countless luncheons, dinners, card parties, receptions, weddings, and balls. "Every afternoon," Gertrude Hubbard noted, "from two to half after five there are calls to be made or received. On Mondays the families and Judges of the Supreme Court receive; on Tuesdays the Senators; on Wednesdays the members of the Cabinet; on Thursdays the members of the House; on Fridays

Army and Navy and on Saturdays residents. Every evening there are large formal receptions, dinners or dances." Gertrude pretended to a certain aloofness—"those who really enter into the gay life must find it most exhausting"—but it was a world in which she wanted to move. She had three unmarried daughters, and after the travails of the past few years and the slurs on her husband's good name, she yearned for social vindication.

Mabel found the rituals strange. "People here seem to spend their time in making calls and in receiving," she wrote her mother-in-law in Canada. But she too wanted to "belong" and quickly slipped into the habit of identifying a woman by her husband's position. "I am just now waiting anxiously for the member from Boston to call on me as she said she would and the Senator from Mass. has been here." She enjoyed the adventure of new acquaintances within the unfamiliar world of politics. One of the first calls she made was on the wife of Republican senator James G. Blaine. "I have seldom seen a more thoroughly handsome suite of rooms than those into which we were ushered," she noted in her journal. "The prevalent tint was a soft, yet rich tint of dead gold, heightened by the glow of a wood fire. . . . Mrs. Blaine herself harmonized well with the surroundings, a tall stately woman of the Martha Washington type." When she and her sisters heard that the wife of Anthony Pollock, one of the Bell Company's lawyers in the patent case, was sending out 250 invitations to a party, they waited anxiously to see if they would receive one. Mabel noted in her journal her relief when, belatedly, the yearned-for card arrived. In the end, however, she missed the party. Alec was in Boston, working with Thomas Watson on various improvements to the telephone, including a new kind of circuit, a call-bell, and two variable-resistance transmitters. "All working satisfactorily," he blithely reassured his wife in a telegram. Yet he failed to return in time for the Pollock soirée and Mabel noted that "I decidedly did not think it satisfactory."

Washington had always been a southern city, and its rigid southern code of morals often appalled Mabel. Coming from progressive

Boston, she didn't agree with the tut-tutting triggered by any woman who stepped out of line. When the famous British actor Sir Henry Irving and his leading lady, Ellen Terry, were scheduled to appear at the National Theater in 1879, Washington's grandes dames decreed that Irving would be "received" but that Terry would be given the cold shoulder. This thin-lipped disapproval was sparked by the diva's unfortunate marital history. After her marriage at age sixteen to the British painter George Frederick Watts, who was thirty years her senior, she had eloped with architect Edward Godwin, by whom she had two children. She had returned to the stage when that liaison failed, and had married a fellow actor, Charles Kelly, who turned out to be a drunk. Since then, she had poured her passions into her children and her stage career. In a private journal, Mabel deplored Washingtonians' hypocrisy: "Because of a fault committed years ago and repaired as far as possible by devoted care of her children, Miss Terry is to be excluded from all good society while Irving, about whose past there are stories and who is certainly divorced from his wife, is received and feted everywhere." But she acknowledged that she did not have the nerve to buck the conventions: "My own position in society is not yet secure and I have no right to injure my husband or children."

Alec had no appetite for afternoon calls, let alone gossip, and often resorted to strategic headaches. "This afternoon I spent making Alec a new cravat," another journal entry reads, "and getting ready to make some calls. My [new] maroon dress came home in time for me to wear it, but when Alec saw me in it he utterly refused to make calls with such a gorgeous person and I had to go down on my knees to him before he would hear of it and then to put on his things myself." More to Alec's taste was an expedition to a new skating rink: "Alec was so fascinated that after a while he got over his dread of being laughed at and joined in. . . . He did cut a ridiculous figure at first. . . . He upset one young lady and wasn't able to help her up and ran into the arms of another beside tumbling down himself."

Washington's social life did offer some tantalizing prospects to Alec. A meeting with some members of an Arctic expeditionary force gave rise to the suggestion that the inventor accompany them. Alec burst through the front door of the Rhode Island Avenue house, afire with enthusiasm: "Oh Mabel! Wouldn't it be lovely to see the sun above the horizon for all the twenty four hours!" As he rattled off the rationale for his going—"Only a few months, perfectly safe, such a chance for making experiments and discoveries determining the influence of the North Pole on the magnets . . ."—Sister joined Mabel in the hall and assumed a look of extreme skepticism. "Yes," she snapped at her brother-in-law. "Go, and leave Edison a clear field in which to steal marches." Alec gave a weary nod of submission and returned to his photophone experiments.

Both Alec and Sumner Tainter felt they were making little headway with the photophone in January 1880. Alec wrote in his notebook that they were having such problems with selenium that "we have both been quite unwell ever since." But the following month, there was a breakthrough. One day when Alec's cousin Charlie Bell dropped by, Alec managed to get the photophone working in the laboratory. He showed Charlie how to operate it, then he disappeared to the basement where he had rigged up a receiver to which photophone sounds could be transmitted on a telephone wire. Charlie began uttering various sounds into the mouthpiece, including Alec's own signature greeting. Alec described in a letter to his father what happened next: "I heard the words Hoy-hoy-hoy in different tones—the vowels being wellmarked. . . . Then came a song 'God save the Queen'—the words in this case seemed perfectly intelligible and plain." Alec could barely contain his excitement as he explored the implications of his invention: "I have heard articulate speech produced by sunlight! I have heard a ray of the sun laugh and cough and sing! . . . I have been able to hear a shadow and I have even perceived by ear the passage of a cloud across the sun's disk." He immediately understood the potential of his discovery. "We can talk by light to any visible distance without any conducting wire."

The photophone would have to await the development of fibre-optic technology, a century later, before it could become a practical substitute for the telephone. In Alec's day, the photophone was limited to line-of-sight transmission on sunny days. It was typical of Alec that he had been exploring an idea that was not profitable in the short term, rather than perfecting the telephone and ensuring its continued development. His reluctance to think commercially about his intellectual preoccupations meant that he could never be the Bill Gates of the nineteenth century.

Alec's excitement over his new experiments almost overwhelmed another event in the Bells' life. In late February Mabel wrote to her mother-in-law, "I am on the sofa for the first time this morning, and must try and give you some account of my little one, as I fear Alec has been far too busy with his baby to talk or write much about mine." On February 15, after only a couple of hours' labor, Mabel had given birth to a second daughter, who weighed six and a half pounds. "Only think!" Alec wrote gleefully. "Two babies in one week! Mabel's baby was light enough at birth, but mine was LIGHT ITSELF! Mabel's baby screamed inarticulately but mine spoke with distinct enunciation from the first!" Mabel managed to dissuade her husband from naming the baby either Electra or Photophone. Instead, the child was given a much more prosaic name—Marian Hubbard, after Mabel's little sister who had died. She would always be known as Daisy.

Cocooned in the gush of telephone revenues, both Alec and Mabel assumed that there would be more profitable inventions and healthy babies in their future. But their comfortable assumptions would soon be shaken.

Chapter 12

SAD LOSSES, FAILED HOPES
1880–1885

Alexander Graham Bell was only thirty-three, and already world-famous. In five short years, he had made both his name and his fortune. The success with the photophone experiments reassured him that there were more adventures ahead in the laboratory. "Can Imagination picture what the future of this invention is to be!" he wrote to his father. No novelist describing a brave new world or poet describing the movement of the planets could have been more prescient or lyrical than Alec as he spoke of his discovery's potential.

> Some of the practical results to be obtained I clearly foresee. When Electric Photophony is practiced in warfare the electric communications of an army could neither be cut nor tapped. On the ocean, communication may be carried on by word of mouth between persons in different vessels when great distances apart—and lighthouses may be identified by the sound of their lights. In general science discoveries will be made by the Photophone that are undreamed of just now.

Every variation of a light will produce a sound. The twinkling
stars may yet be recognized by characteristic sounds, and
storms and sun-spots be detected in the sun.

Alec needed this boost because, like many early achievers, he had
been assailed by self-doubt. Would he ever achieve another great
invention? Was his brilliant future behind him? The endless litigation
battles over his patents gnawed away at his spirits, particularly when his
opponents continued to whisper that it was pure luck that a professor of
elocution had won the race to invent the talking telegraph. "Oh Mabel
dear," he had earlier written in one downhearted missive. "Please, please,
make me describe and publish my ideas that I may at least obtain credit
for them and that people may know that I am still alive and thinking."
He wondered whether he should temporarily abandon inventions and
instead write a book detailing his telephone research.

Now, however, he was riding high, his confidence restored. At first,
the photophone received a much warmer welcome than the telephone
had in its early days. The *Times*, in London, congratulated its inventor "on having made an addition to our scientific knowledge and discovered another possible application of science to practical purpose."
Alec told Mabel, she confided in her mother in July, that he was the
"most lucky fellow alive as he has but to hold out his hand and discoveries drop into it." He immediately sold the photophone prototype and
patents to the new president of the National Bell Telephone Company,
a cautious Boston financier called William H. Forbes.

Alec's elation was not dented by whispers that the photophone, in
the words of the scientific journal *Nature*, might not have a "widely
extended future of usefulness." He ignored Forbes's lukewarm
welcome for a discovery that he suggested might have less "practical
importance" than the telephone. He even shrugged off a nasty little
item in the *New York Times* ridiculing an invention that appeared to
require "a line of sunbeams hung on telegraph posts." For once, he
seemed to take criticism in his stride. Living in his imagination rather

than in the realm of what today we call "mission-driven research," he gave barely a backward glance when the National Bell Telephone Company allowed the photophone to wilt on the vine. He was too fired up by yet another invention: the spectrophone. Fascinated by what he had already discovered about light, he developed an instrument that could detect by sound the invisible colors of the spectrum. He acknowledged that his instrument was a supplement to, rather than a substitute for, the spectroscope, which had been around since the early 1800s. "Of course, the ear cannot compete with the eye in the examination of the visible part of the spectrum," he told the Philosophical Society in Washington in April 1881, "but in the invisible part beyond the red, where the eye is useless, the ear is invaluable." Like the photophone, the spectrophone had no immediate practical application and was soon gathering dust in Alec's workshop.

Mabel, now an attractive and graceful twenty-three-year-old, was enjoying the newfound stability and solidarity of the Hubbard-Bell clan in the early 1880s. She was no longer the only Hubbard daughter to be married. In January 1880, Mabel's oldest sister, Gertrude, had married an ebullient Hungarian actor, Maurice Grossman. Gertrude, in Mabel's words, was "just the daintiest, most fastidious little creature imaginable, the last word in two centuries of New England gentility," so the Hubbard clan were even less impressed with her choice of husband than they had been with Mabel's shabby Scotsman. Maurice, Mabel recalled years later, was "so big and rough and unpolished . . . he wore little gold spectacles and such little sticks of legs supported his huge fat body that I could hardly help laughing." But Gardiner Hubbard gritted his teeth and welcomed his exotic new son-in-law, even though he had no visible means of support, because his beloved eldest daughter was in love. He had to accept that, at thirty-one, Sister probably knew her own mind—and there were no other suitors on the horizon. Gardiner even found Maurice a job, as president of the International Telephone Company, which Gardiner had recently founded in Germany. This breathtaking piece of nepotism did not faze any of the family. "Maurice had of course little acquaintance with the business details of the organization of companies," Mabel wrote to Alec's mother, Eliza, "but he has many influential friends in the chief European capitals and . . . is very much interested in the telephone."

Within weeks of Sister's wedding, Mabel's youngest sister, nineteen-year-old Roberta, started showing an unusual interest in Alec Bell's cousin Charlie Bell. Charlie, son of Melville's brother David, had arrived in Washington to act as Alec's secretary until he too could be slotted into the family business. Mabel warmed to this young man, who had the black hair and dark eyes of all the Bells and who always laughed at her husband's jokes. In her opinion, Charlie was "a good kind gentle boy," while her spirited sister Berta was an incorrigible flirt. Nevertheless, as Mabel noted in the journal she kept after Daisy's birth, the two young people "took advantage of my confinement to fall in love with one another." Gardiner Hubbard, who had had a tough time coming to terms with the idea of an impoverished Hungarian actor marrying his favorite daughter, hit the roof when he heard that a second penniless Bell was wooing another of his daughters. As he peremptorily ordered the young man out of his house, Mabel must have recollected her father's behavior toward Alec only five years earlier. The whole drama played out exactly as it had for Alec and Mabel. Gertrude Hubbard took the young couple's side and calmed her husband down; after much grumbling, Gardiner agreed that he would take Charlie seriously once he had the means of supporting a wife. And once again, nepotism came in handy. With much gentle cajoling from his wife and daughters, Gardiner decided that Charlie should join the Grossmans in Europe in order to promote Bell telephone sales there. Alec assured Mabel, she told her father, that as soon as he was working full-time in his laboratory, he would have "something that will provide work on this side of the ocean for Maurice and Charlie." Soon, wedding plans were being made for Berta to become another Mrs. Bell.

One further event during these years reinforced Mabel's and Alec's sense of being settled in Washington. Alec's parents were now relatively well off, thanks to Alec's gift of the Canadian telephone rights. Melville remained feisty and argumentative, eager to break into a recitation of Robbie Burns or Shakespeare whether or not anyone wanted to listen. But his wife, ten years older, had now lost almost all her hearing (she smiled serenely through her husband's recitations) and ached to spend her final years closer to her only son. The arrival of a second granddaughter reinforced her yearning. So in 1881, after eleven years at Tutelo Heights, Melville and Eliza prepared to leave Canada. Alec went north to help them. "My father has sold off everything," he reported to Mabel, "and the only thing left here is poor Willie the dog. A most forlorn and disreputable beast, but full of love for his master and foreboding for the future." Alec's parents (but not poor Willie) were soon installed in an elegant four-story home at 1527 Thirty-Fifth Street, in Georgetown, a mile away from Alec and Mabel across Rock Creek. The following year, Melville's brother David and his wife and daughter moved into the redbrick house next door to Melville and Eliza, on the pretty street, lined with oak trees, that still retained a village atmosphere. The Bells had cut their ties with Canada and regrouped around their most famous relative.

Phone lines buzzed between all the Bell and Hubbard households. Mabel reveled in the frequent visits from her sisters to admire her babies, her mother's advice on servant problems, and the Washington social engagements she attended with Grace, the only sister still single. ("She is an accomplished caller," Mabel noted. "When she sits up so straight with such cold dignity, I feel quite afraid of her!") Alec's morale was further raised when he heard that the French government had decided to award the prestigious Volta Prize to the inventor of the telephone. Established in 1801 by Napoleon Bonaparte to honor the Italian physicist Alessandro Volta, the prize had been awarded only twice before.

In September 1880, Alec arrived in Paris with his wife, daughters, and cousin Charlie to collect the gold medal and the 50,000-franc ($10,000) prize money that went with it. He was thrilled with the honor, but not with the soft Parisian light. He had hoped to demonstrate his photophone to French scientists, but the misty atmosphere of the French capital prevented him from demonstrating selenium's potential to translate light beams into sound. From Paris, he traveled alone to London for business discussions about the International Telephone Company and for meetings with various scientists. His packed program included experiments at the Royal Institution of Great Britain, several lectures in London and Leeds, visits with eminent scientists (including Charles Darwin), and demonstrations of both the photophone and Visible Speech.

Separated from Mabel on the anniversary of their betrothal, and resentful of business obligations, Alec's spirits sagged and his old hypochondria reasserted itself. "I have been in the blues all day—bad headache," he wrote to Mabel from London. He had a "general feeling of despondence at the lecture engagements I have entered into." He worried that he wouldn't be able to speak without notes at his forthcoming lecture at the Society of Arts, especially because he had an ugly red sore on his eyelid. "I dread having to appear with a blinder over one eye!" He was homesick for America, and he begged Mabel to join him in London so they could all sail home as soon as possible.

Mabel remained in Paris with Charlie Bell, who had fallen ill. She knew her husband too well to be too upset by his complaints. And sure enough, once Alec walked onto the stage for his first lecture, he "dived deep into the middle of my subject, forgot my audience, and went on swimmingly to the end." His experiments at the Royal Institution were received even more enthusiastically. The eminent Irish scientist John Tyndall, who numbered Michael Faraday, Thomas Carlyle, and Alfred Lord Tennyson among his friends, was currently working there on transmission of sound and the diffusion of light in the atmosphere. A tall, thin, jovial man who loved encouraging younger researchers and explaining science (his lecture on the atmosphere was entitled, "Why the Sky Is Blue"), Tyndall welcomed Alec warmly. "It was delightful," Alec told Mabel, "to see another man get excited over my experiments."

Tyndall's support persuaded him to use his Volta Prize money to establish a small private laboratory in Washington, in which he and others might pursue research. On the Bells' return to the United States, Alec bought a modest two-story brick former stable, half-hidden among trees, at 1221 Connecticut Avenue. He named it the Volta Laboratory and fitted it with workbenches and gas lighting (although the smell of horse manure and the saddle posts on the walls revealed its original purpose). Alec seemed all set for more lucrative inventions ahead.

In July 1881, Washington sweltered in a heat wave—the kind of heat wave that brought a warm malarial wind from the undrained Potomac swamps into the city, and drove most Washingtonians away. The Bells had already moved their ménage to the Hubbards' Brattle Street home in Cambridge to escape the oppressive temperatures in the capital. On July 2, President James A. Garfield took a carriage to the station of the Baltimore and Potomac railroad, intending to catch a train to Williamstown, Massachusetts, in order to attend commencement exercises at Williams College. He had been in office barely four months, but in Washington, that was more than enough time to make enemies. As the president, a vigorous fifty-year-old veteran of the Civil War, strode across the waiting room, two shots rang out. A lawyer named Charles Guiteau, angry that his application to be the U.S. ambassador to France had been denied, had followed the president to the station and pulled out a pearl-handled revolver that had cost him fifteen dollars. The broad-browed, stocky Garfield fell in a heap on the waiting room's tiled floor. One bullet merely grazed his arm, but the second lodged in his lower back, only inches from the spine. A crimson pool of blood spread across the pink tiles.

Guiteau was quickly captured, and Garfield hurriedly driven back to the White House. Panic gripped Washington and rippled out across the country. From Cambridge, Mabel wrote to her mother, who was in Europe, "You can have little idea of the state of intense excitement it has thrown everyone in. Cousin Mary says she never saw anything like it since the shooting of Lincoln." Physicians congregated at the White House, debating whether (in the days before X-rays) an attempt should be made to remove the bullet. The debate spread to the newspapers, where correspondents named Medicus, Old Practitioner, and Common Sense championed their own remedies to heal the head of state. Meanwhile, at the White House, cold air was pumped up from the basement to the sick room through the hot air registers, and staff tried to keep their stricken leader comfortable with glasses of iced champagne. The humid atmosphere was thick with fear: that week, four White House employees came down with malaria.

Mabel was late in a third pregnancy, and eager to escape the heat and move to a house they had rented on the Maine coast. But when Alec heard the news of the president's predicament, he immediately caught the streetcar into Boston and hurried over to Charles Williams's workshop. There he started designing a kind of early metal detector that would use an electric current to locate the hidden lead bullet. Sumner Tainter helped him translate his ideas into instruments. Alec's fame meant that his offer to help was taken very seriously by White House physicians. Within days, according to a Washington newspaper, "Professor Graham Bell and Professor Tainter came here from Boston . . . for the purpose of making, under supervision of attending surgeons, a series of experiments intending to test the practicability of ascertaining by electrical means the location of the bullet which lies embedded in the President's body. They were driven at once to the Executive Mansion and are now . . . in the surgeon's room."

For the next few weeks, Alec worked day and night at the Volta Laboratory and in the White House to develop an apparatus that combined elements of two different devices. It consisted of an induction balance, which he had originally developed as a possible way of finding metallic deposits in the ground by means of an electric current, and a telephone receiver, which would signal success by means of sound. A reporter from the *Boston Herald* who visited him at his laboratory described the scene: "In cabinets, on tables, chairs and floor were coils

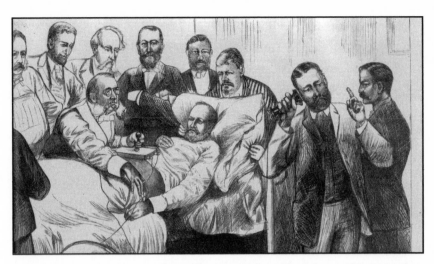

A contemporary engraving of President Garfield's sickbed shows Alec trying to detect the bullet by sound.

of wire, batteries, instruments and electrical apparatus of every sort. The light from the [gas] jets, burning brilliantly in the center of the room, was reflected from a hundred metallic forms." There were no screens on the windows, which were open in the sweltering heat, and the continuous whine of mosquitoes was punctuated by the sound of slaps as Alec and Tainter tried to protect themselves. Alec ordered more instruments from Boston and Baltimore, and consulted widely among scientists with similar ideas. The president's carriage was often seen waiting outside the little Connecticut Avenue building, as the president's physicians watched Tainter hold a bullet in his mouth, clenched hand, or armpit while Alec held the induction balance and telephone receiver in the vicinity of the bullet and tried to detect a sound. "Last night I located successfully a bullet hidden in a bag filled with cotton-waste," Alec wrote to Mabel, who was still in Cambridge. "This looks promising." He and Tainter shot lead bullets into slabs of raw beef from the local butcher, to approximate the conditions in which they had to work.

On July 26, Alec and Tainter brought their apparatus into the White House through a private entrance. Alec tiptoed into the sickroom, where Garfield slept, so he might decide how to set up the equipment. He was shocked by the president's appearance. The once-florid complexion of a man accustomed to open air and good living was now "ashen grey colour," Alec wrote to Mabel, "which makes one feel for a moment that you are not looking upon a living man." When the president awoke, Alec and Tainter proceeded with the experiment. There were five physicians plus several White House staff in the room, and Garfield, fearful of electric shock, stiffened at the sight of connecting wires trailing over his body. His weary eyes never left Alec throughout the test. But the results were inconclusive: "a spluttering sort of sound" was all that Alec could catch. So the experiment was concluded and Alec and Tainter returned to the Volta Laboratory. Meanwhile, the wounded president's condition seemed to improve. He sat up and was able to eat a sandwich and drink a glass of sweet white wine. Prayers for his health, which had been said daily at the Vermont Avenue Christian Church, were suspended.

Mabel missed her husband, and resented the way his efforts were being received. "While he is working the papers are laughing at him and classing him with the crowd of . . . nostrum inventors." But Alec was now totally consumed with perfecting his "induction balance bullet detector." Even his aversion to the Washington heat wave was forgotten, "partly I fancy because he is so busy and excited," Mabel confided to her mother. Mabel worried that "failure would be so mortifying now," but Alec was convinced he was on the right track. After five more days and nights spent testing the equipment, Alec was ready to try again. But the second test was also a failure. Instead of the clear localized signals he had detected in the Volta Laboratory experiments, all he could hear was a faint buzzing over a wide area. The following day, he discovered that part of the reason was that the White House staff had ignored his request to remove the steel-spring mattress from the president's bed. He was furious. "You poor boy," Mabel consoled him when she heard the news. "How sorry I am for you in your disappointment. I can imagine just how chagrined and mortified you must have felt. . . . Never mind, courage; from failure comes success, be worthy of your patient and don't lose heart even if all others are discouraged. I have not the least doubt but that you will eventually succeed. You have never yet failed and will not now. Only I wish I could be with you to help try to cheer you. If it were not for the little ones, I would come right down. I am only impatient to be with you again."

Oblivious to hints in the newspapers that he was a publicity-seeking crackpot, Alec continued to work alongside Sumner Tainter on apparatus that might detect the bullet. But the president's life was ebbing away. On September 19, James Garfield died, and his vice-president, Chester A. Arthur from Vermont, was sworn in as the twenty-first president.

The autopsy on Garfield revealed that the bullet, although too deep for Alec's apparatus to detect, was not the cause of death: it was lodged harmlessly in soft tissue. Garfield's death was the result of infection introduced by all the physicians who, scornful of theories about the importance of antisepsis, had thrust their fingers into his wound. A few weeks later, Alec demonstrated the apparatus he had used with the president to a group of physicians in New York. Although he himself took little further interest in it, an enterprising doctor called Dr. John H. Girdner put it into commercial production. It was used to help locate bullets in military hospitals behind the lines in the 1894–1895 Sino-Japanese War, the 1899–1902 Boer War, and the First World War. Alec himself was surprised and delighted to hear, in 1886, that the University of Heidelberg had awarded him an honorary doctorate in medicine for his contribution to surgical practice.

Alec was in his study in the Bells' Rhode Island Avenue home, scribbling a letter to Mabel, when he heard a newsboy shouting news of the president's death. "Poor Garfield has gone," he wrote. "After seventy-nine days of suffering to be obliged to give up at last. I hope indeed that there may be an immortality for that brave spirit. . . . A few days ago we had sermons preached all over the country claiming the favorable turn in the condition of the President, *as a direct answer to*

the prayers of the nation that were specially offered up the day before
. . . and now that results have shown that the ministers were mistaken,
their belief in the efficacy of prayer will not be shaken in the least. *It
will be the will of God."*

Alec was in a grimly philosophical mood as he wrote these words.
The president's death was not his real preoccupation: he and Mabel
had suffered their own personal tragedy during the president's illness.
While Alec was sweating over his induction balance, and trying to save
Garfield's life, Mabel had given premature birth to a boy, Edward.
"He was a strong and healthy little fellow," according to Mabel, and
"might have pulled through if they could once have established regular
breathing." But the infant lived only a few hours. Edward's death was a
brutal blow to Mabel, who had long nursed hopes for a son who would
share his father's scientific interests and carry the name and fame of
the Bells into the next generation.

Neither parent found it easy to accept their son's death as "the will
of God." Mabel struggled to maintain her health and equanimity, but
was pale, thin, and weak for months after the tragedy. Her mother
visited her every day; her young daughters often caught her weeping
quietly. A year later, when Mabel caught sight of President Garfield's
successor in the casino at Newport, she allowed herself to hint at the
reproach she felt toward her husband: "But for Guiteau [the assassin],
our own lives might have been different. You might not have gone to
Washington, but have stayed with me and all might have been well."

Edward's death was no easier for Alec. He kept telling himself that
Mabel had been well looked after during her pregnancy, and had he
been close during the birth, he probably could not have done anything. But he had to admit that his determination to harness science
into the cause of saving the president's life had taken him away from
his family. He was never much good at expressing any feelings other
than his devotion to Mabel (the closest he ever got to swearing was a
muttered "Shee-sh" through gritted teeth), but he grieved for both his
wife and the dead child. He recalled the photo of his deceased brother

Edward, taken in his coffin, which his own mother treasured, and in a clumsy gesture of contrition he quietly commissioned a photograph of his own deceased son, and then asked the French artist Timoléon Marie Lobrichon to paint a portrait from it. (There is no evidence that Mabel allowed the painting to be hung.) He also started working round the clock on a mechanical device for administering artificial respiration to patients with breathing difficulties. This "vacuum jacket" was a forerunner of the iron lung: it consisted of an airtight iron cylinder surrounding the patient's torso, and a suction pump that forced air in and out. But it was never commercially exploited in Alec's lifetime.

During these months, Alec's health began to deteriorate. Was it exhaustion? Grief? Guilt? Whatever the cause, he began to suffer the familiar pattern of headaches, sleeplessness, and sciatic pain, and his doctors told him that he had to stop working so obsessively. Mabel forgot her own suffering as she worried about her husband. "Entire rest from mental labor," she wrote to Eliza Bell, "is necessary to avert serious constitutional trouble." Nagged by his anxious wife, Alec handed over all his work in the Volta Laboratory to his associates there, who were working on improvements to an Edison invention: a sound-recording device. Alec agreed to take Mabel and their little girls *"wherever in the wide world you want to go."* Mabel knew exactly where she wanted to go: Europe, where Gardiner Hubbard was busy establishing telephone companies with the help of his other two sons-in-law. Both Mabel's sisters were now expecting babies, and Mabel's mother had accompanied her husband so she could be with them. In late October, the Bells arrived in Liverpool. They spent the next six months taking in all the fashionable cities of late-nineteenth-century Europe, including Rome, Nice, Paris, and London, often with Hubbard relatives. It was the first of many such Grand Tours that they would take.

Mabel always found motherhood easiest when Alec was with her. Elsie was now three and a half years old, with wide-spaced gray eyes and long brown hair—"a little round fat thing," in her mother's eyes, "bright and happy and kind to little baby sister." Daisy, at one and a half, had inherited her father's thick dark hair and black eyes, and was "perfectly lovely . . . far more affectionate than Elsie ever was. She talks a great deal more and has such pretty loving ways." Mabel had developed techniques to "hear" her babies when they cried; she checked their room regularly, and there was always a nursemaid in the household. A series of small terriers accompanied Mabel throughout her life, and their changes in posture and visible excitements would alert her to the crash of furniture or a ring at the doorbell. "I may not be with the [babies] very much," she wrote, "but I feel that I can look in on them any moment and be sure they are all right." Nevertheless, all the insecurities of a new mother were compounded by her deafness. As each child mastered speech, she also became aware of her mother's disability, and this further undermined Mabel's confidence as a parent. "I am getting rather anxious about it," Mabel had confided to her own mother in July 1881. "Elsie says so little to me, and so much more to others. She very evidently sees the difference and points out things to me instead of saying them. I fear she may learn to give her confidence and tell her little stories to others until it will be too late for her to care to come to me with them." Mabel struggled to follow her daughter's "childish prattle" but was constantly frustrated: "I see nothing as she talks so indistinctly, hardly moving her lips."

Mabel urged Alec to compensate for her handicap. "Do take care of your precious health," she instructed him, "for the sake of your wife and the helpless little ones who have only half a mother. They need you more and more every day." She was quick to correct him when he referred in a letter to "your children": "our, not *yours*," she replied crisply. And she constantly reminded him that "they so need their mother and their father's eye . . . when their characters are developing, and little faults springing up, that if not checked may grow serious."

Newborn Edward's death increased Mabel's maternal insecurity. While Alec, a self-declared atheist, chose to see the baby's death as a physiological event that could have been prevented by a better medical

Alec rarely left his Washington study before midnight.

device, Mabel sought answers within her Christian faith. She wondered if God was chastising her. One day, when Elsie was being naughty, Mabel withheld candy from her and explained that she was punishing Elsie as God had punished her. She told the little girl, she explained in a letter to Alec, "He had promised me a baby if I would be careful, but I had not been, and He took the little one from me." When Alec tried to brush off such thoughts with cold hard reason, he made his wife even more unhappy. In the end he could only write, "I would not for the world have my skepticism destroy your faith. You do me injustice to suppose that I smile at your ideas and only 'pity your credulity.'"

When the Bells were in Washington, Alec was typical of his era in being a distant father—frequently absent, and not particularly involved with his children when he was at home. There were always nursemaids, servants, sisters, or female cousins around to keep Mabel company and help mind the children. But on the lengthy journeys by steamer, train, and carriage through Europe in 1881–1882, Mabel, Elsie, and Daisy saw far more of him. Not that he got involved in all their activities—in Paris, Mabel found his secretary, William Johnson, a better shopping companion than her husband. "If I took Alec," she told her mother, "he would be sure to have a bad attack of heart trouble in the first shop and make me buy all I didn't want." He was far happier, his wife reported, hunting for caterpillars in the Bois de Boulogne, "as he has an idea that he can find a method of preventing them from climbing trees!" He also complicated their ménage by acquiring two little monkeys, which wrought expensive havoc in their suite of rooms in the Hotel Metropolitan, tearing at the wallpaper, pulling down curtains, and chewing through bell cords. Nevertheless, Alec was there to horseplay in hotel rooms with his little girls and to carry Daisy on his shoulders when they went sightseeing. There is a refreshingly modern ring to Alec's attitude to parenthood during an era when most affluent parents regarded themselves first and foremost as figures of authority. He believed that "play is Nature's method of educating a child" and that a parent's duty is to "aid Nature in the development of her plan."

The renewed closeness helped Mabel and Alec recover from the grief of Edward's death. When Mabel heard that her sister Berta Bell had given birth to a daughter, she wrote to her mother, "I begin to envy . . . Berta, and think it almost time I had a wee baby too. . . . Do you suppose I ever shall?" Two weeks later, she heard that her sister Gertrude Grossman had also had a safe delivery of her first child. "Of course," she wrote, "I am very much disappointed and sorry for you and Papa and Maurice [at the arrival of yet another girl] though if it had been a boy I could not have helped feeling just a tiny little bit jealous when I thought of my own little one." Such thoughts reopened old wounds: "I would like a boy Oh so much." But Mabel's sturdy good sense had reasserted herself, and she determined to enjoy all that Paris could offer, whatever her husband's eccentricities. "I am having a calling dress of ruby silk and velvet for next winter made at Worth's. . . . I feel like having the name pasted on it in a prominent position, I believe in getting my money's worth and what's the use of a dress from Worth's if no one knows it." Within a few days she was writing to her mother, "No little boy can be half as nice as little girls. My own little ones I think grow nicer every day."

In June 1882, the Bell family returned to the United States and purchased their first permanent home in Washington: 1500 Rhode Island Avenue, on Scott Circle. It was an immensely grand three-story, red-brick mansion, with a billiard room, a library, and a music room containing a grand piano for Alec. It was lit by electricity throughout, and the extensive servants' quarters enjoyed a steam-heating system. The house was so splendid that the Bells did not at first realize that it was badly built, with inadequate sewage pipes that put them at risk for typhoid—the scourge of nineteenth-century Washington. Even the stables had pine paneling. They also had some unusual occupants: alongside the horse there was a menagerie of cats and monkeys. The cats were all white with blue eyes—Charles Darwin had asserted that cats with these characteristics were always deaf, and Alec wanted to see if he was right. The monkeys were some of Alec's favorite pets, but Mabel wouldn't let them in the house because the servants objected to cleaning up after them, and Perrin, the coachman, complained that "Mr. Bell's monkeys and white cats were driving the horse out."

Thanks to his fame and to his efforts on behalf of President Garfield, Alec's black eyes and bushy beard were well known by now in Washington. Few realized that he remained a British citizen: he had not pursued the application he made for naturalization papers in 1874, because Gardiner Hubbard had taken on the job of submitting the patent applications. But he had now lived in the United States for over a decade, his own parents were also living here, and he was the father of two small Americans. He decided that it was time to declare his loyalty and become a full-fledged citizen of his adopted country. On November 10, 1882, Alec marched off down Rhode Island Avenue toward the center of the city, and entered the office of a local judge. There, with his right hand on a bible and his left hand making a salute in the air, he took the oath of allegiance to the United States. Mabel probably did not accompany him, but she certainly heard all about it. She complained to friends that her husband was irritatingly proud of his new status. "Yes!" he would proclaim to her. "You are a citizen because you can't help it—you were born one, but I *chose* to be one!"

While Mabel organized the new house, Alec plunged back into the work of the Volta Laboratory. Before his departure for Europe, he had been working with Sumner Tainter and yet another Bell relative—David Bell's elder brother, Chichester (known as Chester)—on improvements to Thomas Edison's sound-recording design, which he had named the "phonograph." Edison's invention, patented in 1877, recorded sound on cylinders covered in metal foil, but it was useless. The cylinders were scratchy, brittle, and unreliable, producing speech or music that was recognizable but not clear. Alec, along with Tainter and Chester Bell, was looking for ways to make the recordings clearly audible and sturdy. In Alec's absence, Tainter and Bell had experimented with different techniques and materials for what they preferred to call the "graphophone." After trying several different systems of recording, including jets of different substances (compressed air, paraffin, and maple syrup), and different materials on which to record, including wax-coated paper, cylinders, and disks, they had come up with a cylinder with a wax surface on which a sharp stylus incised the sound. Then they moved on to flat wax disks, with the idea that verbal messages incised onto them could be sent through the mail, like written letters. Once back in Washington, Alec spent almost every day in the laboratory, discussing his colleagues' advances and making suggestions on different approaches. His participation justified the inclusion of his name on some of the Volta Laboratory patents for the phonograph. (Thomas Edison would have to purchase those patents when he put the phonograph into commercial production in the early 1890s.) But he also worked on his other projects, including the vacuum jacket and induction balance. And he was increasingly drawn back into a world in which he continued to feel a strong sense of mission—the world of the deaf.

The Massachusetts State Board of Health had asked Bell to undertake a statistical study of hereditary deafness, and as he plodded through pages of census data, he became fascinated by the laws of genetic inheritance. In England, Darwin's cousin Sir Francis Galton was analyzing the pedigrees of famous men, and Alec was infected by the rising interest in the study of human heredity. Deaf parents, he found, were more likely than the population at large to have deaf children. Moreover, deaf people were more likely to marry deaf partners.

This research led him, in November 1883, to submit a paper to the National Academy of Sciences under the unfortunate title, "Memoir upon the Formation of a Deaf Variety of the Human Race." In this paper, he discussed his research on family patterns of deafness. He suggested that a congenitally deaf couple should be warned of the risk of having deaf children, and he also urged that deaf individuals should be given more opportunity to form friendships (that might lead to marriage) with hearing persons, in day schools. He even raised the possibility that "intermarriage between deaf-mutes might be forbidden by legislative enactment," although he went on to dismiss such legislation on the grounds that "interference with marriage might only prompt immorality." But his paper triggered a storm. The *New York Times* gave its report the headline, "A Deaf-Mute Community, Prof. Bell Suggests Legislation by Congress." And the deaf community was outraged by the paper's title, since it seemed to imply they belonged to an inferior species.

To a modern reader, who knows to what abuses the science of eugenics would lead, the pamphlet has a chilling ring. But Alexander Graham Bell was only one of many nineteenth-century men of science who pursued ideas that today seem naive or malevolent. There was as much bad science as good science around. President Garfield's death, for instance, had resulted from a belief in the miasma theory of disease—the assumption widespread during this period (Florence Nightingale was among its proponents) that infection was spread by "bad air." It took American physicians years to accept the idea that germs caused infections and to adopt the practice of scrupulous cleanliness that had been advocated since the mid-1860s by Scottish surgeon Joseph Lister. Similarly, after Garfield's assassin, Charles Guiteau, was hanged, his skull was carefully preserved so that another group of nineteenth-century quacks, the phrenologists, might compare its shape and bumps with those of other criminals. Phrenologists believed that the protuberances on the skull provided an accurate index of talents and abilities, such as benevolence or laziness.

Soon after the Bells moved to the house on Rhode Island Avenue, Alec opened a little private day school, on nearby 16th Street. He rented the ground floor to a local kindergarten, and then found a teacher for a class of six deaf children, aged between three and eight, on the second floor. At recess, all the children (including his own two daughters) played together, so that the deaf children might watch the hearing children speak during communal play. The parents of deaf children received instruction in Visible Speech to enable them to continue teaching their children at home.

Though Alec and Mabel loved one another deeply, the stresses on both of them kept mounting during these years. Alec felt strung out between the phonograph, his deaf studies, the little school, and the telephone litigation that periodically sparked commotions in the press about whether or not he was the "real" inventor. He frequently withdrew from family life and retreated to his study or the Volta Laboratory to read scientific journals. He often sat down at the piano and played to himself for half the night. "He played very dramatically, and it seems to me that he particularly liked big, stirring passionate things," his daughter Daisy would recall after his death. "He played with his whole soul and his whole body too, and I liked the way the big curl on top of his head would wobble about."

Meanwhile, Mabel resented how little she saw of her husband and how often he was too preoccupied to pay attention to their daughters. "It does feel awfully lonesome without you, my big burly husband," she wrote to him. "When you are here you are the object around which

Mabel and Alec with their daughters Elsie (left) and Daisy in 1885.

all my life moves, and now you are away it feels empty and objectless." Were all his activities really warranted, she asked? "Why was our wealth given us if not to give *you* time to make up to *your* children what they lose by their mother's loss?" After a family visit to her parents, who had now returned to Washington, she wrote despairingly in her journal, "Alec talked genealogy all the time. He thinks that in the course of a hundred years, material will be gathered through Genealogical Societies from which important deductions can be made affecting the human race. . . . I am afraid I am not particularly interested in investigations that can only be used a hundred years hence."

The most difficult blow for Mabel came on November 17, 1883. She was seven months pregnant, her two little girls both had bad colds, and Alec had disappeared to a meeting of the National Academy of Sciences in Hartford, Connecticut. She began to feel ill, but the doctor reassured her it was just a cold and told her to stay out of drafts. In the middle of the night, a painful cramp awoke and her, and she realized with horror that she had begun contractions. With only Sister and a neighbor in attendance, she gave birth prematurely to another boy. "Poor little one," she recorded sadly in her journal. "It was so pretty and struggled so hard to live, opened his eyes once or twice to the world and then passed away."

Alec arrived home three hours later. He was saddened by the baby's death, but he was particularly upset because he knew how much Mabel longed for a son and he did not know how to comfort her. He berated himself so vehemently for once again being away from home during such a crisis that Mabel had to dry her own tears and look after him. He would brood for years on his lost sons, his helplessness in the face of their deaths, and Mabel's sorrow. "My true sweet wife," he wrote in December 1885, "nothing will ever comfort me for the loss of these two babes for I feel at heart that *I was the cause* . . . when death came and robbed us of the little ones we wanted so much, you forgot your own suffering to try and comfort me." Two years later, the sight of Mabel cradling a doll in her arms (during a sitting with a portrait painter) prompted him to write, "I love you very much my darling little wife, and wish indeed you could be blessed as you desire."

The Bells named their second son Robert. Mabel had little chance to recover her usual stoicism in 1883 before more family tragedies struck. Her brother-in-law Maurice Grossman died the following year, aged forty-one, of a liver tumor. Within a year, in 1885, her sister Roberta, wife of Alec's cousin Charlie Bell, died in childbirth, leaving two little girls, Helen and Gracie. Meanwhile, Sister, Maurice Grossman's widow, was fighting a losing battle with tuberculosis, from which she had suffered for twenty years. She eventually died in 1886, leaving her parents to raise the Grossmans' only child, a daughter nicknamed Gypsy. The Hubbard-Bell clan now appeared much more fragile than it had at the start of the decade. Reflecting on the loss of her exuberant Hungarian brother-in-law, Maurice, Mabel noted how sadness had subdued her family: "We seem such a quite ordinary family now. Alec is a man out of the ordinary certainly but he is quieter in general life. He never shocks and takes away our pride with his overflowing spirits and utter disregard of conventionalities as Maurice sometimes did. We are all so quiet-mannered and self-restrained."

The emotional stamina she had acquired as a child, alone in the school for the deaf in Vienna, gave Mabel reserves of strength with which to deal with death and with dashed hopes. But in Alec, stress induced all of his old hypochondria and feelings of panic. Writing from Boston in 1882, during another telephone patent hearing, he begged Mabel to join him: "I feel very unhappy and anxious about myself . . . going over old telephonic experiments excites me so much that I cannot sleep." He complained constantly of chest pains: "When I feel badly now I find that my heart gives two beats then a stop. . . . I feel that the only chance of my getting over this heart-trouble is that *you should not leave me.*" Even when things were going well, such as at a scientific meeting in Philadelphia in September 1884, the heat often prostrated him. "All my spirit taken out of me," he reported home. "My whole body and arms completely covered by heat eruption. The bath-tub my only refuge." He traveled incessantly, visiting schools that taught lip-reading and lobbying state governments for day schools that taught either the oral method or a mix of signing and lip-reading, rather than boarding schools that taught only sign language. He complained nonstop in letters home. "I am tired out and going to bed," he wrote Mabel from Chicago in 1885. "Even my hand refuses to write properly—feels inclined to cramp. Think I must have caught cold as my head feels full."

The lowest point of Alexander Graham Bell's life came in the fall of 1885. Against the backdrop of the loss of two sons, his wife's grief, and his own frustration in the laboratory came news that the Pan-Electric Telephone Company had persuaded the U.S. government to file suit against the Bell patents. New York newspapers ran editorials suggesting that the Bell patent had been won by dishonesty and that Alec himself was a charlatan. For a man as temperamental and intense as Alec, this was devastating.

Mabel, still recovering from her sister Roberta's death and worried about Sister's failing health, could see her husband's despair. He could not concentrate on any of his projects. He could barely play with his daughters. His thick hair and bushy beard, once so jet black, were flecked with gray, and he complained regularly of indigestion and back pain. He decided to close the little school he had founded for deaf children on Washington's 16th Street, behind his Rhode Island Avenue mansion. After he announced his decision to teachers and parents and exchanged tearful farewells, he "came home to spend the afternoon, as he had spent the morning, on the sofa with a splitting headache and a heartache harder to bear," Mabel noted in her journal. He told his wife that he felt his life had been "shipwrecked," and begged her to travel with him to Boston for the next round of telephone hearings. With a characteristic touch of melodrama in his appeal, he told Mabel that he needed her to "keep him from an accident," because, as she noted in her journal, he "feels as if I am all he has left."

But Mabel was not all that Alexander Graham Bell had left. He was young and rich, and he retained an insatiable scientific curiosity. And by the time the Pan-Electric suit arrived in court, he and Mabel had also discovered a haven thousands of miles away from the heat and hurly-burly of Washington, a place that would be their refuge for the next thirty-seven years.

Chapter 13

ATLANTIC ADVENTURES
1885–1887

The Bras d'Or is the most beautiful salt-water lake I have ever
seen . . . the afternoon sun shining on it, softening the out-
lines of its embracing hills, casting a shadow from its wooded
islands . . . here was an enchanting vision . . . We came into a
straggling village [and] stopped at the door of a very un-hotel-
like appearing hotel [the Telegraph House]. It had in front a
flower garden; it was blazing with welcome lights; it opened
hospitable doors . . . and we enjoyed the luxury of spacious
rooms, an abundant supper, and a friendly welcome. . . .
The reader probably cannot appreciate the delicious sense
of rest and of achievement which we enjoyed in this tidy inn,
nor share the anticipations of undisturbed, luxurious sleep,
in which we indulged as we sat upon the upper balcony after
supper, and saw the moon rise over the glistening Bras d'Or.

Once Alec had read this idyllic description of a tiny village
in a remote eastern corner of North America, he couldn't get it out

of his mind. The book in which it appeared was *Baddeck, and That Sort of Thing: Notes of a Sunny Fortnight in the Provinces,* first published in 1874 and written by Charles Dudley Warner, a Connecticut resident who was editor of the *Hartford Courant.* He was a neighbor of Sam Clemens (Mark Twain), with whom he co-wrote a book with the resonant title *The Gilded Age.* Warner was a frequent visitor to Boston, and Alec and Mabel used to see his spare, genial figure pottering down Brattle Street when he came to visit his Cambridge friends, Henry Wadsworth Longfellow and Professor Charles Eliot Norton. A friendly encounter with the well-known editor prompted Alec to read his description of a journey through the Atlantic provinces of Canada. Warner had traveled through Nova Scotia and crossed the Strait of Canso to Cape Breton Island, an isolated region populated by Gaelic-speaking (and, in Warner's gently satirical view, quaint) Scots. Warner and his traveling companion had taken nearly a week, by train, steamer, and horse-drawn cart, to reach the charming village of Baddeck from Boston. Alec read Warner's gushing prose some time in the mid-1880s, at a point in his own life when he was oppressed by greedy American litigants, family tragedy, and the muggy heat of Washington summers. He found Warner's hyperbole and the promise of seclusion and "undisturbed, luxurious sleep" irresistible.

Earlier in the decade, the Bells had spent summers on Massachusetts or Rhode Island beaches or in the mountains of western Maryland, but they had never found a summer property to which they felt drawn back each year. By 1885, they were eager to go farther afield. Alec's father, Melville, was now sixty-six and had decided he wanted to make a sentimental journey to Newfoundland, the British island colony off Canada's east coast where he had spent four bracing years as a young man in search of better health. Eliza, now seventy-six and lame as well as deaf, had no interest in any adventures, so Melville, who was still inclined to treat his son as if he were an unreliable eighteen-year-old, decided it would be good for Alec, Mabel, and their daughters to explore North America's east coast with him. He found an unlikely ally in Gardiner Hubbard, who had sunk a lot of money into the Caledonia coal mines at Glace Bay, on the eastern tip of Cape Breton Island. If local legend is to be believed, Gardiner had even persuaded his Cambridge neighbor, the poet Longfellow, to invest alongside him. The Caledonia mines had been operating for nearly half a century, and Gardiner wanted his son-in-law to take a look at them. Once Alec had read Warner's euphoric description of a tidy inn overlooking the glistening Bras d'Or Lake, the expedition came together.

So in late August 1885, Alec, Mabel, their two little girls, Melville, and a nursemaid called Nellie, carrying a mountain of bags, steamer trunks, and other luggage, took a steamer from Boston to Halifax, then a train from Halifax via Truro to the Strait of Canso (the narrow channel that separates mainland Nova Scotia from Cape Breton). There they boarded the S.S. *Marion,* a paddle steamer that took them through the narrow southern opening into St. Peter's Inlet and Bras d'Or Lake. Mabel sat on the deck, spellbound by her first sight of one of the largest saltwater lakes in the world. Seventy miles long and ten to twenty miles wide, Bras d'Or Lake covers four hundred square miles in area and has over a thousand miles of interior coastline, bays, and channels. It is almost two lakes, as its eastern and western shorelines nearly touch halfway up its length, at a slender channel called the Barra Strait.

Bras d'Or Lake is the center of an idiosyncratic North American island community initially composed of Catholic Scots who were driven out of the Scottish Highlands by famine, rising rents, and land enclosures from the late eighteenth century onward. Attracted by free land and salmon-rich rivers, they weathered the harsh winters and imported a tradition of ceilidhs (festivals) and fiddle-playing. In 1802, when Presbyterian Scots had joined the exodus from their native land, the island boasted 2,513 inhabitants. By the time the Bells arrived, over eighty years later, the population had soared to over 80,000. What made this particular corner of the British Empire unique was that nearly all Cape Breton immigrants came from the Gaelic-speaking Highlands and islands of Scotland. And geography dictated that they remained undiluted. With no railroad until the 1890s and no causeway to the rest of Nova Scotia and Canada until the mid-twentieth century, the island nurtured a deeply conservative society, clinging to its old clan structure, its belief in fairies, and an addiction to "toddy." A substantial proportion of its population didn't speak English. Long-established Gaelic customs, such as "waulking" or milling the cloth, were occasions for singing old Gaelic songs and composing new ones. Cape Breton was, in fact, a more deeply Gaelic society and culture than the one that many of its members, and certainly Alec Bell himself, had left behind in Scotland.

Mabel had little inkling of this as she sat on the *Marion*'s deck. She was captivated by the dazzlingly white little lighthouses on the shoreline, punctuating the dark backcloth of spruce, fir, and pine. "It is perfectly lovely here, the archipelago of islands and the mainland with its steep hillsides sloping down into the water," she told her mother. "I do not see that there is anything more beautiful on the Corniche between Menton and Nice." Alec and Melville listened with amusement to the Gaelic (which neither of them spoke) and Scots accents of their kilted and cloth-capped fellow passengers. "Everybody here is Mac something," Mabel noted.

The Bell party stopped only briefly at Baddeck, to see whether the Telegraph House lived up to Warner's description (it did). However, the overnight visit did allow Alec to discover that a telephone line ran between the local general store and the local newspaper, the *Cape Breton Island Reporter*. According to Baddeck folklore, Alec was walking past the newspaper office's plate glass window when he caught sight of the newspaper's editor, Arthur McCurdy, trying to make the wall-mounted phone work. The next thing that the young editor knew, a tall, beefy man with bushy black hair and whiskers had entered his office to ask him what he was doing. McCurdy explained that he was trying, without success, to fix the phone. The stranger unscrewed the end of the earpiece, removed the diaphragm, brushed a small fly away, replaced the diaphragm, screwed the end back on, and said, "I think

The crowded wharf at Baddeck.

now you will find it working." McCurdy tried the phone and immediately got through to the store, which was run by his father. Shaking the stranger's hand, he asked him, "How did you happen to know about this?" The stranger smiled back: "Because I am the inventor of that instrument."

The Bells made an even briefer stop at Glace Bay, to check out the Caledonia mines. But Melville Bell was eager to reach Newfoundland, and hustled everybody back to Halifax and onto the S.S. *Hanoverian,* a steamer bound for St. John's, Newfoundland, under the command of a Captain Thompson. There were 280 passengers aboard, 40 of whom (including the Bells) were saloon class and the rest in steerage, plus a cargo of canned meat, tobacco, and pork.

The Bells had made several transatlantic voyages without incident by now, and as dusk fell and the *Hanoverian* weighed anchor, they settled happily in their saloon cabins. But a thick fog rolled in while the steamship made its way up the coast of Nova Scotia toward Newfoundland's Avalon Peninsula. The captain insisted he knew exactly where he was, and despite the fog (now as dense as pea soup) and the concern expressed by some passengers, he refused to check his position with a couple of fishermen whose boat appeared out of the fog and who waved vigorously at them. In the laconic account that appeared in the St. John's newspaper, the *Evening Telegram,* "Expecting to be around Cape Race, he took Cape Mutton to be Cape Ballard." It was a fatal miscalculation: the *Hanoverian* had not yet rounded the vicious rocks of Cape Race, at the southeastern tip of the Avalon Peninsula. Just before ten o'clock on the third morning, all hell broke loose. Mabel felt the boat come to a shuddering, grating halt; Alec heard panicky shouts from the crew. The steamer had gone aground on a reef off Portugal Cove South. Melville Bell came roaring out of his cabin, Nellie the nursemaid fell to her knees and started praying, and Mabel quickly dressed her little girls in their warmest clothes. Alec was already on deck, trying to find out the extent of the damage.

Captain Thompson, according to the *Evening Telegram,* did not rise to the occasion. When two "respectable planters [settlers] . . . launched their skiffs and offered assistance . . . the captain declined, and referred to them as savages, this unwarranted expression coupled with a strong seaman's adjective." But Captain Thompson had his reasons for such behavior: Newfoundlanders from Cape Race had a fearsome reputation as wreckers. And Captain Thompson's situation was not eased by the conduct of his own crew, who (again according to the admittedly partial *Evening Telegram*) "behaved disgracefully, exhibiting incompetency, drunkenness and insubordination."

Luckily, among the *Hanoverian's* passengers were some British sailors, who soon started lowering lifeboats and organizing people onto them. Meanwhile, Alec had returned to the cabin to tell Mabel—who could not hear the shouted orders to leave the ship—what was going on. Then he took five-year-old Daisy in his arms, grabbed seven-year-old Elsie's hand, and chivied his wife and father ahead of him. Daisy would never forget watching her Grandfather Bell clamber laboriously down a rope into the dinghy while a sailor called out, "Here! Wait for this big fat man!" Despite the panic around her, Mabel displayed her usual sangfroid as her husband got them all seated in a lifeboat. But as she admitted to her mother in a letter, "Just fancy! Alec left his valise with his precious deaf-mute books in his cabin. He did not forget them, but actually preferred leaving them to run their chance to letting his wife go on the boat without him. I felt quite elated!"

The passengers and the crew of the S.S. *Hanoverian* all reached the little village of Portugal Cove South safely. Things didn't go so well after that. The *Evening Telegram's* correspondent claimed that everyone was "comfortably sheltered in the houses of the hospitable fishermen in the Cove," but that wasn't Mabel's experience. "If the wreck was a gentle one," she reported to her mother, "the rescue was a hard one badly managed or rather not managed at all. A more solid inhospitable set of people than these of Portugal Cove I hope never to see. Not a thing would they do except for pay and that as badly and grudgingly as possible." Since there was no road along the rocky coast to St. John's, the passengers had to wait for the British naval vessel H.M.S. *Tenedos* to pick them up. The following day, the seas were too rough for open boats to ferry them to the British vessel, which steamed away to Trepassey, a village eight miles along the coast with a better harbor. The *Hanoverian's* passengers were told to make their own way there overland. As they straggled down the rutted coastal path, past peat bogs and rock fences, they saw coming toward them a menacing spectacle: bands of men with boat hooks on their shoulders and knives in their belts—the ill-famed wreckers, looking for salvage, or, as Mabel put it, "vultures intent on their prey."

Alec proved a tower of strength in the melee. He paid for tea and coffee to be served to the bedraggled passengers, then strode off beyond Trepassey to find some means of transport for children, the elderly, and the sick. He returned with horse-drawn wagons, brought his own family to a Trepassey inn, and installed them in a private set of rooms there. Mabel wrote admiringly of his efforts to her mother but noted wryly, "Tomorrow I fancy the collapse will come, but he feels happy for he thinks yesterday has proved that he has not heart disease."

The *Hanoverian's* passengers finally sailed off to St. John's aboard the *Tenedos,* with the ship's captain carefully taking depth soundings throughout the voyage up the Avalon Peninsula's craggy shores. By now, the calamity had become an adventure. A band played in the ship's salon for the rescued passengers. The British naval officers charmed Mabel and had an endless supply of candy for Elsie and Daisy. A couple of days later, the vessel sailed through the Narrows into St. John's harbor, and nosed its way between the numerous tall-masted schooners, bobbing up and down at their moorings, toward one of the narrow finger wharves that jutted into the harbor waters. "We found all St. John's on the wharf," Mabel reported to her mother, "waiting to see the shipwrecked passengers land, but they must have been disappointed for [although] nearly every one owned only the clothes they stood up in, they looked very respectable—Alec especially, who is fast developing into a dandy in his old age." Most of those on the wharf were there to greet the wife and daughter of the bishop of Nova Scotia, en route to join the bishop in England. There was no mention in the local newspaper that the American millionaire Alexander Graham Bell, famous inventor of the telephone, was among the shipwrecked, since the party was listed on the ship's manifest as "Mrs. G. Bell, two children and maidservant . . . Messrs. Bell, G. Bell" from Halifax. But Alec was delighted to find a welcoming party of telegraph men, who had booked rooms for his entourage at the posh new Atlantic Hotel, on Duckworth Street. While Mabel sorted out their soggy luggage ("I shall save my underclothing but everything that won't stand thorough cleaning is ruined"), Melville Bell went off to find the Macphersons, a family from Glasgow who had befriended him in Newfoundland half a century earlier.

Alec emerged as leader and advocate for all the *Hanoverian's* passengers. The sight of her terminally undomesticated husband being consulted by so many women amused Mabel: "All the ladies depend on Alec for everything. He must advise about their trunks, about drying their clothes, prayer books and photographs, it is really absurd." He began a campaign to find funds and clothing for their more destitute fellow passengers. "We hope that the people of St. John's will

The view from Baddeck of Red Head, which Alec would rename Beinn Bhreagh.

make ample amends for the inhospitality of those of Portugal Cove," commented his wife.

The Bells remained in St. John's for about a week. They were fascinated by its dramatic beauty and its idiosyncrasies; since much of the population was illiterate, the stores on Water Street and Duckworth Street followed the age-old practice of advertising their contents by hanging out signs such as a giant pocket watch for a jeweler, a boot for a shoe mender, and a red and white pole for a barber. On the fish-drying sheds, noted Mabel in her journal, "lie the whitening carcasses of more fish than I ever saw in my life before." But Newfoundland was too isolated, too harsh, too *Irish;* Alec hankered after the gentler Scottish charms of Cape Breton. So the party boarded a ship in St. John's harbor, bound for Nova Scotia.

Three days later, they were steaming back across the Bras d'Or Lake to Baddeck and the spool beds, old-fashioned crockery, and creaking stairs of the Telegraph House. Years later, Maud Dunlop, daughter of the hotel's owner, recalled seeing the family on their return in September 1885: "Mrs. Bell, at this time in her late twenties, [was] a slender person with the gentlest manners, her sweet sympathetic face framed in the most beautiful soft brown hair. She and her dark-eyed little girls made a lovely picture; they were most devoted to her. . . . I can see her [now], sitting on the upper verandah reading aloud to them while their governess was out."

In the soft September sunshine, with puffy white clouds reflected in the lake's calm water, Baddeck was as tranquil and rejuvenating as Charles Dudley Warner had promised. During the next few weeks, Alec and Mabel walked, drove, and explored. They swam in the invigorating salt water; they rowed along the rocky shore. They sailed up the bay, and had to be rescued by a thoughtful Cape Bretoner when the breeze died. At dawn, they watched the sun rise behind the Kempt Head headland opposite the village, and in the evening they watched its last scarlet rays fall on Red Head, a headland just beyond Baddeck harbor. Alec luxuriated in the fresh breezes off the water and played chess with Arthur McCurdy, the local newspaperman whom he had met a month earlier. Mabel savored the precious family intimacy. "I think we would be content to stay here many weeks just enjoying the lights and shade on all the hills and isles and lakes," she wrote in her journal. "I cannot see why C. D. Warner should be possessed of such a mad desire to come here that he must travel so uncomfortably day and night . . . to spend two days here. No wonder the good people here took him for insane." By late September, when it was time for the Bells to return to Washington, Alec had decided they must come back the following year and find a cottage by a running brook where they and their children might enjoy simple pleasures, far beyond the reach of the Pan-Electric Telephone Company's lawyers. Baddeck had revived the dream that had lain dormant since he and Mabel had spent their romantic (albeit uncomfortable) week in the cottage in Covesea, Scotland, just after their marriage.

Sure enough, the Bells were back in Cape Breton in July 1886. Arthur McCurdy had found a property for them to rent: a bare four-room cottage ("at best it is a regular shanty," according to Mabel) a mile outside the village, overlooking Baddeck Bay. They named the cottage Crescent Grove, and these American millionaires set about devising ways to improve on its scarce furnishings. A wooden armchair, a box, and a hay-filled mattress became a chaise lounge; a hammock was improvised from barrel staves held together with rope; fabric flung over boxes artfully transformed them into shelves and tables. It was all very make-do-and-mend compared to the splendors of Washington, but Baddeck fulfilled the fantasy shared by Alec and Mabel that they wanted only "the simple life" and that their wealth and privilege were of no consequence to them. (Nevertheless, Alec was soon the proud owner of a stylish steam launch, which was certainly beyond the means of most year-round Baddeck residents.) "Alec's patience with all his following of wife, children, nurse and two dogs, butterfly net, lunch basket etc. is something wonderful," Mabel wrote to her mother-in-law.

Mabel had decided that it was stupid for her little girls to be climbing trees and running through the long grass in petticoats and skirts. So in Halifax, en route to Cape Breton, she had stopped at a men's store and ordered trousers for boys "about the same size as my daughters." (An open admission that she intended dressing her girls in boys' outfits would have triggered shock and horror.) Once in Baddeck, Elsie and Daisy climbed trees, acquired a pet lamb called Minnie, and lost their hair ribbons. Mabel made raspberry jam and blackberry jelly and watched the nursemaid Nellie teach their newly acquired cook some fancy dishes. Alec pitched a tent on the lawn and insisted on spending the night there. The Bells were so determined to achieve simplicity that they acquired a cow, "Miss Miggs," and a wooden churn so they could make their own butter: "We all worked, Alec, the children and I, and after a long hard bout we had the delight of seeing butter form. It did look like tightly scrambled eggs at first but Nellie . . . soon brought it into shape. . . . But we found the churning process less delightful in practice than in theory and Alec is trying to invent a windmill to do our churning for us." In the meantime, Elsie, Daisy, and Mabel churned while Alec played "Onward Christian Soldiers" on a portable organ to cheer them on.

Crescent Grove, after the Bells had jacked up the Baddeck cottage and added another story.

Alec purchased Crescent Grove and enlarged it by jacking it up and building another story underneath it. The locals, who watched Bell family activities with amused skepticism, dubbed it "the cottage on stilts." However, Alec already had his eye on another property: the headland called Red Head, across the bay from Baddeck. One bright August day, when the sun was sparkling on the lake and there was not a cloud in the deep blue sky, he and Mabel hired a wagon and drove over to Red Head, then found a rough trail there that took them to the other side. "Fancy driving over the crest of a mountain, the highest for many miles," Mabel told her mother, "and seeing the land stretched out on every side of you like a map. Hills and valleys, water and islands are all around, the hills of Sydney on one side, Boularderie Island right at your feet with the Big and Little Bras d'Or entrances on either side. . . . Immediately beneath, peeping through its fringing trees with sharp contrasts of white plaster cliffs, lay two lovely little harbors."

Both Alec and Mabel were captivated by the extraordinary beauty around them. One look at her husband's face convinced Mabel that this was a place he could be happy. Alec was already planning how to take possession of this magical headland, and over the next seven years, he would acquire it, a few acres at a time. He took psychological possession of it, too: he would soon rename it "Beinn Bhreagh" (pronounced "ben vreeah"), which is Gaelic for "Beautiful Mountain." Beinn Bhreagh and Baddeck provided the kind of therapeutic release for which Alec's tightly wound temperament yearned. The solitude and natural splendor also gave him a new lease on life: they would stimulate the kind of frenzy of invention he had enjoyed during the 1870s.

Mabel confided to her mother that her husband "is quite a different person from what he is in Washington. . . . Here he is the life and soul of the party. . . . He is forever on the go; at night when all are sleeping, he is paddling about but he is up again at his usual hour." At Baddeck, both Mabel and Alec lived in the present. Whenever they were far from Cape Breton, their thoughts would wander back to the sound of the wind in the fir trees and the long view down the vast inland seas. "It is strange what a hold Baddeck has already on Alec," Mabel wrote to Melville Bell in June 1887 from Washington. "The other night when Alec was in pain and fever, he bade me remember if anything happened that he would be buried nowhere else except at the top of Red Head."

Chapter 14

A SHIFTING BALANCE
1887–1889

Each fall, the Bells were pulled back to Washington. In 1885, Alec had to return to face the Pan-Electric Telephone Company's assault on his patent and his reputation. He was outraged by the Pan-Electric suit, and sat up for several nights writing a stinging letter to the U.S attorney general. "Poor Alec," Mabel wrote in her journal that November, "all this week he has been working hard and drawing immense drafts on the reserve of health and strength laid up this summer. . . . He has not taken a single meal with us all week."

The case was still grinding on in January 1887 when the Bells faced another Washington crisis. Early one freezing-cold morning, a policeman noticed flames creeping along the mansard roof of 1500 Rhode Island Avenue. The fire had begun in a grate, then traveled into the supporting timbers in the roof. The policeman gave the alarm, which woke the Bells' cook, maid, and nursemaid. In a panic, they rushed up and down stairs, shrieking "Fire," shaking the children awake, and banging on Mabel's bedroom door. But Alec was out of town and Mabel slept on, unconscious of the danger, until Sheila, her Yorkshire terrier, jumped onto her bed. When Mabel realized what was happening, she displayed her usual presence of mind: her priorities were her children and Alec's books. First, she told Nellie to dress Elsie and Daisy and take them and the dog across the road to a neighbor's house. Then, pulling on her new white cashmere coat and leaving her brown hair loose on her shoulders, she went off to find the firemen and instruct them that their most persistent efforts must be directed to saving her husband's study and library, on the third floor. She made sure they did what she asked, pretending she couldn't understand their pleas to leave the burning building until she was finally driven out by the smoke. By then, the interior of the mansion was drenched with water that quickly froze solid: the exterior of the house was encased in an armor of ice, and icicles as large as a man's leg hung from the gutters. Meanwhile, across the road, Elsie and Daisy watched the "magnificent sight of the flames leaping into the sky from the roof." Daisy never forgot the "beautiful . . . way they swirled around in a little turret, which was my father's study."

Early the next morning, Alec was awakened in his hotel room by a bellboy with a telegram from Gardiner Hubbard. It read, "Your house is on fire, Mabel and the children are safe." Never at his best when woken early, Alec sleepily asked for a train timetable. Informed that there was no train to Washington for four hours, he turned over and went back to sleep. When he finally arrived in the capital, he found his wife infuriated by newspaper reports that described her as rushing out of the house in her nightgown. But he also found most of his library intact.

Alec's third-floor study, however, was a mess. After negotiating the charred staircase and the squelching carpet, the inventor stood in the doorway, staring in dismay at the litter of sodden notebooks, scraps of paper, and old letters covered in a layer of ice three inches thick. Mrs. Sears, the housekeeper, had hired some men from a nearby rooming house to rip up carpets and clear the debris. Among them was a tall, good-looking African American from Virginia, an eighteen-year-old

called Charles Thompson, whom she had put in charge of the study
because he could read and write. Half a century later, Thompson
remembered this turning point in his life: "During the late afternoon
the door of the study was pushed open; turning around to see who
had entered, I faced a tall, heavily built man with black hair and black
beard mixed with gray. 'Ah ha, here you are; is this Charles?' 'Yes, Sir.'
'Well, Charles, I am Mr. Bell, how are you?' extending his hand with a
genial smile, he shook my hand as if he had known me for years." To a
young man accustomed to being treated as a second-class citizen, this
courtesy was extraordinary: "I loved Mr. Bell from that moment, and
if I had left that house that day never to see him again I never would
have forgotten that handshake; it electrified my whole being."

Alec quickly ascertained that Charles could read his own untidy
scrawl. "Ah ha!" he said, when Charles read the notes without hesita-
tion. "Now young man, you have the most important work of all the
people employed in this house. This room contains a lot of important papers, in fact, all of my important papers bearing on the work in
which I am now engaged is in this room." Standing in the middle of
the chaos, Alec fixed his intense eyes on his new employee, and issued
precise instructions: "Don't throw away any scrap of paper, however
small, that has any figures, writing or drawings on it, put everything in
a basket, or box, and when you are leaving in the evening bring them
to my room." The young man carried out these instructions so con-
scientiously that Alec decided to employ him permanently. Charles
would remain with the Bells for the next thirty-five years, looking after
everything from travel arrangements to Alec's wardrobe, and always
greeting his employer with the same words that Alec used as a telephone greeting: "Hoy! Hoy!"

Charles's most significant contribution to the household, however,
was that he lifted from Mabel's shoulders the full weight of dealing with
her husband's eccentricities and obsessive observance of rituals—rituals that became more entrenched each passing year. "His daily routine
or 'schedule' as he used to call it," observed Charles, "was adhered to

in the most minutest details." Within months of Charles's arrival in the household, as he recalled in a memoir he wrote in 1922, he learned that "any changes by accident, or intent, would upset [Mr. Bell] for the entire day sometimes mentally as well as physically. To call him before nine o'clock in the morning usually gave him a bad headache. But not to call him at nine, and to tell him the exact time was even worse, as he often explained to me—because it left him in doubt as to the exact time that I did call him, and would cause him to lose confidence in trusting me to help keep up his schedule." Charles would ensure that when Alec did finally rise, his breakfast tray, the morning mail, and the newspapers were waiting for him on a table in his study. Alec insisted on remaining undisturbed while he ate his cold oatmeal (never rolled oats) served with brown sugar and cream on plain white (never patterned) china, and read the papers. If someone did come in and start talking to him, he was likely to go straight back to bed. "He told me many times that he never felt really awake until he had finished his breakfast, read the papers and lighted his cigar or pipe."

Once these morning rituals were complete, Alec would press three times the bell that communicated with the kitchen three floors below; this was a signal that Charles should remove his breakfast tray and tell Alec's secretary to go up. After Alec had dealt with his mail, he would dictate an account of the events of the previous twenty-four hours. Then he would get back to his researches on deaf issues, or he would disappear to the Volta Laboratory. "He never liked for any one to knock on his door before entering the room," Charles always remembered. "If he was following a train of thought and there was a tap on his door, his attention being diverted to the noise, he very often lost the thread and for days would not be able to pick it up again." Alec once told Charles, "Thoughts, my dear sir, are like the precious moments that fly past; once gone they can never be caught again." Charles often saw his employer work for up to twenty-two hours without sleep, ignoring his wife's request that he join her and any guests (usually Hubbard relatives) for dinner.

Any disturbance was such anathema to the telephone's inventor that he never had a telephone installed in his own study—the bell might have triggered a migraine. He could not stand chiming clocks either. His obsession with his own routine made him maddening to live with. One morning, Mabel asked why every single clock in their mansion was stopped. Her husband told her, "I had an awful time with those clocks last night. . . . I could not work for the noise, so I went around and stopped them all." Oblivious to ticks and chimes, Mabel needed to be able to tell the time. She found a clock mender to remove all chiming mechanisms.

After the fire, Mabel refurnished the Rhode Island Avenue mansion, and the family returned there. But the fire had spooked Mabel. "Mother decided it was too big a house for her anyway," her daughter Daisy wrote later. So in 1889, it was sold to Levi P. Morton, vice-president of the United States under newly elected Republican president Benjamin Harrison. The Bell entourage spent the next two years in rented accommodation on nearby 19th Street while they watched their new house being built at 1331 Connecticut Avenue. The new house was designed by the fashionable firm of architects Hornblower and Marshall, who were responsible for changing the streetscapes of Washington in the 1880s and 1890s with their designs for dozens of residences for prominent Washingtonians, including the Smithsonian's Natural History Museum and the mansion that today houses the Phillips Collection. Mabel loved the location of the new house: her sister Grace lived next door and her parents' townhouse was across the road. Mabel and Alec finally moved into their rather plain, four-story dwelling, built of red brick and stone, in 1892. Soon their little girls were sliding down the banisters of the spiraling oval staircase and Alec was enjoying musical evenings in the parlor. This was their Washington home for the rest of their lives.

By 1888, it was evident that Alec would emerge unscathed from the Pan-Electric suit. Yet frustration and dismay continued to gnaw away at him, as he felt time passing without his making any further contributions to science. It was now fourteen years since he had conceived

*Gertrude Hubbard (left),
with Alec and Mabel,
outside her Washington
mansion, Twin Oaks.*

of the telephone, and eight years since he had dreamed up the photophone. Had his brain slowed down, his imagination ground to a halt? Admittedly, deaf studies preoccupied him, but he was still scribbling ideas in his notebooks. The trouble was, he kept telling Mabel, he simply didn't have the time or the qualified assistance to develop them properly.

But this was only part of the problem. A larger part was his own nature. He did not have the self-discipline to adopt a methodical strategy for testing out his intuitions in practice. He had knuckled down to his experiments when he was in his twenties because he was driven by the need to make money to marry Mabel and to prove to his father that the telephone was as important as Visible Speech. But once he had achieved these goals, he spread his energies too thin. He himself was rarely prepared to do the painstaking testing that inventions like the photophone or the vacuum jacket required. He preferred leaps of the imagination, during which he might experience a eureka moment, to the dull slog of careful calibrations. Soon after they became engaged, Mabel had commented that "you like to fly around like a butterfly sipping honey, more or less from a flower here or another flower there," and Alec's intellectual capriciousness had only increased since then. He *so* wanted to keep on inventing, but he allowed himself to be pulled in too many directions at once.

Moreover, he lacked the knowledge and training that were increasingly required for scientific innovation. As Robert V. Bruce has pointed out, Bell's understanding of mathematics was limited. Too often he relied on analogy. This had proved to be a stroke of genius in the invention of the telephone, when his firm grasp of the principles of human speech and hearing allowed him to speculate that an electric current could mirror sound waves. It was also useful for the photophone. But when dealing with challenges outside his knowledge base, it could lead him up intellectual alleys that were little short of ridiculous. At one point, as he recorded in a notebook, he wondered, "Are odours vibrations? If so, may they not be vibrations between sound & heat?" A few years later, he wondered whether "thought transference" was possible. He coiled wire around his own head, then connected it to a coil around his assistant's head, to see if he could transmit ideas without speaking. All that was shared was a headache.

But a larger issue was the changing nature of the business environment in the nineteenth century. Alexander Graham Bell was increasingly out of step with his times.

From today's vantage point, the United States from 1865 to 1900 seems compellingly exuberant, particularly as it is often called, in the phrase coined by Mark Twain and Charles Dudley Warner, "the Gilded Age." But the Bells' world of big houses, confident, bearded men, and a settled pace of life is deceptive. Post–Civil War America was in a state of transition, as a motley collection of antagonistic states became a nation. The speed of that shift to nationhood was breathtaking, matched only by the speed of the technological revolution

One of Alec's more bizarre ideas involved an apparatus for transferring brain-waves between individuals.

that accompanied it. Alec himself had contributed enormously to the latter with his invention of the telephone: by 1898, the United States, with a population of 76 million, would have 800,000 telephone sets. This was twice as many as in all of Europe, which had a considerably larger population. Equally important was the rapid spread of railroads, knitting distant regions together. The first transcontinental railroad was completed in 1869; by 1880, there were 93,000 miles of track in service across the continent, and that number jumped to 165,000 miles by 1890.

The rail network created national markets both for industrial products, such as the steel required for rails, and for consumer goods, such as household furnishings. Everything was going farther and farther, faster and faster—and the accelerating economy was changing the way that innovations went from the inventor's brain to mass production. Scientific knowledge was crucial for improvements in how everything was produced, from textiles and domestic appliances to steel, oil, and railroad equipment. Smart businessmen no longer simply waited for the next good idea to come along: they actively sought out technologies that would increase output and, of course, profits. And savvy inventors recognized this appetite for their innovations. Before 1850, there were rarely more than five hundred inventors eager to patent their inventions each year in the United States. Thirty years later, the figure was twenty thousand and growing.

One of the canniest of such inventors was Alec's rival Thomas Edison. Edison recognized, in the words of his biographer Neil Baldwin, that "progress and competition were identical." Edison was not content with dreaming up new innovations, such as the phonograph, the incandescent lightbulb, and the carbon microphone for telephones. He was an entrepreneur: he also wanted to push his ideas through development and patenting to commercial use. He would be granted 1,093 patents during his lifetime, compared to Alexander Graham Bell's 31 patents (some of which were in association with colleagues). Edison was single-minded: he allowed nothing to get in his way. (He deserted his first wife, Mary, on their wedding night, scurrying back to his workshop to check a new device. She died when she was only twenty-nine, after a breakdown, leaving three children Edison had little to do with.) But Edison was prepared to do the laborious detailed work that Alec either ducked or delegated to others. It was Edison who gave us the aphorism, "Genius is 1 percent inspiration and 99 percent perspiration." At Menlo Park, twenty-five miles east of New York City and strategically situated on the Pennsylvania Railroad line, he constructed both a laboratory and a machine shop in the late 1870s. He hired experts, such as John Kruuesi, a Swiss clockmaker, Charles Batchelor, a British machinist, and Ed Johnson, an engineer, who together could translate his brainwaves into models and patents. And he was, in the words of author Harold Evans, "an impresario of invention." As soon as Edison saw a glimmer of commercial potential in any of his inventions, he trumpeted his success near and far, in the interests of raising the public interest and capital he would need to bring it to market. "Anything that won't sell," he once announced, "I don't want to invent." The buzz he created attracted the attention of the major investors of the Gilded Age (or "robber barons," as they are more often known), most notably Jay Gould and J. P. Morgan.

Alec was a throwback to a different age, when a lone inventor waited for inspiration to strike and then watched to see if others might exploit his invention. Alec didn't have an entrepreneurial bone in his body. For example, in the mid-1890s, he devoted a lot of time to perfecting an automated central switchboard that would do the work of twenty telephone operators, yet he never bothered to file a patent application. "Mrs. Bell was shocked," Charles Thompson recalled, and she asked her husband, "Why not?" Alec replied, "It would turn all of those poor girls out of their jobs."

Alec pursued his fascination in science for science's sake at the Volta Laboratory. In 1889, he moved his laboratory from Connecticut Avenue to another former stable—this time in Georgetown, behind his father's house. After his breakfast ritual each morning, he would disappear down M Street, to his office in the little redbrick building. In the evenings, as he strode home, his fellow Washingtonians would tip their hats at the distinctive figure with a thick beard and well-padded paunch. Sometimes Mabel would walk west along M Street to meet him and enjoy his company by street lamp. "I do not very often feel that I have so much of my husband's attention as I did tonight," she noted in her journal late in October one year. "He is usually so full of other things, but tonight it was too dark for Alec to talk to me so I had things all my own way."

There were two other outlets for his scientific curiosity in Washington. The first was the weekly journal *Science*, founded in 1880 and self-described as "the medium of communication among the Scientists of America." Thomas Edison had been an early backer of this publication, but he quickly lost interest in swallowing its deficits and felt no obligation to play a larger role in the scientific community. In 1882, Alec and his father-in-law, Gardiner Hubbard, became the journal's major investors. Alec invested as a way of keeping abreast of science; Gardiner

Hubbard was looking for innovations that might have commercial application. Year after year, the journal continued to lose money as subscriptions fell far short of the six thousand necessary to break even. "We are feeling very short," Mabel noted in February 1885, "*Science* having robbed us of so much." Two years later Alec noted gratefully, "There are few wives as self-sacrificing as you are, very few that would allow their husbands to take thousands of dollars from their income to invest in unprofitable enterprises like *Science*." The journal would cost Alec and his father-in-law some $80,000 by 1891 ($1.5 million in today's currency), of which about $60,000 ($1.2 million) came from the Bells. But Alec enjoyed the scientific camaraderie. Describing to Mabel an editorial meeting held in Philadelphia in September 1884, he wrote, "It was a grand thing to see all these splendid men there." Eventually, the American Association for the Advancement of Science took over *Science* as its official journal in 1900.

The second Washington opportunity for Alec to stay in the scientific swim was an institution of his own invention: his Wednesday-evening get-togethers. Architect Joseph Hornblower designed at 1331 Connecticut Avenue a special one-story wing, topped by a stone balustrade, for these gatherings. The large room was lined with books of travel, biography, and general literature, the woodwork was of carved teak, and there was an elaborate backlit stained-glass window that depicted the Temple of Isis at Philae. The guest lists for these weekly meetings were a who's who of late-nineteenth-century scientists: they included John Wesley Powell, director of the U.S. Geological Survey; Edward Morse, an eminent zoologist and orientalist; William H. Brewer, professor of agriculture at Yale University; Thomas Mendenhall, president of Worcester Polytechnic Institute; Simon Newcomb, director of the American Nautical Almanac office and professor of mathematics and astronomy at Johns Hopkins University; Edward Drinker Cope, professor of geology and mineralogy at the University of Pennsylvania; Dr. John Shaw Billings, first director of the New York Public Library and designer of the Johns Hopkins medical school; Alpheus Hyatt, professor of zoology and paleontology at Boston University; and Samuel Scudder, editor of *Science*. There was probably no other household in America that could claim so many remarkable thinkers in regular attendance— or so many highbrow debates, gray beards, and well-chewed pipes. But one personality dominated the room.

As Dr. L.O. Howard, a renowned entomologist and Wednesday regular, would recollect in later years, "What *interested* [others], *delighted* him. I never knew a man with so many enthusiasms, I never knew a man who was so instantly and truly responsive to an interesting or quaint turn of thought or to a fine new idea. He was as like a child, to whom his little world is a wonderland, as it is possible for a great man with a great mind to be." David Fairchild, a plant biologist who would become Alec's son-in-law, put it another way: "Mr. Bell was tall and handsome with an indefinable sense of largeness about him. He always made you feel that there was so much of interest in the universe, so many fascinating things to observe and to think about, that it was a criminal waste of time to indulge in gossip or trivial discussion."

The evening would start with general conversation, then Alec would turn to one of his distinguished guests and invite him to speak about his own particular interest. Since Alec had found out the activities of his guests in advance, the speaker usually had a paper prepared. By the end of the evening, as Charles helped Alec's guests into their cloaks and overcoats, the room would be thick with tobacco smoke and Alec himself would be afire with new knowledge or ideas. On one occasion there was a heated discussion about the health risks attached to the widespread habit of spitting on Washington's sidewalks. Alec's guests decided to form a committee to educate the population about the dangers of spreading diseases like tuberculosis in this fashion, and they elected Alec chairman. (This particular campaign achieved little because the first meeting was scheduled for ten o'clock one morning, far too early for the nocturnal Dr. Bell. When Charles tried to wake him for the meeting, Alec growled, "Let them spit all they mind to," rolled over, and went back to sleep.)

The Wednesday-evening soirées gradually became a well-known and prestigious Washington institution. But news of colleagues' investigations and scientific breakthroughs reinforced Alec's uneasy sense that instead of piling up new achievements, he was simply coasting on his reputation, if not sliding downhill.

By the late 1880s, the Bells were well established in Washington. Washingtonians were proud to have Alexander Graham Bell in their midst: gleaming white invitation cards to balls and social functions adorned the mantelpiece in the Rhode Island Avenue parlor. Although Alec did not really enjoy the social round, he was always affable and charming in public. Mabel commented how, at parties, the "listlessly polite face of the hostess [would] change and light up when you come, and the look of interest [would] deepen when your name is announced." After she attended a reception at the Smithsonian Institution with her father, she noted, "I felt what a very distinguished man my husband was that they should all be so attentive to his wife."

With her parents and her only remaining sister, Grace, living close by, Mabel felt safely cocooned by family. She herself was gradually emerging as the matriarch of the clan, replacing her mother, who was losing her sight. Hubbard-Bell ties became even closer in 1887 when Alec's cousin Charlie, whose wife, Roberta Hubbard, had died in 1885, married his former sister-in-law Grace. Gardiner Hubbard must have pondered the irony that, despite his initial doubts about Bell marriages, three of his four daughters had become "Mrs. Bell."

The balance of Mabel and Alec's marriage shifted with the passing years: as Mabel matured, the ten-year age gap between her and her husband shrank in significance. Cool-headed in a crisis and the source of inexhaustible reassurance to her volatile husband, Mabel kept her husband grounded. When Alec erupted with exasperation about the fire going out in his study, Mabel soothed him and suggested that, instead of blaming the staff, he might even relight it himself. When Alec withdrew into his researches, Mabel coaxed him back into the family circle with maternal solicitude. "Please darling, make an effort to be more sociable," she urged him, "and go to see people and have them come to see you. . . . [P]lease try and come out of your hermit cell." When Alec overtaxed himself by working all night, she scolded him: "I am frightened to think where you will be when sickness comes, as come it must sometime if you continue your present irregular life."

Mabel knew that the private Alec was a much more insecure and anxious man than his public persona at white-tie dinners or Wednesday-evening get-togethers suggested. "If anything went wrong," their daughter Elsie recalled years later, "Father had to be looked after first, while Mother took charge of the situation." Mabel was aware that he hungered to make more contributions to scientific knowledge and that his endless health complaints were often related to his frustrations. He *always* got a headache when he had to give a speech or meet new people. Alec acknowledged his emotional dependence on his wife in a letter he wrote to her in June 1889:

> Oh Mabel Dear, I love you more than you can ever know,
> and the very thought of anything the matter with you
> unnerves me and renders me unhappy. My dear little wife,
> I feel I have neglected you. Deaf-mutes, gravitation or any
> other hobby has been too apt to take the first place in my
> thoughts, and yet all the time, my heart was yours alone. . . .
> I want to show you that I really can be a good husband and a
> good father, as well as a solitary selfish thinker. . . . You have
> grown into my heart my darling and taken root there—and
> you cannot be plucked out without tearing it to pieces.

A year later, when he was forty-three and Mabel thirty-three, Alec deplored his own tendency to bury himself in whatever project obsessed him but took refuge in self-pity to excuse his conduct: "Our minds are apart and it is my fault. I remain solitary and alone—and every year takes me farther from you and my friends and the world—and I seem

Alec and Mabel Bell in the late 1880s.

powerless to help. But for you I should lead the life of a hermit, alone with my thoughts. I hang like a dead weight on your young life, and crush you. What can I do? Is it in me to be young again? I fear that I am older than my years, while you my dear are young. What can I do to now make you happy?"

Charles understood the dynamics of the marriage perhaps better than anybody. "Only those who were closely associated with the daily life of Mr. Bell," he wrote after the death of his employer, "knew how much his success, in the things he was interested in, depended on the never failing care, constant watchfulness over his health, his recreation and his hours of study, by Mrs. Bell. They were indeed united for life."

Most of the time Mabel played the traditional role of the dutiful and admiring wife—the role her mother had played for her father. She made sure that the household revolved around Alec's needs, even if this meant that everyone had to tiptoe past his bedroom until the early afternoon or that he kept everyone except her awake past midnight by playing Beethoven vehemently on the grand piano. She tried to allay his anxieties by reassuring him that rivals like Edison and Gray were just "inventors, and you are a scientific man." But sometimes she could not suppress her resentment at the way that Alec left so much to her and took her and their daughters for granted. When he was pursuing a new idea or caught up in his latest projects, everything else went by the board, including sleep and children.

"My dear Alec, I do think you are behaving shabbily all round," reads one undated letter, written during one of their frequent separations.

> Why will you always break your word to your wife at the slightest provocation. . . . I really need you badly here and have not been feeling quite well for some days, and had nightmares every night so that I am frightened being alone. . . . They are all laughing at me here for believing you would keep your promise [to join the family]. They think it a good joke, and oh Alec, it hurts. I never dare say a word to you for you always say I am scolding and so sulk and will not hear. . . . Your loving but distressed wife.
>
> P.S. It is so nice being rich, please don't spoil it for me by feeling that perhaps it is a curse instead of a blessing because if we were poor you would have to work hard and regularly.

Mabel couldn't help contrasting the way that Alec treated her with the way that her father treated her mother, or Alec's cousin Charlie treated her sister. On a later occasion, she wrote reproachfully, "I realize as I see Mama and Papa, Grace and Charlie together, how little you give me of your time and thoughts, how unwilling you are to enter into the little things, which yet make up the sum of our lives. There are so many more of those things than you realize, which I cannot decide alone. I feel as if I were giving more and more to others the dependence for help and advice that should be yours."

There were undeniably times when Mabel's spirits drooped, although she was an optimist by temperament. Her sisters had always been her closest friends, and she was often lonely now that two of her sisters were dead and her husband was frequently absent. The women in Washington society admired the way she had overcome the handicap that deafness presented, but they did not befriend her. Not only was her speech hard to understand, but her shyness and Boston aloofness meant that she wouldn't (or couldn't) play their social games—comparing invitations, fashions, and houses.

And there was another issue, only hinted at in Mabel's and Alec's letters to each other. Mabel was still young, young enough to try for the longed-for son. But she had suffered two premature births and the loss of both infants, and her physician had warned her after Robert's death that she should not yet attempt a fifth pregnancy. In the Bells' correspondence, there are veiled references to the contraceptives (pessaries, abstinence) on which she relied until her doctor told her she was strong enough to try again. There are also hints of Alec's frustration.

"Dear, dear Mabel," Alec wrote to her in 1885. "My true sweet wife— nothing will ever comfort me for the loss of those two babes for I feel at heart that *I was the cause.* I do not grieve *because they were boys,* but because I believe that my ignorance and selfishness caused their deaths and injured you. . . . After [the first child's] death *I prevented you from fully recovering* and gave you another child before you were well. You have not even yet fully recovered and I believe you never will until you have had a complete and prolonged rest."

Volatile and needy, Alec hungered for his adored wife at the same time as he feared another conception that might put her health at risk. Yet celibacy appears to have been beyond him. Perhaps distance was the only remedy? "I pause here in New York," he wrote, "and ask myself whether it was best for you that I should return just now. Think seriously of it my darling wife." Only separation, he suggested, would give Mabel the chance to regain her strength and then "have another sweet baby-face smiling in your arms."

Allusions to Mabel's longing for another baby and to Alec's fear that sexual intercourse might damage her health appear in their letters for the next decade. In a pre-contraception era, Alec's fear was probably justified. But her loneliness and his fear of intimacy created tensions, especially when they were apart. In 1888 Mabel wrote to Alec, "I wonder do you think of me in the midst of that work of yours of which I am so proud and yet so jealous, for I know it has stolen from me part of my husband's heart, for where his thoughts and interests lie, there must his heart be."

Despite Mabel's hopes, despite some mysterious surgery that she underwent in 1891, despite assignations with Alec when she was convinced she was ready to conceive, Mabel would never have another child.

Summers in Baddeck gave the Bells the opportunity to shuck off the social pressures and marital stress that Washington induced. Memories of the lake's misty shoreline or the sun's last rays on Red Head were a psychological balm during winters stuffed with social engagements, speeches, interviews with lawyers or journalists, children's illnesses, domestic crises, family obligations, disappointments. In 1888, even as Mabel reproached Alec for, once again, accepting too many out-of-town speaking engagements at schools for the deaf, she reminded him that soon they and their daughters would be at their island refuge: "I live in hope that you will not quite forget me and that we may pass many another summer like the last [in Baddeck] when we had thoughts and interests in common."

There was, however, one subject to which Alec was devoted heart and soul, and for which Mabel had little sympathy.

Chapter 15

HELEN KELLER AND THE POLITICS OF DEAFNESS 1886–1896

One day in 1886, there was a knock on the door of the Bells' Washington home. The housekeeper opened the door to find a well-dressed gentleman clutching the hand of a small girl. The six-year-old child was neat and pretty, with a chestnut cloud of corkscrew curls. But there was a slight bulge to one eye, and the other eye wandered strangely. She appeared unaware of the stranger in front of her, or of the invitation to enter the house.

Alexander Graham Bell, summoned from his third-floor study, joined the housekeeper at the door. The gentleman was Captain Arthur H. Keller, a former Confederate officer who owned a run-down cotton plantation in Alabama and edited the local newspaper. Captain Keller had dropped a note at the Bells' house the previous day, asking if he could bring his daughter to meet Dr. Bell. Alec, who never refused to see a deaf child, had immediately sent a note of assent. Now he politely shook Captain Keller's hand, then knelt on the floor so he could say hello to Helen. Her busy, curious hands rapidly explored the territory of his face—the bushy beard, strong nose, and thick mustache. Her small fingers rested on his lips as he said her name, then touched his stiff collar, soft tie, tight waistcoat, and watch chain. Her responsiveness did not register in her face; he later described it as "chillingly empty." The party moved into the parlor, with Helen holding Alec's hand. As soon as he sat in his big wing chair, she scrambled onto his lap and her fingers, like little mice, resumed their exploration. She found Alec's pocket watch, which chimed on demand, and when he set it to chime, the vibrations finally elicited the ghost of a smile.

This meeting was an epiphany for Helen Keller, who would become one of the most admired celebrities of the twentieth century. "Child as I was," she would write in her 1903 memoir, *The Story of My Life,* "I at once felt the tenderness and sympathy which endeared Dr. Bell to so many hearts, as his wonderful achievements enlist their admiration. He held me on his knee while I examined his watch, and he made it strike for me. He understood my signs, and I knew it and loved him at once. But I did not dream that interview would be the door through which I should pass from darkness into light, from isolation to friendship, companionship, knowledge, love."

At this point in her young life, Helen was untutored, unmanageable, and probably very unhappy. Born in the remote Alabama town of Tuscumbia, on the Tennessee River, she had developed at nineteen months what doctors at the time called "brain fever" (probably scarlet fever or meningitis) and was expected to die. She recovered, but she could no longer see or hear anything. Helen had only the senses of smell, taste, and touch with which to explore her environment, and with every passing year she grew more screamingly frustrated with her limitations. Her parents were deeply distressed by their "little bronco," as they called her, whose behavior and understanding of the world steadily deteriorated.

Then Helen's mother, Kate Keller (who by now had a second daughter), read in Charles Dickens's *American Notes* of Boston-based Dr. Samuel Gridley Howe's success with Laura Bridgman. Dr. Howe had died in 1876, but Kate Keller was as determined to help her disabled daughter as the Hubbards had been twenty-three years earlier to help their daughter Mabel. Just as the Hubbards had hoped that Dr. Howe might provide a key to their deaf daughter's education, so Mrs. Keller hoped that the answer to Helen's problems might be found at Boston's Perkins Institution for the Blind, founded by Dr. Howe, where Laura Bridgman still lived. Kate Keller also heard of an eye surgeon in Baltimore who was said to have helped blind children recover their sight. So she sent her husband, Arthur, to find help for Helen in the distant (and still not particularly welcoming) northeast. The Baltimore surgeon could do nothing for the little girl, but he recommended that the Kellers consult the famous inventor Alexander Graham Bell, who had a particular concern for the education of the deaf. Captain Keller and Helen, along with Arthur Keller's sister, trekked on to Washington, in the hope that Alec might recommend a teacher for the girl.

At their first meeting, Alec could see that this child, despite her blank face, nursed a huge curiosity about the world. As soon as she tired of his watch, she scrambled off his knee and started a restless tour of the library, touching and patting every object and piece of furniture she encountered. Alec watched her with both paternal sympathy and scientific detachment. This was a fascinating example of an individual learning new ways to negotiate her world. He decided that he should consult the other great Washington expert on deafness, Edward Miner Gallaudet, on the right course of action for her.

Gallaudet was superintendent of Washington's Columbia Institution for the Instruction of the Deaf and Dumb, founded in 1864. Soon after Alec had moved to Boston in 1871, he had encountered Dr. Gallaudet at the prestigious Clarke Institution at Hartford, where Alec had given a demonstration of the Visible Speech system. Gallaudet, ten years older than Alec and with appointments at both these institutions, was a firm supporter of sign language, which his father, Thomas Hopkins Gallaudet, had introduced into the United States. He insisted on the intellectual superiority of signs over words, though he did allow both sign language and lip-reading (or "oralism," as it was known) to be taught at the Columbia Institution. In 1872, he was so impressed by the young Scots immigrant's teaching methods that he invited Alec to come and teach in Washington at the Columbia Institution. Seven years later, when Alec and Mabel established their home in Washington, Alec wrote to Gallaudet to renew their acquaintance. Gallaudet sent a warm reply: "I should be very happy to have you connected with the work of this college should you take up your residence in Washington."

During the Bells' early years in Washington, the two men's relationship was friendly and respectful. In 1880, the Columbia Institution awarded an honorary degree to Alec. He and Gallaudet had much in common. Both had deaf mothers, and both were motivated by the need for more educational facilities for the deaf. The Gallaudets dined with the Bells from time to time, and Gallaudet attended Alec's early Wednesday-evening soirées.

Alec dashed off a note to Gallaudet: "Mr. A. H. Keller of Alabama will dine with me this evening and bring with him his little daughter . . . who is deaf and blind and has been so from nearly infancy. He is in search of light regarding methods of education. The little girl is evidently an intelligent child and altogether this is such an interesting case that I thought you would like to know about it . . . and [I] hope you may be able to look in in the course of the evening." Neither Gallaudet nor Mabel, who was probably present at the dinner, left any record of what they thought of this "interesting case." Since Helen was blind as well as deaf, sign language would have had no place in the two men's conversation about how to help her. But on the strength of his encounter with Helen, Alec judged her educable. He suggested to Captain Keller that he ask Michael Anagnos, now director of Boston's Perkins Institution, to recommend a teacher for Helen. Anagnos recommended a remarkable twenty-year-old former Perkins pupil called Annie Mansfield Sullivan, who had spent four traumatic years in a Massachusetts poorhouse as a child and had been near-blind until recently because of trachoma. Annie had become a new person at Perkins, thanks to surgery to improve her vision. She learned Braille and the manual alphabet, and spent a great deal of time with Laura Bridgman, studying the instruction methods that the late Dr. Howe had used to teach Laura. She had the skills required to educate Helen.

Annie Sullivan agreed to move to Tuscumbia and live with the Keller family as Helen's full-time teacher. At first she was almost defeated by a child who threw cutlery, pinched, grabbed food off dinner plates, and hit Annie herself so hard that one of her front teeth was knocked out. Yet within weeks, Annie had engaged Helen's intellect. An intense, symbiotic relationship blossomed between stubborn, emotionally starved Annie and the bright little girl. To Helen Keller, Annie was her extraordinary "Teacher," who would "reveal all things to me, and, more than all things else . . . love me." The rest of the world would soon know Annie, in Mark Twain's phrase, as "the miracle worker."

Annie started by teaching Helen the one-handed manual alphabet that Annie had learnt at the Perkins Institution. This alphabet, in which letters are denoted by the different position of fingers, is different from both the English double-handed manual alphabet, with which Alec communicated with his mother, and the "glove" alphabet that Alec had taught George Sanders in Boston. However, all three of these manual alphabets entail spelling out words letter by letter, and so (unlike sign language) belong to writing rather than to speech.

Eight months after Annie had arrived in Tuscumbia, Helen's father, Captain Keller, wrote to Alec,

> It affords me great pleasure to report that her progress in learning is phenomenal and the report of it almost staggers one's credulity who has not seen it. . . . [Within a month of Annie's arrival] the little girl learned to spell about four hundred words and in less than three months could write a letter, unaided by anyone. In six months she mastered the "Braille" system which is a cipher for the blind enabling them to

read what they have written. She has also mastered addition, multiplication and subtraction and is progressing finely with Geography. . . . I send you a picture of Helen and her teacher and also a specimen of her writing believing you will be glad to hear again from the dear little treasure.

Helen's ability to learn was far in advance of anything that anybody had seen before in someone without sight or hearing. From the start, Helen was a writer: she relied on a symbolic knowledge of all physical reality that she acquired through Annie's spelled-out words. She learned oral speech only painstakingly years later, and she was never easily understood. But she began writing almost as soon as she had an elementary vocabulary.

By 1890, Helen Keller and Annie Sullivan were living in Boston, at the Perkins Institution, and for the first time in her life Helen found herself among children with disabilities similar to her own. The sense of belonging was a revelation to her. "I was delighted to find that nearly all my new friends could spell with their fingers," she wrote in an article entitled "My Story," for the January 1894 issue of the magazine *The Youth's Companion.* "Oh, what happiness! To talk freely with other children! To feel at home in the great world! Until then, I had been a little foreigner, speaking through an interpreter, but in Boston . . . I was no longer a stranger!" In the early 1890s, Helen made a huge leap forward in communication when she learned to "read" what people were saying by laying her fingertips on their lips as they formed words. With her left hand, she would feel the lip movements of her interlocutor; with her right hand, she would spell out her side of the conversation in one-handed manual language.

Annie Sullivan was the most important person in Helen's life, but Alexander Graham Bell was a close second. He followed Helen's intellectual development with deep interest, and was a constant source of support for both teacher and student. The affection was mutual. One of the first letters Helen wrote was addressed to "Dear Dr. Bell." The letter demonstrates Helen's brilliant intelligence: it was a remarkable achievement for a child who, only nine months earlier, had absolutely no vocabulary or linguistic ability. "I am glad to write you a letter. Father will send you picture. I and father and aunt did go to see you in Washington. I did play with your watch. I do love you. . . . I can write and spell and count. Good girl. . . ." Alec could scarcely believe his eyes when he read it: Helen proved all his theories about the capabilities of disabled children.

At one level, Alec's interest in Helen was warm and paternal. As Annie Sullivan unlocked Helen's intelligence, Alec took an almost personal pride in Helen's achievements. He never personally taught her, but he learned the one-handed manual alphabet and Braille so that he could communicate with her. When he was with her, he treated her as first a child, and later as a young woman; he never patronized her or behaved, as so many of his contemporaries did, as though her disabilities made her less than human. He put a lot of effort into raising funds for her schooling. He also helped Helen herself financially on several occasions, sending her $400 when her father died in 1896, $100 toward a country holiday in 1899, and $194 in 1905 so that she could buy a wedding gift for Annie Sullivan.

Alec's support for Helen Keller was unwavering, particularly during a nasty incident that occurred in 1891. Eleven-year-old Helen had written a short story, "The Frost King," as a birthday present for Dr. Anagnos, principal of the Perkins Institution. Anagnos was so impressed with the story that he arranged for it to be published. But it turned out that Helen had unwittingly taken her inspiration from a story by Margaret Canby, published twenty years earlier, that had been read to her when she was eight. Mrs. Canby was not troubled by Helen's inadvertent error: she sent Helen her love and called her version "a wonderful feat of memory." But Anagnos and his colleagues at the Perkins Institution reacted as though the child had been a thief. A committee of trustees ("a collection of decayed turnips," in Mark Twain's phrase) cross-questioned a frightened and humiliated Helen about how she had developed the plot. Alec assured Helen that she had done nothing wrong: "[T]he child is quite incapable of falsehood about the matter," he wrote to Mabel. Anagnos had behaved "in a most unjust and outrageous manner." Alec suggested that "we all do what Helen did," because "our most original compositions are composed exclusively of expressions derived from others."

At another level, however, Alec saw Helen as "an interesting case"—a real asset to his crusade to improve education for the deaf and integrate them into the speaking world. In the late 1880s, he was very much caught up in a statistical study of deafness. He traveled to Augusta, Maine, to trace the history of a family called Lovejoy in which there were deaf members of every generation. He spent several weeks on the island of Martha's Vineyard, where he had discovered an isolated community, with the glorious name Squibnocket, in which one-quarter of the population was deaf. He was able to track the course of deafness from an early settler on the island to Squibnocket's living residents. He was also successful in persuading the U.S. Census Bureau to use the term "deaf" rather than "deaf-mute" on its 1890 census forms, so that more accurate figures on the incidence of deafness could be constructed.

Genealogies, census returns, family histories, and notebooks piled up in the Volta Laboratory in his father's backyard in Georgetown. Alec divided the laboratory building into two sections, one for the scientific experiments for which it had originally been intended, and the second, to be known as the Volta Bureau, for "the increase and diffusion of knowledge relating to the deaf." (He reminded his assistant, John Hitz, that the phrase "increase and diffusion" was the kind of language that all great research institutions, such as the Smithsonian, had in their charters.) He accepted every invitation to speak about deafness or to visit schools for the deaf, long after he had given up speaking about the invention of the telephone. "I never knew him to refuse to lecture for the benefit of the deaf, or on anything related to the deaf," his butler, Charles Thompson, recalled after his death. "There certainly was a

*Helen Keller (left) with Annie
Sullivan and Alexander
Graham Bell.*

tender chord in Mr. Bell's heart for the deaf people of the human race. He seemed never so happy as when in company with the men and women identified with the teaching and training of the deaf." And he never stopped campaigning for education in lip-reading rather than sign language, so that the deaf would not, as he saw it, be marginalized into their own language ghetto.

Helen Keller was a wonderful figurehead for Alec's campaign to raise awareness of the causes and consequences of deafness. The entrepreneurial spirit that was so absent in Alec's scientific work blossomed in his efforts on behalf of the hearing-impaired. He did not hesitate to feed the public hunger for information about Helen by talking about her and Annie whenever opportunity arose. As early as 1888, he had given a copy of one of Helen's letters and a photograph of her to a New York newspaper, and this made the eight-year-old a national celebrity. "The public have already become interested in Helen Keller," he wrote in 1891, "and through her, may perhaps be led to take an interest in the more general subject of the Education of the Deaf." While others described Annie Sullivan's success with Helen as "a miracle," Alec insisted that it was the product of educational method—of the way Annie constantly spelled out idiomatic English into Helen's hand, allowing the child to learn meanings from the context.

When Helen was only twelve, Alec introduced her to his friends at one of his Wednesday-evening get-togethers. Helen sat in the middle of the erudite gathering, which included Professor Newcomb, Professor Langley, Gallaudet, and Alec's father-in-law, Gardiner Hubbard. Despite being treated as a scientific specimen (Alec recorded that she was grilled on "the rotundity of the earth [and] her conception of geometrical relations"), the self-possessed youngster charmed her cigar-puffing audience. Her pièce de résistance came when she stood up, smoothed her skirts, and recited, in her strangely nasal tones, a poem familiar to all those present: Longfellow's "Psalm of Life." How could such an eminent assembly resist such lines as

> Lives of great men all remind us
> We can make our lives sublime,
> And, departing, leave behind us
> Footprints on the sands of time.

She had the "great men" in her protector's parlor eating out of her hand. Dr. Daniel Gilman, the founding president of Johns Hopkins University, told the Washington *Evening Star* that "her story is remarkable, and the skill of her teacher, Miss Sullivan, is admirable in the highest degree."

Thanks to Alexander Graham Bell, Helen Keller was regularly in the public eye. In early 1893, Washington newspaper reporters watched her turn the first spadeful of earth for a new building to house the Volta Bureau and its library of information about the hearing-impaired. The building, a handsome neoclassical yellow brick and sandstone structure, was located at 1537 35th Street NW, in Georgetown, next to Melville Bell's house. A few days later, Alec took Helen and Annie with him to a day school for deaf children in Rochester, New York. Alec vehemently argued that day schools were preferable to boarding schools for deaf children, because the children returned to their families at the end of each day. In Rochester, he spent a couple of days raising money for the school and promoting the superiority of teaching lip-reading over sign language. "The teachers and pupils have been profoundly impressed with Helen and much encouraged in their work," he wrote triumphantly to Mabel, "for the method of instruction pursued here is the same as that adopted by Helen's education. The elder pupils wept over Helen and there was not a dry eye among the teachers either." Later the same year, he escorted Helen and Annie around the Chicago World's Fair, and introduced Helen to teachers of the deaf who "saw enough to remove all their doubts."

Alec also used Helen in demonstrations to prove the effectiveness of oralism. Helen described their stage act some years later, in a chapter entitled "My Oldest Friend" in her 1930 book, *Midstream: My Later Life:* "After he had talked awhile, he would touch my arm, I would rise and place my hand on his lips to show the audience how I could read what he was saying. I wish words could portray him as I saw him in those exalted moods—the majesty of his presence, the noble and spirited poise and action of his head, the strong features partly masked by a beautiful beard that rippled and curled beneath my fingers. . . . No one can resist so much energy, such power." Who would not be moved by the sight of the slender young woman, dressed in a demure white muslin gown and with an expression of rapture on her face, reaching up to touch the lips of one of America's best-known men?

Despite her disabilities, Helen was a natural performer—pretty, outgoing, and eager to capture public attention. She loved being a star. Over the next few years, she totally eclipsed the aging Laura Bridgman as a fascinating specimen of human achievement. With her severe dress, prim manners, and eagerness to please, Laura had satisfied the mid-nineteenth-century appetite for a frail victim-heroine whose rescue from spiritual imprisonment was a Christian morality tale. Helen, who was truly brilliant, embodied the new robust "anything is possible" ethic. In time, she would master French, German, Latin, and Greek (although only her close friends learned to understand her speech in any language). Urged on by Alec, she insisted on attending a school for normal students rather than a special school for the deaf and blind. She would attend Radcliffe College, and she would be published widely. She read Roman history, German philosophy, and English literature, and embraced radical socialism and Swedenborgian beliefs. In 1903, the Harvard philosopher William James (brother of the novelist Henry James) noted that, while Laura was "almost a theological phenomenon," Helen was "primarily a phenomenon of vital exuberance. Life for her is a series of adventures, rushed at with enthusiasm and fun."

If it was Alec who put Helen Keller on display, she herself was happy to publicize her relationship with the world-famous inventor. It is doubtful whether she would have achieved her celebrity status without Alexander Graham Bell's intervention. His whole-hearted devotion to the interests of the deaf as well as his warmth toward her were crucial to her successes. She dedicated *The Story of My Life* to "Alexander Graham Bell, Who has taught the deaf to speak and enabled the listening ear to hear speech from the Atlantic to the Rockies." Years after his death, when she met one of his granddaughters, she confided, "I am still hungry for the touch of his dear hand."

While Alec's attitude to Helen Keller was a mix of warm friendship and scientific detachment, his immediate family's attitude to his protégée was far more complicated. Alec could be as insensitive to their feelings as, in earlier years, his father Melville had been to his.

Mabel kept her distance from Helen Keller, despite everything they had in common—disability, Alec's support, an appetite for life. Mabel was not present in 1893 when Helen appeared at the Wednesday-night get-together in their Washington home. When Alec escorted Helen to the Chicago World's Fair the same year, Mabel and her children did not join the party, although Daisy was the same age as Helen, and Elsie only two years older.

This was partly because the Bell girls were in school and Mabel had her hands full running households in both Washington and Nova Scotia. But there was something else going on too. Mabel was, in the best Boston tradition, accustomed to being the self-effacing wife of a Great Man, but Helen Keller consumed Alec's attention and affection to a remarkable degree. At heart, Helen Keller, like her mother, was a Southern belle, whatever her disabilities. She cultivated her looks, well aware that both the "vital exuberance" noted by William James and her sweet expression won men's hearts. She went through the agony of the removal of her own eyes (which bulged) and the insertion of sparkling blue glass eyes, widely admired for their living beauty and humane depth. Because she could neither see nor hear, Helen and Alec could communicate only through the most intimate of senses: touch. How did Mabel feel as she watched the pretty young woman twenty-three years her junior laying her fingers on Alec's lips and looking up at him through those eerie azure eyes?

Mabel professed to admire Alec's efforts on Helen's part. "My own darling," she wrote to him in March 1892, when she was in Europe and he was in Boston with Helen and Annie. "You are a perfect knight errant going about to succor distressed damsels, and I love you." She often wrote a note at the end of her husband's letters to Helen, seconding his invitations to visit the Bells. But she also sympathized with the less visible member of the Helen-and-Annie double act. "I know just how that poor Miss Sullivan must feel." A week later, she scolded him for not writing to her: "I do think Helen Keller and the summer meeting might rest for one night. . . . I tell you I am a jealous woman, and that is at the bottom of what you consider my want of sympathy with your work."

Mabel wasn't the only person suffering flashes of resentment. "Both children are a little jealous of your appreciation for Helen Keller," Mabel informed her husband in 1895. Elsie and Daisy Bell often found themselves outside the closed door of the drawing room, listening to the sound of Helen's laughter within as she talked to Alec. They watched their father set off to tour Washington Zoo with Helen, and without them. They saw the yellow cockatoo he presented to her on her fourteenth birthday, which she called Jonquil. Only rarely were they allowed to play with Helen. When they did, they were deliberately careless of her handicaps.

Years later, Elsie admitted to Helen E. Waite, author of the 1961 authorized biography of the Bells, *Make a Joyful Sound,* that they resented their exclusion from their father's special relationship with another child their own age. On one occasion, she and her sister were told to entertain Helen Keller, and they decided to show her their secret hiding place. "Later," Waite recorded, "strolling outside in search of the three children, Dr. Bell heard a cheerful call. 'Here we are, Papa! Look up here on the roof!' Raising his eyes, the horrified man beheld his two little girls with Helen between them happily perched on the roof of Grandfather Hubbard's stable! His gaze went to the ladder. How had a blind child made the ascent, and then taken the perilous walk along the roof?" Alec carefully guided Helen down the ladder and gave his daughters a stern rebuke. Elsie insisted that "Helen certainly had a good time!" but there is a hint of revenge in the episode.

However, there was a more subtle dimension to Mabel's wariness of Helen Keller that had nothing to do with Helen herself. Throughout the Bells' marriage, Mabel nursed a deep-seated ambivalence toward her husband's concern for the deaf, and it was a source of tension in their relationship. In 1888, Alec reproached her: "[Y]ou haven't a particle of sympathy *for my work*—there I am fated ever to be alone." For Mabel, the fundamental issue was that she felt devalued when she was identified as deaf, so she shunned not just the label but also the company of the hearing-impaired. "It has been my life-long desire to forget, or at least ignore, the fact that I was not quite as other people," she admitted to her son-in-law Gilbert Grosvenor at the end of her life.

"I have striven in every possible way to have that fact forgotten, and to so completely be normal that I would pass as one." For this reason, she preferred to keep her distance from Helen, no matter how appealing the child was or how close she grew to Alec. "To say a child was deaf was enough to make me refuse to take any notice of it. If help had to be given, given it was from a distance." Mabel was particularly wary of the teachers of the deaf. "Of all the people I hated [to meet] a teacher of the deaf [was foremost]. I was always on the lookout for a little difference in their manner of addressing me, which would reveal the fact that I was a 'case' in their eyes. . . . Above all things I antagonized my husband's efforts to keep up his association with them and to continue his teaching of them. Well, this is a confession of great selfishness. The only excuse is that it is just the spirit enlightened teachers of the deaf wish to see manifested in their pupils. They don't want them to herd together and become a 'peculiar people.'"

Mabel knew that strangers often found her hard to understand, but nothing exasperated her more than being treated as, in her expression, a "Barnum monstrosity." Her own daughters were taught that it was only courteous to face anybody they were talking to, and that it was rude to shout from another room; as a consequence, they were slow to realize that their mother could not hear. Daisy learned the hard way. When she was only about three, her mother came into her room one night to check on her health. At first Daisy pretended to be asleep, then she spoke to her mother as Mabel was leaving the room. When her mother apparently ignored her, left the room, and closed the door behind her, Daisy was frightened and began to bawl. "Her father heard her," according to Helen E. Waite, "came in, listened to the story, and quietly explained." Later, Elsie and Daisy often acted as Mabel's "ears" during telephone conversations, but they were usually oblivious to their mother's condition because they could all communicate with each other easily. That was exactly the way their mother wanted it. She strove to create the kind of self-sufficient little family unit that, as a child, she had known within the Hubbard family.

Mabel's reluctance to be identified as deaf, combined with her husband's single-minded support for the assimilation of the hearing-impaired into mainstream society, began to affect Alec's relationship with the single most important champion (other than Alec) of deaf education: Edward Miner Gallaudet. The cordiality of that 1886 dinner, when Gallaudet joined Alec to meet Captain Keller and his daughter, would gradually dissolve in the next decade.

Perhaps the deterioration in relations was inevitable, considering the men's different philosophies. Courtesy and good manners could not smother the rivalry between their two systems of education. As superintendent of Washington's Columbia Institution for the Instruction of the Deaf and Dumb, Gallaudet endorsed the "Combined Method": students in his institution were instructed in both sign language and lip-reading. But Gallaudet never hid his conviction that sign language was intellectually superior for deaf people than any form of the English language. "So far as motions or actions addressed to the senses are concerned," he would write in 1899, "[sign] language is superior in accuracy and force of delineation to that in which words spelt of the fingers, spoken, written or printed, are employed." According to Richard Winefield, author of a study on the Gallaudet-Bell communication debate, Gallaudet believed (like the Abbé de L'Épée and his own father before him) that "sign language was the natural language of deaf people, as natural as standard English was for children in Britain or the United States. The objective of the oral method was communication in standard English, something Gallaudet considered artificial." Gallaudet felt that deaf children had a right to use this "natural language," and he considered himself the champion of that right. He lobbied for the right of deaf people to speak for themselves at conferences and conventions and to teach fellow members of the deaf community.

Alec couldn't accept Edward Gallaudet's conviction that sign language was "natural" for the deaf, or that acquisition of the English language, for signing, speech, or the manual alphabet, was "artificial." Alec also believed that if deaf students were grouped in a residential institution with deaf teachers and allowed to rely on sign language, they would never be able to function in the world outside their institution. "The great object of the education of the deaf," he wrote in 1889, "is to enable them to communicate readily and easily with hearing persons."

Alexander Graham Bell's faith in the ability of almost all deaf people to learn both lip-reading and some form of speech was sustained both by Mabel's extraordinary skill and by Gardiner Hubbard's long-term support for the cause. As he put it to Gallaudet, "I wish you could realize, as I do, how important even imperfect articulation is to a deaf person. I have daily—hourly—experience of its value in my own home." It wasn't just Mabel's skill that shaped his certainty that most hearing-impaired people could master lip-reading and mix easily with hearing people. He was also influenced by Mabel's own attitude to her disability, by her determination to live a "normal" life. He was apparently oblivious to the happiness that many deaf people found in the company of others like themselves, where, as Helen Keller herself had put it, they were no longer "foreigners." He was convinced that the deaf population far preferred lip-reading to sign language and that it was only because people like Gallaudet had so much professionally invested in sign language that signing continued to dominate North American deaf education.

Through the 1870s and 1880s, the two men agreed to disagree. But there was always a danger that the simmering disagreement might burst into open flames. The first warning sign came with Gallaudet's proposal to start training teachers for the deaf at the Columbia Institution. In 1890, Gallaudet asked Congress for $5,000 to establish the program, and in pursuit of his subsidy he quickly learned his way around Congress. A good-looking man who waxed his mustache and sported pince-nez and immaculate white linen suits, Gallaudet was a real operator on Capitol Hill. A former student at the Columbia Institution described how, when the superintendent appeared before a congressional committee, Gallaudet would always "make a graceful bow to all the members, and then, in a clear cultured voice, explain the needs of the college, its need for expansion. . . . Should he be the butt of a joke by a congressman who had an abundant sense of humor, he would always laugh with them and prove that he was a good fellow." His unctuous social manner had helped him obtain grants for new dormitories and other buildings on his school's campus and money for scholarships for the deaf. Now he put the same charms to work on the congressional committee that was considering the budget request for his teacher-training program. He haunted the U.S. Congress's marble corridors and invited leading committee members to lavish dinners.

Alec was very suspicious of Gallaudet's plans for a teacher-training program because he thought Gallaudet was going to employ deaf teachers. This, he argued, would guarantee that deaf students would never learn to speak or lip-read. Although Gallaudet assured him that deaf people would not be eligible for his new program, Alec didn't believe him. He went public with his opposition to the proposal. "I consider this a backward step," he told Congress's Appropriations Committee, "and not a step in advance. . . . The employment of deaf teachers is absolutely detrimental to oral instruction."

The two men were quite a contrast. On one side stood Gallaudet, slender and smartly dressed, the single-minded spokesperson for his institution and a political wheeler-dealer. On the other side stood the world-famous millionaire Alexander Graham Bell, a big bear of a man who spoke with passion on the many issues that were close to his heart. The debate on the subsidy could have gone either way. Gallaudet's students told reporters that they would prefer to be taught by deaf teachers, who might be less likely to make them feel "strange." In the nineteenth century, as in this one, many deaf people thought deaf was a perfectly good way to be—as good as hearing, perhaps better.

Alec was blithely convinced that the correctness of his case was self-evident. But some of his Wednesday-evening friends were worried, so they suggested that he might take a leaf out of Gallaudet's book and give a little dinner to the committee members, to butter them up. Charles Thompson, the butler, was in the room, and remembered his employer's reaction: "Mr. Bell resented this suggestion at once, and raising his right arm, his eyes flashing fire, his hair seeming to raise on his head, face aflame, he said, 'If the facts presented to these gentlemen do not convince them of the merits of this case, they can go to blazes.' With this, his clenched fist came down upon the table with a bang."

To Gallaudet's fury, Alec's arguments prevailed and Congress refused Gallaudet's request for a grant. The refusal prompted the first open clash between America's two leading educators of the deaf. Unaccustomed to failure, Gallaudet exploded. He accused Alec in the press of "meddling in affairs that did not concern him." Alec tried to take the high road, simply sending a chilly letter to Gallaudet asking whether he had perhaps been misquoted.

Mabel watched from the sidelines the clash between her husband and Gallaudet, and made little comment on the debate. Gallaudet was no friend of hers. At one meeting, she confided to her mother, Gallaudet had urged her to acquire either the manual alphabet or sign language and to encourage her family to do likewise, so they could all use it together, and "then I should understand more of general conversation." Mabel was furious: "I hate that man," she wrote to Gertrude. She was adamantly opposed to Gallaudet's campaign for special institutions for the hearing-impaired, because of her conviction that this would lead to the deaf being treated as a "peculiar people." But she never publicly voiced these views, or her conviction that it was cruel to send deaf children to residential institutions like the Hartford School ("a virtual prison" in her private opinion) to learn sign language, because then they would be unable to communicate with people who did not know it. In the interests of passing, in her words, as "normal," she hadn't allowed Gallaudet's arguments or pretensions to get under her skin. Now she winced as she saw that her husband was hurt by Gallaudet's accusation, as well by Gallaudet's more rabid supporters, who attacked Alec's views as "pestilential," "ranting," and even the "vaporings of an idle brain." On this occasion, knowing Alec's propensity to explode with rage, she urged him to stay calm: "I am very pleased with your letter to Mr. Gallaudet. It is just exactly what it should be, to the point, dignified, and wasting no words." But her own fury began to build.

In 1890, the two men managed to turn down the heat of their disagreement and dampen the flames of personal invective—although not without a great deal of spluttering. Gallaudet accused Alec of conducting the row "more in the spirit and attitude of a lawyer. . . . bent on winning his case, than as a philanthropist . . . striving to advance a worthy cause." Alec reproached Gallaudet: "You have impugned my motives . . . and you have publically [*sic*] discredited me before young men and women of your college, whose interests I have at heart."

A few years later, however, the hostility erupted into a firestorm. It was sparked by the breakdown of efforts to merge two teaching associations: the American Association to Promote the Teaching of Speech to the Deaf (AAPTSD), which Alec had founded and now financed, and Edward Gallaudet's Convention of American Instructors of the Deaf (CAID).

This time, the breach was brutal. It occurred at an August 1895 CAID conference of teachers for the deaf in Flint, Michigan, at which many Bell supporters were also present. The meeting was held in an unventilated wooden hall, where the temperature hovered near 100 degrees Fahrenheit during that steamy week. Despite the heat, delegates had packed the hall in anticipation of Edward Gallaudet's keynote speech. Members of the different associations fanned themselves, mopped their brows, and glowered at their opponents. Gallaudet, as he confided to an ally, had decided it was his moral duty "to 'do-up' Professor Bell." He stalked up to the podium and proceeded to describe his valiant fight to establish the training college for teachers for the deaf—and Alec's allegedly unprincipled opposition to it. He then mocked the rival teachers' association, drawling out the cumbersome name "American Association to Promote the Teaching of

Speech to the Deaf." The interpreter made the name seem even more ridiculous by spelling the whole thing out, letter by letter, in manual language. Gallaudet ridiculed Bell (who was not present) personally, and emerged, in his own eyes, as the champion of the deaf.

Gallaudet's supporters applauded wildly. They saw Alec as blindly prejudiced against them: they had not forgotten the paper Alec had published in 1883, with the unfortunate title, "Memoir upon the Formation of a Deaf Variety of the Human Race." But Alec had friends in the audience too. They saw the speech as a spiteful attack on their hero, designed to present him as a narrow-minded, unprofessional egomaniac, and they lost no time in describing it to Alec. Then they sat back and waited for his reaction. It was disappointingly mild. When Alec arrived in Flint a couple of days later, he made a speech in which he calmly rebutted the charges. Although his fans insisted that Alec's speech was "inspired," "a model for the highest type of a Christian gentleman," his critics dismissed it as "lame." But there was no mistaking the state of affairs between the two men. At one point, they were invited to come to the front of the hall and make peace. Both men hesitated: Dr. Gallaudet's thin nostrils flared, while Alec frowned so fiercely that his bushy eyebrows almost met over his nose. Finally, each reluctantly made his way forward and, without smiling, extended his right hand. As Richard Winefield describes the encounter, "The tips of their fingers barely touched in the most frigid of handshakes. . . . The last chance to unify the field had been lost, resulting in many more years of antagonism and frustration for teachers, parents and deaf children."

Writing to Mabel after the Flint fracas, Alec shrugged off the frigid handshake: "Had a lively time in Flint—Gallaudet having made a most outrageous personal attack upon me and my work for the deaf. . . . The address was simply 'bosh'—to excite the passions of the deaf. . . . I am seriously troubled about Gallaudet. . . . Don't think any man in full possession of his senses would have written that address." Alec insisted that the mood of the convention swung in his favor because it was dominated by "adult deaf-mutes from Illinois, Ohio and Michigan. . . . Enough to swamp the votes of all the Sup[erintenden]ts and Prin[cipal]s present." Gallaudet, in Alec's opinion, was "not quite sane upon the subject of Bell."

This time, Mabel was not prepared to stay on the sidelines, watching the two men paper over the breach. Gallaudet's personal attack on her beloved husband enraged her. She immediately wrote to Gallaudet, severing all social ties between the two families. The following March, at a reception at the National Geographic Society, she realized that the superintendent of the Columbia Institution (recently renamed Gallaudet College, after Edward's father) was approaching: "I was sitting talking and looking straight ahead when Dr. Gallaudet walked slowly past looking me full and steadily in the eyes, just as a dead man might and almost as white. His face never changed, his eyes never left mine until the slow walk had carried him past me and I, well, I went on laughing and talking." In Washington society circles, such a public "cut" was a dramatic gesture. In vain, Edward Gallaudet wrote to Mabel, assuring her that "I must, of course, accept your decision as to our future relations, but I am unwilling to have our friendship come to an end." Mabel was unrelenting.

Even Alec was shocked by Mabel's intransigence. "I don't approve of your carrying your resentment so far at all," he wrote to her. "I am very seriously of the opinion that he is only partly responsible for his actions—and that the future will reveal the fact of insanity taking the form of mono-mania against myself." In fact, the ferocity of the disagreement did subside: Alexander Graham Bell was too large-spirited to harbor grudges. In later years, he would occasionally join forces with Gallaudet to protect the rights of deaf people. But the deep split that first cracked open in Flint, Michigan, between deaf educators who endorse lip-reading and those who promote sign language gapes as widely today as it did in 1895. It remains a battle between those, like Alexander Graham Bell, who aggressively push for the assimilation of deaf people into the mainstream and those, like Edward Gallaudet, who communicate through American Sign Language and regard the deaf subculture as a valuable minority, like any ethnic or linguistic minority—one that deserves an independent status within the larger culture.

Through all the *Sturm und Drang* of clashes between the two schools of deaf education, the friendship between Alexander Graham Bell and Helen Keller flourished. Helen made several visits to Beinn Bhreagh, where she enjoyed midnight swims in the lake with Daisy and Elsie: "The air was quite cold, but the water was deliciously warm and our joy knew no bounds!" She knew she could always call on Alec for assistance. In January 1907, she sent him a telegram asking him to accompany her on stage and translate her speech at a meeting at New York City's Waldorf Astoria Hotel three days later. From Washington, Alec telegraphed back, "My dear little girl, I will come and help you."

Yet Alec's attitude to Helen Keller reflects both his large-hearted generosity to humanity as a whole and his insensitivity to individuals. One evening in 1902, when Alec was fifty-five and Helen was twenty, they sat alone on the creaking wicker chairs outside Beinn Bhreagh's drawing room, enjoying the cool breezes off Bras d'Or Lake, which glinted below them in the moonlight. Alec mused on the unexpected turns his own life had taken, then turned to the young friend sitting close beside him and, as she recounted in her memoir *Midstream*, remarked, "There are unique tasks waiting for you, a unique woman. . . . The more you accomplish, the more you will help the deaf everywhere."

But then Alec took the conversation in an unexpected direction. He spelled out on his young guest's hand, "It seems to me, Helen, a day must come when love, which is more than friendship, will knock at the door of your heart and demand to be let in."

Startled, Helen replied, "What made you think of that?" Alec continued, "Oh, I often think of your future. To me you are a sweet, desirable young girl, and it is natural to think about love and happiness when we are young."

"I do think of love sometimes," Helen admitted, "but it is like a beautiful flower which I may not touch, but whose fragrance makes the garden a place of delight just the same."

The conversation, as Helen described it twenty-eight years later, was suffused with Alec's heartfelt belief that disability should not prevent anybody from enjoying the full range of human experience. Happily married himself, he could not bear the idea that a woman whose attractions he thought obvious might not enjoy a similar relationship. "He sat silent for a minute or two, thought-troubled, I fancied. Then his dear fingers touched my hand again like a tender breath, and he said, 'Do not think because you cannot see or hear, you are debarred from the supreme happiness of woman. Heredity is not involved in your case . . .'"

Helen, however, was all too aware of the stigma attached to someone with her disabilities. Where would she find a partner who would match her intellectually, desire her physically, *and* have the endless patience of Teacher? Even if she found him, would he respond to her? Smiling in Alec's direction, she said quietly, "I cannot imagine a man wanting to marry me. I should think it would seem like marrying a statue."

But Alec was too wrapped up in his own romantic fantasy of Helen's future to listen to her doubts. "You are very young," he told Helen, patting her hand tenderly, "and it is natural that you shouldn't take what I have said seriously now: but I have long wanted to tell you how I felt about your marrying, should you ever wish to. If a good man should desire to make you his wife, don't let anyone persuade you to forego that happiness because of your peculiar handicap."

What prompted Alec to speak like this to Helen? Probably a mix of paternal affection and an undercurrent of sexual attraction that he himself would not have recognized (let alone acknowledged). Yet the conversation reveals so much, particularly Alec's lack of insight into both his own emotions and Helen. As an inventor he lived in his imagination, yet he was unable to empathize with someone whose life was very different from his own. Helen continued to sit alone on the porch with Alec, and continued to leave her hand in his, but she felt increasingly uncomfortable with the intimacy of the conversation. "I was glad," she wrote later, "when Mrs. Bell and Miss Sullivan joined us and the talk became less personal."

Part 3

MONSTER KITES AND FLYING MACHINES
1889–1923

I never saw anybody who threw his whole body and mind and heart into all that interested him in a hundred different directions, like the waves beating on the shore [that] fling seaweed on the sand and then retreat to fling more seaweed in some other wildly separated place. Papa flings ideas, suggestions, accomplishments upon all recklessly and leaves them lying there to fertilize other minds, instead of gathering them all together to form creations to his own honor and glory.

Mabel Bell to Daisy Fairchild, 1906

Chapter 16

ESCAPE TO CAPE BRETON
1889–1895

In late April 1889, a sharp wind blew off Bras d'Or Lake, and patches of snow lingered on the shady side of the headland. But sap was rising in the dark woods, and the sound of hammers and saws drifted across the water to the village of Baddeck. A large bearded figure, in a baggy tweed suit complete with watch chain, chuckled as he watched a handful of men frame what would be a very special building—the Bells' first home on their own headland, Beinn Bhreagh.

"I am very much pleased with the site of the new house," Alec wrote to Mabel in Washington when he returned to the comfort of the Telegraph House that evening. "The more I see of it, the more I like it. . . . As you row towards Beinn Bhreagh it seems as though the house were built on the very edge of the cliff, and is in danger of sliding down. When you climb up there, you find there is room for a good broad road between the house and the cliff."

The foundations for the dwelling had been laid the previous autumn, and all winter Alec had chafed to leave Washington and travel north to check on the builders' progress. Finally, in April, he heard that passage to Cape Breton was possible. He caught the train north to Canada, via Boston—the first stage of the complicated three-day journey to Baddeck. When the train reached Truro, Nova Scotia, he probably took a slow train east across the province toward the Strait of Canso, which lies between the mainland and Cape Breton, and then crossed the strait on a vessel that steamed through St. Peter's canal, the gateway opened twenty years earlier from the open ocean into Bras d'Or Lake. As the steamer butted its way past shadowy, silent islands, any passengers brave enough to stand on deck in the cutting wind would see seals occasionally surfacing in the chilly waters of the huge inland sea. A few hours later, the steamer arrived at the village of Grand Narrows, where the Barra Strait links the southern half of the lake with the northern half. In the northern half, three deep channels splay out like fingers. Once through the straits, Alec could see north up St. Andrew's Channel toward Red Head, looming in the distance against the dove gray winter sky.

As soon as he set foot in Baddeck, where the Dunlops welcomed him back to the Telegraph House, he felt as though he had come home. He called in at Crescent Grove (the "cottage on stilts" where the Bells had spent the previous three summers) and was greeted by a procession of sheep. ("First there was Minnie, followed by her daughter Nellie, followed by her son, what's his name, and a little baby lamb born only a few days ago. Minnie also has a lamb born today so there will be five in the procession when the children arrive.") And when he had found a rowboat and rowed himself across Baddeck Bay to his beloved headland, he found his team of local workmen hard at work: "The tree trunks that are to form the verandah poles are cut and ready to be placed in position. . . . There is a great deal more stonework about the house than I had any idea of. . . . The different parts of the house are being riveted together with iron. . . . It looks therefore as if it will require a two-fold Samoan hurricane to move the building from its place."

This new house, which had been sketched out by Arthur McCurdy and Alec and named "the Lodge" by Mabel, was intended only as a temporary residence. Still flush with more capital and income than

Alec and Mabel rowing in Beinn Bhreagh bay, below the Lodge.

they had ever imagined, the Bells were already thinking much bigger. Once Alec had managed to buy all the land on the headland, they planned to build a grand mansion, designed by a Boston architect, on the tip of the headland. The Lodge was therefore modest—too modest, it quickly emerged, given the Bells' love of guests. It originally included only four bedrooms, with two-level rustic verandas on all four sides. Within months, as friends and family members began arriving, the verandas were enclosed to provide extra sleeping accommodation. One of those visitors was acerbic cousin Mary Blatchford from Boston, who sniffed that the extra rooms "hurt the architecture."

That first April, Alec was thrilled to be back in Baddeck, despite the icy winds and unexpected snow squalls. He would stand on Beinn Bhreagh, taking great gulps of Cape Breton air and letting his gaze skim down the ten-mile length of St. Andrew's Channel to Grand Narrows. The frustrations and worries that haunted his Washington life—his inability to get launched on a new invention, Mabel's yearning for another child, his conflict with Edward Gallaudet—fell away. "For the first time in months, I have been able to take exercise without perspiration. I have certainly walked not less than six miles a day since I came to the island," he bragged to Mabel that month. "I never felt better in my life." Alec's enthusiasm for cold, damp weather knew no bounds. While Cousin Mary was staying, a violent storm broke one night: "[T]he rain it rained," she wrote to a friend, "and the wind it blew, the trees slapped and shivered and bowed like mad things . . . and all Nature was in a frenzy." She was horrified when Alec announced he was going to walk up the mountain during this downpour, and even more appalled when he donned his slippers and swimsuit ("a closefitting dark blue jersey, *very* becoming") for the outing. By three o'clock in the morning, everybody was downstairs. Mabel watched lightning zigzag across the lake; her guests jumped each time they heard thunder roar or branches splinter; the driving rain leaked through the window frames. *Where* was Alec? Suddenly, there was a tap on the window. "There stood Mr. Bell," Mary Blatchford recorded, "dripping like a merman, and looking as handsome as an Apollo, with his gray curls wet and shining, and his white arms and legs."

There was a new energy in Alec's step as he strode around the headland, planning roads and buildings and convincing Cousin Mary that "all men are boys." In Cape Breton, he could be himself in a way he could never be in Washington, with its snobbery, racism, and unbearably humid summers. He was invigorated not only by the bracing Atlantic air but also by the locals' amused tolerance of the Great Man who had turned up in their midst.

Baddeck's taciturn Scots residents, most of whose families had lived in Cape Breton for at least three generations, found Alec an exotic creature, with his British accent, American wife, international reputation, and untold wealth. But by 1889, after four summers, the Bells had proved they were not just another set of American tourists, come to stare at the natives. They had endeared themselves to their neighbors and had begun to create their own world. Alec shed the Washington uniform of morning coat and stiff collar and wore comfortable tweed knickerbockers and a flat cap. As a child in Edinburgh, he had been ignorant of the Highland customs and mythology that Scots emigrants had brought to Cape Breton half a century earlier; he had never worn a kilt or played the fiddle or the bagpipes. But Cape Bretoners' dry humor and love of music made him feel at home. The men of Baddeck often gathered on Sunday evenings to sing Scottish ballads, and he regularly added his rich baritone to the chorus. Alec's imagination, no longer smothered by the hauteur of Washington, was fired up. He started thinking about the kind of scientific experiments he could pursue on Beinn Bhreagh's gently sloping acres. His next great passion was unfolding.

This passion grew out of his research on the family histories of deaf people. Genetics was still a little-understood science, although the Austrian monk Gregor Mendel had published evidence of inherited traits, based on his experiments growing peas, as early as 1865. The word "genetics" would not even be coined until 1905. So Alexander Graham Bell was groping in the dark as he tried to understand the role played by genetic heritage. His study of the transmission of deafness through several generations of Martha's Vineyard families in the village of Squibnocket had alerted him to patterns of inherited characteristics. He had stumbled on the existence of dominant and recessive genes, even though he never knew those labels. And he had the perfect open-air laboratory in Cape Breton: his first scientific adventure involved selective sheep breeding.

A flock of sheep had come with a parcel of land he purchased in 1888, and while the Lodge was being completed, he ordered the flock to be turned loose on Beinn Bhreagh's hilltop. "We have about fifteen lambs so far," he reported with excitement to Mabel in April, "and expect many more about May 1st." When one ewe lost her lamb, he designed a pen that would force the ewe to adopt one of the twins born to another ewe. "The experiment has been a grand success," he wrote home. "Today they seem to be fast friends." He was particularly interested in exploring why some sheep had a rudimentary set

Alec studied genetics through sheep-breeding experiments, which provided another excuse to remain in Nova Scotia.

of extra nipples, besides the functioning pair, and whether these extra sets were related to a ewe's propensity to multiple births. His goal was to breed ewes that would regularly produce twins or triplets and then provide milk through several nipples simultaneously, as a sow does for the several young piglets that are the norm in every pig litter. In other words, Alec wanted to breed a superflock of sheep.

Alec loved the idea that, in pursuit of his own scientific interest, he might benefit the people among whom he was now making his summer home. His specially bred sheep, he reasoned, would double a farmer's income without materially increasing his labor. He started to dream about fecund ewes filling the pastures of Cape Breton with crowds of gamboling lambs and swelling the emaciated wallets of the island's weather-beaten farmers. By the time Mabel arrived in June, a series of sheep pens, named "Sheepville," had been constructed at the end of the headland, to house all the lambs that were twins or had extra nipples.

Though Mabel had remained in Washington, Beinn Bhreagh was never far from her thoughts. "Do order trees planted immediately," she asked Alec. "Please order a lot of white pine (pinus strobes), white ash, sugar maple, locust (robinia pseudacacia)." Cape Breton assumed almost as much psychic importance for her as for Alec. Like Alec, Mabel was in love not just with its scenic grandeur but also with the sense of release it gave her. Her deafness constrained her less there than anywhere else in the world, because it seemed less of a stigma in Baddeck. She was already regarded as a bit of a hothouse flower by Cape Bretoners, with her fashionable gowns and sweet manner, so that her odd speech and her habit of watching their lips closely were just more aspects of her strangeness. The rural community was more tolerant of idiosyncrasy than the American elite. Mabel felt very comfortable in Baddeck. Like Alec, she was stirred to take initiatives that would benefit those who were helping them make their summer home in Cape Breton.

Mabel's first scheme was to organize a sewing school, in the hope of giving local girls skills that would allow them to earn their independence. She found the traditional weaving, crochet, and embroidery work that she saw in some of the cottages she visited so impressive that she started marketing it to relatives and friends in the States. She also established a monthly discussion club for local women, based on a similar club to which she had been invited in Washington at the home of Senator Eugene Hale. The Washington club met weekly, and participants discussed current events, books, and projects that interested them. In Mrs. Hale's elegant Washington parlor, where the women paid as much attention to each others' coiffures as to their words, Mabel struggled to follow the conversation and was often tongue-tied with shyness. But back on Beinn Bhreagh, she raised with Alec the idea of a similar club for Baddeck. If well-connected Washington women, with all the resources available to them, enjoyed such get-togethers, how much more would they appeal to women in Baddeck, where there still wasn't even a public library?

Alec was quick to encourage her, since he was eager to keep his family in Nova Scotia after their summer visitors had gone. One warm October evening in 1891, Mabel sent a rowboat over to Baddeck to collect her guests for the first meeting of a club that, she explained, would be both educational and social. It was an instant success, and soon Mabel was writing to her daughter Elsie, "Yesterday was lovely. The Ladies' Club Board came over for a meeting at four o'clock. We talked until five-thirty and then got out our needlework or knitting and gossiped until six, had a jolly dinner and then Papa showed us lantern slides. . . . There is nothing like real country life when you know how to manage it so that you have real sociability. I have more of this here than I do in Washington." By year's end, the club had forty-one members. One of the more impressive aspects of the club was that Mabel, the guileless outsider, had done what few Baddeck residents would even dare attempt: she had Catholics and Protestants sitting together in the same room.

Much as both Bells loved Cape Breton, Beinn Bhreagh created new stress between them. This idyllic corner of Canada was miles from anywhere. Mabel kept asking herself, was it a suitable place for their growing daughters to spend months in?

Mabel had good cause to worry. There was something obviously wrong with Elsie, aged ten. Five years earlier, the little girl had given her mother a severe shock when she suddenly went into a convulsion. Since then there had been no more convulsions, but Elsie's behavior was increasingly erratic: she fidgeted, twitched, waved her hands about, and could not control these involuntary body movements. Eventually a Washington doctor diagnosed chorea, often called St. Vitus' Dance.

Today, this neurological disorder is called Sydenham chorea, and is known to result from a streptococcal infection such as a sore throat, which can be treated with antibiotics. But there was no straightforward, empirical diagnosis for such a condition in the 1890s. A frantic Mabel was particularly exasperated by Alec's vague conviction that somehow

Elsie would simply grow out of it. Mabel decided to take her daughter, over Alec's objections, to Philadelphia to consult Dr. S. Weir Mitchell. Dr. Mitchell, who had made his reputation treating Civil War gunshot wounds and phantom limb pain, was now the most eminent "nerve doctor" in America, famous for advocating the "rest cure" for women suffering from that vague Victorian diagnosis, "hysteria." When Mabel and Elsie Bell arrived in his consulting room, the thin-lipped and self-assured physician examined the child carefully. Then, seating himself at his desk and fixing a harsh gaze on Mabel, he announced that the only treatment for Elsie was complete quiet and rest, far away from her mother. Elsie was moved to Dr. Mitchell's nursing home in Philadelphia, where she spent most of the next two years under the strict (and expensive) supervision of a nurse companion.

From today's perspective, Dr. Mitchell's approach to women patients was relentlessly misogynist. Even in his own day, his practice was suspect. The American author Charlotte Perkins Gilman was among those obliged by Dr. Mitchell to spend weeks, during a bout of depression, incarcerated in her bedroom undergoing the rest cure. She channeled her disgust with his theories into her famous short story "The Yellow Wallpaper," published in 1892. Dr. Mitchell certainly did nothing for Mabel Bell's already shaky confidence in her own capacity to be a mother. "He said he objected to . . . the atmosphere of our home, that Elsie boasted of how when she went home she would sit up all hours and read all kinds of books," Mabel told her mother. "So my own child has helped to strengthen the doctor's unfavorable opinion of me."

On her regular visits to her daughter in Philadelphia, Mabel wrote Alec, "I am so afraid of doing Elsie harm." And when Alec proposed to collect Elsie from Dr. Mitchell's care and take her on a brief trip to see his mother, Mabel was in anguish. "Please don't take Elsie to Washington with you. I am willing to take more risks with my children for your mother than for any one else because they are her only grandchildren, but think how long it has taken Elsie to get as well as she is . . . and how dreadful it would be for her to have a relapse now.

Your mother would not thank you for making her the cause of such a disaster. . . . Think of having to pay for it with another year's separation from our child." Elsie stayed in Philadelphia, and Mabel continued to blame herself for Elsie's condition, which Dr. Mitchell sternly described as "one of the worst cases of chorea he ever saw."

Alec could see his wife's distress, as Mabel brooded on both her inability to conceive and Dr. Mitchell's disdain for her maternal skills. He tried to reassure her: "There never was a kinder or better or more loving mother than you have been, my dear." He was dismissive of the august Dr. Mitchell's pronouncements and remained convinced that "hygiene and Swedish movements," preferably undertaken at Beinn Bhreagh, would cure his daughter. (His remedy, which amounted to nothing more than fresh air and exercise, would have worked. Modern textbooks state that the condition is "self-limiting": Elsie would have grown out of it, as he had said all along.) To be sure, there was an element of self-interest in his prescription—he resented anything that took his wife and children back to the United States. In his view, their world could come to them, and once they were settled in the Lodge, it did. Each successive summer was more lively and crowded. Hordes of relatives would arrive—Mabel's parents, Alec's parents, Mabel's sister, Grace, and her husband, Charlie Bell (who was also Alec's cousin), and assorted nieces and nephews and uncles and aunts. Some stayed at the Lodge; others took up residence in cottages across the bay in Baddeck village. And each successive summer, there were more children—not just Charlie's two daughters, Gracie and Helen, by his first marriage to Berta Hubbard, but also his and Grace's two sons, who would play with their cousins and the children of various Bell employees.

The long summer days were filled with extended clan picnics, excursions down the lake, and walks to Sheepville, followed by evening games of whist at the Lodge. And there was a new addition to the headland's attractions: Alec commissioned a houseboat from a local boatbuilder and named it *Mabel of Beinn Bhreagh*. Every few days a handful of guests would chug off on overnight excursions, to fish and to explore the endless bays and inlets of Bras d'Or Lake. Meanwhile, Alec threw himself gleefully into family activities. He told his daughters and their cousins Gracie and Helen Bell wonderful stories about a character named "the Great Imagination." He decided they must all learn to swim, so every morning, like a jovial Pied Piper, he donned a woolen swimsuit that covered his bulky torso from shoulders to knees and led his retinue of excited little girls down to the dock. They would all pile into a rowboat, and then, one by one, each child would have a rope tied around her waist and be told to plunge into ten or fifteen feet of water. "The rope pulls them up immediately," Mabel told her mother, "and then they cling to the boat while Alec gets the life preserver under their arms, and then off they go kicking and swimming like little frogs."

"Alec is so happy in this place," Mabel wrote to her mother. Only in his own beloved Beinn Bhreagh, his daughter Daisy would recall years later, could he "let himself go and be as elemental as he pleased." Sometimes this took the form of an urge to "float about in the water by the hour, particularly at night, smoking a cigar and looking up at the stars." On another occasion, he decided to find out how a scientific man, cast ashore with only a penknife and watch on an uninhabited island, could stay alive. He strode off, stark naked, in search of food. "The most promising things he found," according to Daisy, "were the heads of ripe seeds of one of the big weeds . . . along with some partridge berries, but [he] didn't find them very sustaining." His search for clothing was equally unsuccessful: he quickly discovered that he would have to collect a lot of sphagnum moss before he was decently covered. Moreover, "the moss was damp," so he abandoned the experiment, retrieved his clothes, and went off to the Lodge's kitchen to forage for buns, sausages, and hot tea.

After the second summer of such excitements, Alec announced that he wanted to stay in Cape Breton until after Christmas 1890, and thus avoid the elaborate and expensive socializing and gift-giving of the American capital. Once the summer visitors had gone, he returned to his sheep research with a vengeance, spending every afternoon with the sheep and shepherds, and looking, according to his wife, like "a big school-boy among the Lilliputians." His latest brainwave was to design covered sheep pens on wheels, so he could move them around the hillside. He had also invented a "non-pollutable, non-freezable drinking trough for the sheep," Mabel told her father. That fall, Mabel allowed herself to be persuaded to remain with him. "You will be surprised at my enthusiasm for the snow," she wrote to her mother, "but Beinn Bhreagh snow is a very different article from Washington or even Cambridge snow. There it is damp and chilly, a nuisance in every way; here it is dry and crisp and nice to walk in . . . I expect to be quite ready to go after Christmas, but I am more glad all the time that we have decided to remain." By December 17, boys were fishing for trout and smelts through holes in the ice, and "sleds and horses have been using the harbor as a thoroughfare."

A Cape Breton Christmas was Alec's idea of heaven. He and Mabel played in the snow like children, along with Daisy (Elsie was still in Philadelphia). They tobogganed down the headland's slopes onto the frozen lake, snowshoed all around the Lodge, and skated from Beinn Bhreagh to Baddeck—a distance of about a mile. Years later, Daisy would recall how Arthur McCurdy, who was now Alec's full-time assistant, gave her parents "the kind of *young* friendship that they never had with anyone else. They had jolly carefree times with him out of doors." On crisp December days, the party would set out to blaze new trails through the woods. McCurdy, tough and sinewy, was the leader, followed by Mabel, in elbow-length woolen gloves and a sealskin cap, with a little hatchet that McCurdy had given her tucked into the belt of her long, tight-fitting skirt. "How she ever ploughed through the snow in a skirt to her ankles seems astonishing," reflected Daisy, who remembered vividly how beautiful her thirty-two-year-old mother looked that year, with her "exquisite skin . . . gray-blue eyes and masses of soft brown hair." Mabel could not hear the crunch of snow underfoot or the excited cries of her companions when they spied a grouse or deer;

but she marveled at the soft light of the snow-shrouded woods and at the brilliant sheen of the lake ice. Behind Mabel tramped Daisy, and Alec brought up the rear. Large and unkempt, with his iron-gray hair long and curly, Alec was always worried that Mabel might fall, because she had no sense of balance. But McCurdy "made light of everything [and] took it for granted she could go anywhere anyone else could, and yet seemed to know intuitively when she needed a helping hand."

The Lodge, built for summer, was freezing in winter; a blazing fire in the ground floor hearth did not heat the thin-walled bedrooms. It was a struggle for Daisy to drag herself away from the roaring fire each night and scamper up to her bedroom, where the small fireplace scarcely took the chill off the air. Half-awake and clutching a stoneware hot-water bottle, she would look at the flicker of firelight on the ceiling and listen to the sounds of the piano downstairs, "for almost every night after the others had all gone to bed he would sit down and play for hours and hours." Alec's spirits soared in Cape Breton's remote magnificence, and his mind raced with ideas for future research and inventions. "He played vehemently—passionately—pouring, pouring his soul out, and upstairs I would wake up at intervals and listen."

During these months, Mabel watched her husband throw himself into "these breeding experiments with all his characteristic interest and absorption and thoroughness of detail," numbering, weighing, and measuring every sheep. She applauded his efforts when he exulted, "Twins and extra nipples abound. One of the extra-nippled ewes gives milk *from all four nipples*—promising." But much as she loved to see Alec so happy, she could not embrace life on their windy headland with such fervor once summer had gone. She did not "think it very much fun to walk silently along while Alec discussed the relative merits of different sheep with the men, [or] to remain idle in his little house on the mountain top while he counted, weighed and examined them the whole afternoon."

The next crisis in their rural idyll slowly built as she thought about their future. "All my ambition now is for my children," she told her mother. "Our daughters must have suitable friends." Mabel knew she could not launch her daughters into Washington society from Cape Breton, let alone prepare them for an education at Wellesley College, where she was planning to send them. And her daughters' interests were not the only concerns eating away at Mabel Bell: she herself longed for the quiet support of her parents and Grace, her only remaining sister, who were in Washington. It was exhausting being married to an erratic genius and housekeeping in the woods. She finally put her foot down at the end of January and ordered Baddeck's sturdiest horse-drawn sleigh to come to the Lodge and pick up Daisy, herself, and a very reluctant Alec. It was the dead of winter, and the snowdrifts were so deep that it took the little party, swathed in buffalo robes, three days to traverse the island and reach Port Hastings, from where they could cross the Strait of Canso to the mainland and the train south.

In Washington, Alec caught up with his other projects: work with the U.S. Census Bureau, papers on deaf education, correspondence with Helen Keller, the shaky finances of *Science* magazine, his Wednesday soirées. There was *so much* to do, in addition to the social calls that eminent Washingtonians were expected to make. But after the weeks of relaxation in Baddeck, some of the exuberance of earlier years was back in his marriage. One afternoon, Mabel reminded him that he was expected for a formal tea party at Twin Oaks, the home of the Hubbards. She described in a letter to Elsie the fun that followed:

> He tried flatly refusing to go, begging off, appealing to my sympathies, and finally [he] ran away. I chased him downstairs through the parlor and dining room and up the stairs again to the bathroom which he gained first. There he locked himself in. Finding [that] rattling the door knob and calls [were] of no avail, Beckie [Mabel's terrier] and I went to the study, and with the long Mississippi mule whip I commenced striking the bathroom window through the study window

much to the edification I doubt not of our neighbors. After
a while I tired of this amusement and just then I observed
Beckie making for the door, I followed her, opened the entry
door just in time to see a pair of boots disappearing upstairs.
I gave chase, but no Papa was to be seen through the third
story, so I came back, found the bathroom door still locked.
Again I rattled the door knob and poured forth such a series
of commands and entreaties and promises as I was sure must
have brought Papa from his retreat if he had heard me. As
however he gave no sign, I decided that he could not be
there, so I started on another game of hide and seek, and
this time went upstairs to the attic where I finally found Papa
crouching behind rolls of carpet! I got him out, helped him
dress and carried him in triumph to Twin Oaks. By that time
he was in a very penitent mood, ready to promise to be very
nice to the people. Of course we had been in fun all along.

But Alec's heart was in Cape Breton: he insisted on getting back for lambing season. So within months, he was again climbing over Beinn Bhreagh's hillside in his knickerbockers, supervising his shepherds, sleeping in a tent right in the middle of Sheepville, and constantly being woken by sixty newborn lambs "calling out for their Maaa at all hours of the night and morning." Mabel worried about her forty-four-year-old husband "living in tents while the bay is full of ice," but she knew he was happier there. So she let him be and accepted the fact that, to all intents and purposes, she was a single parent; she alone must ensure her children's social success. Elsie had finally been discharged by Dr. Mitchell, so Mabel decided to give her daughters, now thirteen and eleven, the kind of European experiences that she had enjoyed when she was their age, and that were now de rigueur for well-to-do American families.

Managing a Grand Tour was challenge enough—to have done so alone, when she was deaf, demonstrates Mabel's pluck. She packed up her trunks and booked passage across the Atlantic for herself and her girls. Her husband's contribution was to allow her to take Charles Thompson, the Bells' major-domo in both Washington and Baddeck, with her. On the transatlantic crossing, Charles had to sleep in steerage while Mabel relied on her daughters to help her navigate through the finer points of first-class travel. "Daisy is a nice little one to translate between me and the Captain and to keep me informed as well as she can of what is being said at table," she assured Alec. "Elsie is a great relief and manages far better than I dared hope." Once in Europe, however, Charles took all the responsibility for seeing that the Bells' tickets were in order and that the mountain of steamer trunks was conveyed from hotel to hotel. In Rome, Mabel noted wryly, "Charles, my colored serving man, [is] learning faster and profiting more by this journey than my children."

In Baddeck, Alec missed his family, but his imagination worked overtime. He had announced to Mabel in 1887 that he found it hard to explore his scientific pursuits in Washington, a city geared to politics rather than technology. "The fact is that Washington is no place in which to carry out inventions," he wrote. "In a small workshop with one workman, it takes forever to have the slightest thing done and ideas cool before anything is accomplished." Baddeck's distance from other scientific men did not seem to bother him, and he was able to work much more easily on Beinn Bhreagh than in Georgetown. Just below the Lodge, he built a little wooden shed, grandly named "the laboratory." Here, he installed Arthur McCurdy and a wiry middle-aged mechanic he had brought from Washington, William Ellis, and began exploring some new ideas. Some of his brainwaves (all of which were carefully recorded in his "Lab. Notes") went nowhere. He had high hopes of a plan to take a printing impression from photographs using a ceramic material that he called "agate cement." "Here at last is a subject at which we can work together," he wrote Mabel. "This invention . . . will bring us together in a common interest. . . . The first print will be of Jacob's Dream, a stairway to heaven and angels passing up and down." But the scheme was abandoned because the materials were too expensive. He then returned to fiddling with the phonograph, thinking to make cylinders on which sound could be recorded in a mineral wax called ozokerite. But the cylinders were always pockmarked with air bubbles when taken out of the mold, so this idea was also dropped.

Failure in some experiments didn't dampen Alec's enthusiastic efforts in other directions. He developed an apparatus for heating water using an electrical filament in a glass tube of sperm whale oil. "Grand success," he recorded. "Just as Electric Lighting must gradually take the place of gas, so I believe Electric Heating will take the place of dirty coal." And his more general musings were extraordinarily prescient. "Thought," he jotted in his notebook in December 1891, while bemoaning the fact that letters from Mabel in Italy took so long to arrive. "Correspondence between distant places will in future be carried on electrically instead of by mail."

At one level, this was Alexander Graham Bell the polymath, whose brilliant mind roamed unfettered across the scientific landscape. This, of course, was the essence of his genius: he could make surprising and original connections between unrelated fields, as when he tried to save President Garfield's life by combining the telephone receiver, then regarded as the latest development in telegraphy, with an induction balance he had designed for the mining industry. But at another level—the level that exasperated Mabel—his approach seems hopelessly scattered, with wild enthusiasms adopted and abandoned. And during the early years at Beinn Bhreagh, before he had a specific project, his attention seems particularly fragmented. His bedtime reading was the *Encyclopaedia Britannica*: "I have read tonight articles on Fog-signals, Force, Fortification, Fort Sumter, Fossil foot prints . . . ," he told Mabel in one letter. His letters were usually written after midnight, and sometimes he was so tired he could do little more than list the subjects he wanted to cover. A catalog he compiled one evening, as he sat in his study with an oil lamp illuminating the thick sheet of paper on his desk, indicates the range of his interests: "Wild flowers of Canada. Mrs. Hobart has a baby . . . Argon and helium . . . Lambs all in. 24 (four-nippled) 2 (three-nippled) 0 (2-nippled). Sir Francis Galton's enquiry. Centre of gravity experiments. With same rotations greater mass gives greater lift . . . Spherical telescope—dome floating on water. Meteoric ring of Saturn. Identity of level of Baltic, Black and Azore seas . . . Oysters and typhoid. Artificial wood for British men of war . . . Have no time to expand these topics."

In Mabel's absence, Alec's appearance became even more eccentric; his bushy hair and shaggy beard remained uncombed and untrimmed, and his baggy tweeds became even shabbier. He could have been mistaken for Leo Tolstoy in pre-revolutionary Russia, as he organized his rural fiefdom and galvanized the locals into unfamiliar activities. Yet a momentous new project was gradually emerging from all the different experiments that he, McCurdy, and Ellis were performing in the little wooden laboratory below the Lodge. Though he did not abandon his sheep or his frequent trips to schools for the deaf, a new pursuit had begun to preoccupy him—an idea far beyond the imaginations of most of the stolid citizens of Cape Breton.

As the century neared its close, the pace of invention had accelerated. Rudolf Diesel would patent his engine in 1892; Henry Ford would build his first car in 1893, and Guglielmo Marconi was already playing around with radio telegraphy. With the widespread availability of electricity, a new torrent of products, including domestic lighting and appliances such as primitive vacuum cleaners, was transforming the lives of ordinary citizens. The first steel-skeleton skyscraper had been erected in New York City in 1889; the might of Niagara Falls would be harnessed to provide power for Buffalo's electric streetcars in 1895. As astonishing as the torrent of innovations was people's capacity to absorb them. There was, however, one particular dream that was common to many of the era's inventors, even as most people ridiculed it as fantasy: the dream of flight. And Alexander Graham Bell had begun to suffer, as he put it to Mabel, from "a bad attack of flying machines."

Inventors from the beginning of time had been fascinated by the idea of flying machines; the notebooks of Leonardo da Vinci contain numerous designs for machines modeled on avian anatomy. When Alec began thinking seriously about powered flight, such machines seemed more possible thanks to the nineteenth century's advances in both steam power and electricity. The conquest of the skies had already begun: as early as 1783, the French Montgolfier brothers had sent a man aloft in the basket of a silk and paper hot air balloon and launched a ballooning craze throughout Europe and North America. Balloons, however, were entirely dependent on the weather and offered their "aeronauts" little chance of controlling the direction in which they wafted. In 1875, a balloon launched from Toronto dumped its hapless crew and passengers into Lake Ontario. Serious inventors were already thinking that if man was to take to the air, he needed wings.

Bell's ability to tackle challenges from unexpected angles led him into creative ways of thinking about how to achieve flight. Most of the other scientists building prototypes for flying machines were preoccupied with achieving speed and lift in the air. As Alec walked around Beinn Bhreagh, checking his sheep and stopping to watch bald eagles soaring effortlessly in the updrafts created by the cliffs, he decided that those prototypes would be horribly dangerous for the test pilots at liftoff and landing. So he began his new enthusiasm by constructing helicopter-like devices. One day that winter, he made his way down to the laboratory and explained to William Ellis that he had designed a miniature tin boiler that they were going to try to get airborne. Ellis set to work to construct the boiler, which was no wider than a man's hand and weighed less than a pound. Alec's idea was to heat water in this boiler, to which a two-bladed cloth propeller would be attached, and then release the steam so that the propeller would rotate. The theory, as Alec explained in a letter to Mabel, was "that an upright tubular boiler could be made to lift itself in the air—fuel and all—by fan wheel arrangement worked by a simple jet of steam."

The temperature in the little hut steadily rose on January 6, 1892, as the two men labored over their contraption. While Alec held the boiler steady, Ellis carefully soldered a vertical pipe into it. It was fiddly work; Ellis regularly straightened up, adjusted his cloth cap, then bent to the task again. Once the two men had completed the device, Alec began a series of experiments, using different mixes of water and alcohol in the boiler, then measuring how fast the propeller over the steam outlet revolved. Over the next few days, frustration mounted. It was difficult to build enough pressure in the boiler. When he did achieve sufficient pressure, the solder on the boiler's seams burst. The propeller's wings were inclined to catch fire. But at one point, the steam-powered helicopter took flight—when steam started escaping from two pinholes, it "shot over to the other side of the laboratory," an excited Alec wrote to Mabel.

The machine may have belly flopped, but Alec's imagination took wing. "I have the feeling," he wrote to Mabel a few days later, "that this machine may possibly be the father of a long line of vigorous descendants that will plough the air from Beinn Bhreagh to Washington! And perhaps revolutionize the world! Who can tell? Think of the telephone!"

Days after Alec had completed this series of experiments, he joined his family in Italy. Within months, however, he, his wife, and their daughters had all returned to Baddeck. By now, Alec had finally purchased the remaining parcels of land on the headland, and the dream that both he and Mabel had nursed for years was about to be realized. They were going to watch the foundations laid for the magnificent mansion that still stands on the end of headland, with water on three sides of it and a wooded hillside on the fourth. A firm of Boston architects had drawn up the plans, and a Nova Scotia contractor had been hired to construct the building.

At a time when most Baddeck cottages changed hands for a couple of hundred dollars, the cost of the new Bell home was reportedly $22,000

Beinn Bhreagh Point, completed in 1893, was "the finest mansion in Eastern Canada."

(construction costs today would be well over $2 million). And the Bells were not the only wealthy people who had succumbed to the charms of Atlantic Canada. The portly railroad baron Sir William Van Horne, president of the Canadian Pacific Railway, was busy constructing a lavish summer cottage called Covenhoven, near St. Andrew's, New Brunswick. A few years later, Alexander McDonald, John D. Rockefeller's partner in Standard Oil, bought a large piece of seashore in Prince Edward Island and built an elaborate mansion called Dalvay, rumored to cost $50,000. Neither of these plutocrats, however, took as great a personal interest in their properties as the Bells. According to family tradition, Alec and Mabel had chosen not only the exact location for their dream home but also the outlook for each room by traveling in a hay wagon to the end of the headland and then clambering up a rickety two-story structure erected on the wagon so they could assess the view. As a result, the prospect from every window that faces the lake is extraordinary.

From the sunporch, the dining room, and most of the bedrooms, the Bells and their guests could gaze out at Boularderie Island to the east, Baddeck to the west, or the Washabuck lighthouse straight ahead. From the kitchen window, the cook could see right down St. Andrew Channel to Grand Narrows, where the local steamer picked up passengers who were arriving on the Cape Breton railroad. (She could also keep an eye on any boat that was carrying guests to the Beinn Bhreagh wharf, and gauge when it was time to heat the soup.) After the house had been completed in November 1893, the *Halifax Chronicle* announced, "The Bell Palace at Baddeck . . . Said to be the Finest Mansion in Eastern Canada."

Beinn Bhreagh Point (to give it its full name) might be a mansion, but architecturally it is a mishmash of styles. There are elements of a French chateau (two cone-roofed towers and a steep cedar-shingle roof), but bulky chimneys, dormer windows, and balconies spoil the roofline. To the right of the main frontage there is a domestic wing at an awkward angle, and the main entrance is through a dark porch at the back of the house. Moreover, Mabel soon faced a major design fault. "The house would be about perfect but for two things," she wrote to her mother in January 1894. "The great smoky fireplace and the horrid furnaces, and the soft coal which makes my pretty clean new house in the clean country as dirty and sooted as the dirtiest sootiest house in the City of London. There's something radically wrong with the furnaces, the smoke leaks through into the heat pipes and fills the whole house with smoke and gas." These heating problems, in which Alec took no interest at all, took several months to solve.

Nevertheless, for the first time in their lives, Alec and Mabel Bell had a home that truly reflected their different personalities and suited most of their needs. In Washington, Mabel was often lonely while her world-famous husband traveled to distant cities to attend meetings, speak at conferences, or pursue research. She once wrote wistfully, "My dear Alec, I miss you very much, the house feels very quiet and lonely without you. You cannot realize it, but you are the pivot around which the household revolves." In Cape Breton, Alec was always home, and his rhythms dominated the rambling, hospitable two-story mansion.

It could sleep up to twenty-six people, and its large living room often resounded with hymns, spirituals, and Scottish songs when Alec sat down at the grand piano. A portrait of the Scottish poet Robbie Burns hung over the fireplace of Alec's ground-floor library, while upstairs, next to her bedroom, Mabel had her own pretty little study, in which she kept the household accounts and tracked their investments. The kitchen, where there were always a cook and at least two maids, was large and practical, with a wood-burning stove and a scrubbed wooden table at which the children drew pictures, made models with candle wax, and ate most of their meals. Alec's study was on the second floor, because, as Mabel explained in a letter to her father, "it must be as inaccessible as possible as the trouble now is that servants will go to him in any difficulty—consequently driving him from the house as the slightest interruption will upset him for hours." His bed, which was housed on the open porch above the main entrance, was encircled by tightly woven curtains that would block all light from reaching his sensitive eyes.

Mabel took far more pleasure in supervising the decor of the Point than that of their residence on Connecticut Avenue, in style-conscious Washington. In Cape Breton, her originality was given full rein. She bought curtains "embroidered in the Norwegian fashion" from a local sewing woman, and she commissioned a frieze for one of the bedrooms that consisted of mussel shells arranged in an attractive design and embedded in white plaster. Cousin Mary Blatchford was quite captivated. "There are two thousand and seven shells, for I counted them," she wrote to a friend. "The delicate color of the inside of the shells makes a charming decoration." Family festivities became even more elaborate once the Point was complete. In 1894, Chinese lanterns glowed from the trees and bonfires crackled as Alec celebrated his parents' golden wedding anniversary. For Melville and Eliza Bell, who had once feared that all three of their sons would predecease them, it was a magic moment. Melville nearly burst with pride at his son's achievements, and now that there was a comfortable house to

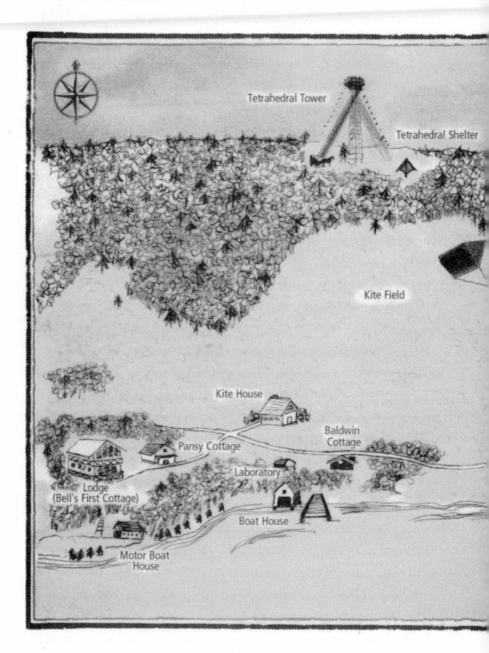

Ice House

Warehouse

Sheep House

Gardener's
Cottage

Observatory

Beinn Bhreagh Point

Poultry House

Barn

Bell's Houseboat

Office

Bras.d'Or Lake

stay in, he would become a regular visitor to Cape Breton. That night, local farmers and fishermen were mesmerized by the brilliance of the electric light that blazed out of the mansion's huge windows and by the military band music that drifted down the lake. Two years later, Gardiner and Gertrude Hubbard enjoyed the same gala treatment when they too celebrated their fiftieth wedding anniversary.

It was almost impossible to dislodge Alec from his scientific pursuits in Cape Breton, so a pattern was established: once family ties had been strengthened by the exuberant summer months in Baddeck, Mabel attended to her daughters' needs. These included both extended trips to Europe, for more of the continental polish Mabel wanted them to acquire, and appearances at fashionable events in Washington. The annual routine also allowed Mabel to spend time with her beloved parents and sister. Alec stretched out his Beinn Bhreagh sojourns as long as he could, rarely leaving before December and rushing back as early as April to resume his sheep-breeding and flying experiments.

The spring of 1895 was typical. Mabel, Elsie, and Daisy had once again set off without Alec to Europe. In Paris, the three Bells installed themselves in a convent on the Rue Nitot, which had a girls' school attached. Each morning, Elsie, now seventeen, and Daisy, now fifteen, donned the blue tunics that were the uniform for the *mademoiselles* enrolled in the elite school, and disappeared to their classrooms. Each afternoon, the two young American heiresses, demurely dressed in straw hats, white gloves, and ankle-length dresses with tightly corseted waists, set off with their mother to explore the Paris of La Belle Epoque. They admired the newly erected Eiffel Tower ("hardly noticeable after the first shock of surprise") and stared at the "bicycle women" who paraded up and down the Champs Elysées in tight white veils and brown bloomers ("a deplorable sight").

One of Mabel's greatest pleasures in Paris was to settle at the table in her little parlor, clad in her pale blue silk *robe de chambre,* after the girls had disappeared to their lessons. She would open the overstuffed envelopes that arrived from across the Atlantic, and unfold the sheets of paper covered with Alec's spidery black handwriting. A typical letter from this period reads,

> I think I explained to you the construction of the new fan-winged machine, a sort of fan made of strips of tin. Having found that it seemed to propel my whirling-table, I determined to try it in the air with three wings. Wings made of strips of tin ¼ inch wide—angle at tip 20°—at centre 90°. When we came to try this with our rotator pulled by hand, it rose so readily that we could not get a good pull. I had to lengthen the pins of the rotator to make it stay on long enough to get any sort of spin. . . . Today we tried it again with a heavy ring of solid brass on the ends of the wings. Weight 460 grammes more than one pound. *The addition of weight made it go higher.* Don't know how high it went—certainly over 250 feet.

At first, Mabel appreciated the regular bulletins on helicopter experiments from Beinn Bhreagh. In June 1895 she replied, "I hope that you have succeeded in definitely fixing the centers of gravity in the arms and in determining whether putting the center of gravity further out on the arms by means of weights you do increase the velocity per second, and thereby cause a greater lift. . . . I do so want your name associated with successful experiments in flying machines. . . . I wish I could be in two places at the same time, with you at Beinn Bhreagh and here with the children . . . You have always been very good to me, my husband, and I love you."

But as the weeks passed, Mabel began to worry. The letters read more like lab reports. Alec's missives became increasingly staccato—he cataloged his activities and forgot to ask about her well-being. Flying-machine experiments were taking over Alec's life: he was sleeping in the laboratory, working round the clock, and grabbing something to eat only when he remembered. In June, Mabel reproached him, "No one can stand such irregularity of life very long. You know you came to me nearly wrecked by just such living some eighteen years ago and if you could not stand it then you certainly can't now." Memories of the haggard young man who had chased her all the way to Nantucket and poured out his hysterical distress in note after note flooded back. By July, the tone of her letters had changed: "I am very much interested in your water experiments and in the center of gravity load shifting, but I am much distressed to think of the lonely solitary life you have been leading. Please darling make an effort to be more sociable and go to see people. . . . The very nuns here are not leading as solitary narrow a life as you. Please try and come out of your hermit cell."

Alec did manage to extract himself from his work long enough to join Mabel, Elsie, and Daisy in France that August. He chafed to return to Cape Breton, but the break did him good. He allowed Mabel to impose some regularity on his life, and to cut his hair and trim his beard. But by the fall, they were all back in Cape Breton. After feeling unable for so many years to find a focus for his inventive mind, Alec was now consumed by the race to build a heavier-than-air flying machine—and he was determined to do it in Baddeck. His pursuit of this goal would consume him—and a great deal of Mabel's money—for much of the next two decades.

Chapter 17

MONSTER KITES
1895–1900

T he two portly, bearded men puffed their cigars in companionable silence as they sat on the deck of *Mabel of Beinn Bhreagh,* the Bell houseboat. The August sun was hot, and they were both dressed in flannel shirts, tweed trousers, and black leather boots. But a canvas awning shaded them from the glaring rays, and a lake breeze cooled them.

There was the occasional grunt as each man studied the seagulls soaring over the water. Finally, one of them grabbed the cigar out of his mouth, shook his head, and barked, "Isn't that maddening!"

"What's maddening?" asked his host, Alexander Graham Bell.

Dr. Samuel Pierpont Langley, the secretary of the Smithsonian Institution and a regular participant at Alec's Wednesday-evening get-togethers in Washington, replied sharply, "The gulls."

Alec chuckled. "I was thinking they were very beautiful."

Each eyed the other for a moment, recalled the Bells' manservant Charles Thompson years later, then both burst into roars of laughter. But the exchange caught the character of each man. They had much in common: both were self-taught scientists who had never graduated from university (Langley, from an old-money Massachusetts background, had read omnivorously in the Boston Public Library in his youth, and emerged a hugely knowledgeable astronomer). Each had chalked up a momentous scientific achievement. In 1878, two years after Alec had invented the telephone, Langley had invented the bolometer, a highly sensitive instrument for measuring solar radiation. The two men even looked alike, with their bushy white beards and jovial expressions. But unlike Alec Bell, Samuel Pierpont Langley's invention was not so lucrative that it had liberated him from the need to work for his living. He was firmly embedded in the American scientific establishment: as secretary of the Smithsonian Institution, he held the most prestigious and influential scientific position in the United States. Before his appointment to the big brownstone building in 1887, he had been director of the Allegheny Observatory and professor of physics and astronomy at the Western University of Pennsylvania for twenty years, and had received countless honorary degrees and international awards. Despite his affable expression, he had an abrasive and competitive nature. Based year-round in Washington, the confirmed bachelor followed the scientific literature on aeronautics with hawk-like attention, undistracted by family or competing interests.

Langley, like Alec Bell, had been watching birds as long as he could remember, and dreaming of manned flights. As soon as he had been appointed to the Smithsonian, he had started trying to launch from a spinning table some of the stuffed birds in the Smithsonian's extensive collection but rarely got the carcasses off the ground. He was trying to understand, he once told a reporter, how birds could "move about the air, rising and falling, soaring up and sailing down without any motion of the wings." The same questions had fascinated Alec, who throughout his life filled notebooks with sketches of bird skeletons or notes about the feasibility of various airborne contraptions. Sam Langley had spurred Alec's interest when he asserted in an 1891 publication on experiments in aerodynamics, part of the *Smithsonian Contributions to Knowledge,* that "[t]he mechanical suspension of heavy bodies in the air, combined with very great speeds, is not only possible, but within reach of mechanical means we actually possess." When his Smithsonian experiments with stuffed birds failed, he moved on to other possibilities for launching a flying machine, or an "aerodrome" as he called it. Between 1887 and 1903, he would transform the Smithsonian's carpentry and machine shops into a research and development facility aimed at the production of a successful flying machine.

Alec followed his friend's efforts closely, frequently asking him to present his ideas at the Wednesday-evening gatherings. The experiments with steam-powered helicopters that Alec had begun conducting at Beinn Bhreagh in 1891 were inspired by similar experiments conducted by Langley at the Smithsonian. In subsequent experiments, Alec had tinkered with propeller blades, gunpowder rockets, and spring-powered helicopters. He made "whirligigs"—toy airplanes powered with rubber bands. He speculated on whether gunpowder, dynamite, or nitroglycerine was the airplane fuel of the future, and on the method by which passengers might exit a hovering machine. (His solution was that they "could be let down by rope.") He kept careful records of all his experiments, but he also continued with his other interests. In addition to sheep-breeding, deaf education, and ideas for improving the phonograph, he made sketches in his notebook of an idea for an automatic telephone switchboard that would eliminate the need for human operators. (His heart was not in the effort, however, and he did nothing to develop his idea. This was his last foray into telephone research.) During these years, both he and Mabel had also developed an interest in photography: Arthur McCurdy had acquired one of the first Eastman Kodak cameras, and the Bells soon followed suit. Mabel set up a darkroom in the basement of the Point. Despite these distractions and despite repeated setbacks on the flying front, by the mid-1890s Alec was obsessed with theories of mechanical flight. "The more I experiment," he wrote, "the more convinced I become that flying machines are practical."

Both Langley and Alec also followed the exploits of the German civil engineer Otto Lilienthal. While the two Americans pursued the dream of flying machines by focusing on powered models, Lilienthal was coming at the dream from a different angle—that of manned gliders. In 1891, he had started jumping off hills near Berlin while attached to the undercarriage of a set of wings, like a modern hang glider. He maintained some control over the machine through movements of his body. What distinguished Lilienthal's experiments was that he understood that maneuvering through air was a different proposition from maneuvering across land. There were plenty of other eager airmen trying to invent flying machines in the 1890s, including Lawrence Hargrave in Australia, Hiram Maxim in England, and Octave Chanute in the United States. Most, however, thought of themselves (in the phrase coined by historian Charles Gibbs-Smith) as "chauffeurs of the air." They assumed that flying a plane would be like driving a carriage, and they concentrated on getting up enough speed. Lilienthal understood that the fluid medium of air made flight a quite different proposition, to which issues like lift, thrust, and drag were crucial. Like all his contemporaries, he did not yet understand the laws of aerodynamics that would govern the stability of a flying machine; but he recognized that such laws might exist—and Alec Bell and Sam Langley grasped that Lilienthal's exploits had moved flight research into a new phase.

Langley's flight experiments were further advanced than Alec's. In 1892, he had purchased a houseboat and towed it to Quantico, a wide, quiet stretch of the Potomac River in Virginia, thirty miles south of the capital. The river was so shallow in this area that a plane could easily be recovered from the water after the test flight. It was also out of the public eye: Langley was paranoid about press attention. The flying-machine business was attracting a fair number of cranks by now, and their antics—outrageous claims, ridiculous contraptions, phony financing—were earning the delighted derision of newspapers. The secretary of the Smithsonian Institution couldn't stand the idea that his steam-powered, propeller-driven aerodromes might meet with ridicule.

But Langley's loyal supporter Alexander Graham Bell had stayed up all night in 1895 to watch the first flight of Aerodrome No. 5, a sixteen-foot-long unmanned biplane launched by a catapult mechanism from the roof of the houseboat on the Potomac River. When the machine stayed aloft for six seconds, the inventor exploded with excitement. "I shall count this day as one of the most memorable of my life," Alec wrote to his friend. On May 9, 1896, Alec delayed his return to Nova Scotia to witness Langley's latest prototype take wing. Langley himself was a bag of nerves, and could hardly bear to watch the test. But Alec found a boy to row him out to the middle of the river so he could photograph the steam-powered aerodrome in flight. He roared with delight when he saw the huge machine, with its fragile wings constructed of spruce ribs, pine spars, and white silk, make not one but *two* successful test flights. On the first, it stayed aloft for one minute twenty seconds and covered an estimated 3,300 feet. On the second flight, it traveled some 2,300 feet before gliding slowly to the water.

Alec Bell immediately recognized his friend's monumental achievement. For the first time, a large flying model with its own source of power had remained in the air for a substantial length of time. It was still unmanned, and its inventor had no control over its flight path and no idea what made it take a particular direction, horizontally or vertically. But it had flown. He urged his friend to release news of his success, and with Langley's permission wrote a lengthy letter to the editor of *Science* describing the test flights. The flying machine, he wrote, "resembled an enormous bird, soaring in the air with extreme regularity in large curves, sweeping steadily upward in a spiral path, the spirals with a diameter of perhaps 100 yards, until it reached a height of about 100 feet in the air at the end of the course of about half a mile, when the steam gave out, the propellers which had moved it stopped, and then, to my further surprise, the whole, instead of tumbling down settled as slowly and gracefully as it is possible for any bird to do." There was now no doubt, insisted Alec, "that the practicability of mechanical flight had been demonstrated." At the Wednesday-evening soirée a week later, members of Washington's scientific elite were almost incoherent with excitement at their colleague's achievement.

Within the next few months, American newspapers were filled with reports of further accomplishments in flights, usually by manned gliders rather than powered machines. Alec followed the different reports in the *New York Times* avidly. In February 1896, under the headline "When We Take to Flying," there was a long article about recent advances, including sketches of Otto Lilienthal's latest gliders, Hiram Maxim's aeronautical machines, and the box kites developed by Lawrence Hargrave. The newspaper quoted Lilienthal as assuring readers that in aeronautical adventures, "the risk is but slight. . . . One can fly long distances with quite simple apparatus without taxing one's strength at all, and this kind of free and safe motion through the air affords greater pleasure than any other kind of sport." It was all too inspiring for words.

But in August the same year, Alec got a nasty shock. He was sitting in his study at Beinn Bhreagh when Charles brought him a bundle of issues of the *New York Times* that had been sent over by the Baddeck post office. Alec sorted the various copies into chronological order, and began reading. When he came to the copy for August 12, he let out a groan. On the front page, datelined Berlin, August 11, he saw the headline, "Killed on His Flying Machine" over the following report: "Herr Lilienthal, an engineer, who for many years was experimenting in the building of flying machines, met with an accident today that resulted in his death. He started with one of his machines to fly from a hilltop at Rhinow, near Berlin. The apparatus worked well for a few minutes, and Lilienthal flew quite a distance, when suddenly the machinery of the apparatus got out of order and man and machine fell to the ground. Lilienthal was so badly injured that he died in the hospital to which he was removed."

The death of the great German aviation pioneer, aged only forty-eight, shocked Alec. Even though he had never met Lilienthal, the abrupt finish to so many plans, so much promise, must have triggered all those painful memories of the untimely deaths of his own brothers. "A dead man tells no tales," he noted sadly. "He advances no further." But how should other pioneers continue the work? "How can ideas be tested without actually going into the air and risking one's life on what may be an erroneous judgment?" Lilienthal's tragedy convinced him that safety should be a priority in flying-machine experiments.

The aeronautics craze continued unabated, prompting the *New York Times,* only a month later, to comment, "The fate of so many predecessors, from Icarus to Lilienthal, does not seem to damp the ardor of those enthusiasts who aspire 'to essay the airy void with wings not given to man.'" In the words of Tom D. Crouch, former curator of aeronautics at the National Air and Space Museum, "even the most skeptical members of a generation that had witnessed the advent of the telegraph, the telephone and the horseless carriage were slowly coming to realize that they stood on the threshold of the air age."

Yet Sam Langley was slow to follow up on his success. Like his fellow flying-machine enthusiasts, he wasn't sure where to go next, and he dreaded the ridicule that failure might bring. During his visits to Cape Breton over the next few summers, he and Alec spent hours on the Beinn Bhreagh porch or the deck of the houseboat discussing the next step—manned flight in a powered machine. Alec continued to pursue his own research, filling his workshop with propellers, wings, and motors. Encouraged by Langley's success in 1896, he spent months testing differently angled propellers and various wing shapes, trying to see which produced the most lift. "I am finding out in the laboratory that a great deal has yet to be learned concerning the best way to combine aero-planes or aero-curves, so as to gain the full benefit of the surfaces," he noted in October 1896. It all made his mother very nervous, but Mabel reassured Eliza Bell that his contraptions were still inside the laboratory. "All Alec tries to do is to see how much lighter the machine becomes in the different arrangements of motors and wings. He will not be ready to construct an outdoor machine for many months yet."

But Alec was smitten. Despite the regularity with which his rockets misfired, despite his ignorance of the mathematics required to calculate wing ratios and propeller angles, despite his lack of a systematic strategy for research, the idea of manned flight obsessed him, much as the telephone and the photophone once had. This left Mabel in the same old quandary: torn between her husband's and her daughters' needs. "He is intensely interested in his flying machine and very hopeful of success," she wrote in despair from Beinn Bhreagh to her mother in 1896. "I cannot believe it right to take him away unless there is no other means of educating our children properly." The following year, it was the same story. Alec had told Mabel that "he would not leave his flying machine experiments until they were in a more satisfactory shape than now," she reported to her mother. "They are killing him, but he won't leave them and he won't stop, it is cruel of me to try and make him leave them." Mabel was chafing to embark on several ventures. She was eager for more travel: invitations to her world-famous husband were multiplying—most recently he had been invited to visit the emperor of Japan in Tokyo. She knew that her parents were right when they suggested that Alec was missing valuable contacts and opportunities when he hid himself away in remote Cape Breton: "I have much the same feeling that you have," she wrote, "that Alec ought to be more in the city with other scientific men, but I cannot get him away." She would have preferred Alec to focus on an invention that had more immediate commercial potential. Although the Bells continued to be extremely wealthy, they were living on capital, and Beinn Bhreagh's heavy expenses far outstripped their annual income. Most important, she knew it was time to launch Elsie into Washington society.

The Bells' two daughters were, by now, accomplished and good-looking young women. Elsie, who was nineteen in 1897, was tall and well built, with her mother's gray eyes, thick, soft brown hair, and gentle manner. Shy and cautious in public (perhaps as a result of the enforced seclusion while she suffered from chorea), she stayed close to her mother and

Mabel Bell handled all the family's domestic and financial affairs.

often let her younger sister take the lead on social occasions. Seventeen-year-old Daisy was smaller and more agile—a tomboy who often challenged authority. "She is a little inclined to be what she calls 'fresh,'" Mabel reported to Alec, "and I [call] a 'little impertinent.'" Daisy had her father's dark hair and black, deep-set eyes, and her mother's artistic tastes. In Paris, she had particularly enjoyed the visits to the Louvre and the Salon des Artistes. Frequent visits to Europe meant that both girls were fluent in French and at ease with people from different cultures, and barefoot summers in Cape Breton had given them a carefree indifference to convention. On one occasion, they were among the very few people invited to a huge Washington charity ball to brave a savage blizzard in order to attend. Julia B. Foraker, a Washington grande dame of the period, recalled in her memoirs that "[t]he two daughters of Alexander Bell waded through the snow in boots and knickers, carrying their ball dresses in their arms."

In the 1890s, according to Mrs. Foraker, Washington society was "as brilliant as any that ever America has produced." Mrs. Foraker's husband was Senator Joseph Benson Foraker, the Republican boss of Ohio and a close friend of William McKinley, who would serve as president between 1897 and 1901. This meant that Mrs. Foraker had a front-row seat on events and could see how everything was getting more lavish—the ceremonies at presidential inaugurations, the hospitality offered by an expanding diplomatic corps, the size of government. The formidably well-upholstered Mrs. Foraker noted with pleasure in her memoirs how "the rich, spectacular New York-crowd-with-the-names came over, took big houses, gave extravagant parties and exotically quickened the pace. . . . Never again shall any of us see such abundance and cheapness, such luxurious well-being as prosperous Americans then enjoyed." Mabel wanted Elsie and Daisy to be part of this Gilded Age world. She worried that, stuck in Baddeck, they would never find suitable husbands. But unlike the well-connected Forakers, the Bells' tenuous connections to the administration and frequent absences from the city made them marginal to the social swim. And without Alec, Mabel felt doubly handicapped as she tried to infiltrate Mrs. Foraker's brilliant circles. Not only was Mabel deaf, she was also too diffident to be a grande dame.

In March 1896, Mabel gave a ladies' luncheon party in the Bells' Connecticut Avenue home, with guests seated at five tables, each with different-colored flowers, lamps, cloth, and centerpiece. The social effort was a strain. "I felt so incapable of taking hold of the crowd and with a smile and strong cheery word directing them and commanding them as a true hostess should!" she reported to Alec, who was puttering away in his Beinn Bhreagh laboratory. But Mabel had been determined to make it work: her purpose was to make the contacts necessary for Elsie's launch in the coming season. "The lunch was very handsome and yes, very expensive," she continued. "It will be the last and I think it will help me. Almost all the ladies were Society People and mothers of next year's debutantes, people whom it was necessary for me to know."

As usual, the Bells spent the summer together at Beinn Bhreagh. In the fall of 1896, Mabel and Elsie returned to Washington for a social offensive. "Debutante and tea and reception cards are beginning to come in thick and fast," Mabel wrote to Alec. Five hundred and fifty people were invited to Elsie's coming-out ball, held at 1331 Connecticut Avenue on December 7. Mabel pleaded with Alec, "You must come in plenty of time." On the big night, at least 450 guests arrived at the house, which, with her mother's help, Mabel had transformed into a fairyland. Despite the month, huge bouquets of delphiniums and chrysanthemums (from the greenhouses at her parents' Washington home, Twin Oaks) decorated the entrance hall and dining room. On the dining tables, silver epergnes overflowing with grapes and oranges gleamed in the flickering light of tall candles in elaborate candelabra. The library's Turkish carpets had been rolled up and the hardwood floor polished to a high gloss for the dancing. Best of all, Alec had arrived two days earlier. Resplendent in white tie and tails, he hailed each new arrival with a cheerful and irresistible warmth. As usual, welcome from such a famous man melted the snobbish crust of even the grandest Washington family. "The party seems to have been a success," Mabel wrote with relief the next day to Daisy, who was at boarding school. "Elsie looked very pretty indeed. . . . Dancing began about twelve and continued until two."

With Alec in Washington, the pace of the Bells' social life intensified further. Despite Alec's professed dislike of the capital, the capital loved the inventor of the telephone. Alec was invited to dinners at the Cosmos Club, the gentlemen's club on Lafayette Square where his flying crony Samuel Langley was a permanent resident. He revived his Wednesday-evening get-togethers with the city's men of science. And in April 1897, the Bells struck up a friendship that would prove momentous. Dr. Edwin Augustus Grosvenor was an acquaintance of Mabel's father who had recently addressed the National Geographic Society. This was a rather stuffy organization that Gardiner Hubbard had established in 1888 for the "increase and diffusion of geographical knowledge." It held meetings every other Friday during the winter about anything that caught its august members' fancy—geology, the weather, anthropology, mineralogy—and also published a journal that was, in its early issues, "often tedious and sometimes somnolent," in the words of its historian Robert M. Poole. Bound in muddy brown and filled with close-set gray text, it was an irregularly published compilation of papers by society members.

Professor Grosvenor had been invited to tell the National Geographic Society's members about Constantinople. He and his family had lived in that city for twenty years, and he had taught history at Robert College, an American school there, before returning to the United States and a teaching post at Amherst College. A few days after his talk, he called at 1331 Connecticut Avenue to meet the famous Dr. Bell, and stayed on for dinner. The neat, weather-beaten academic, with his trim goatee and precise manner, enthralled the Bells with his stories of Greek-Turkish clashes and Eastern customs. He was so interesting, Mabel told her daughter Daisy, "that Papa has forgotten deaf-mutes etc. for awhile." Professor Grosvenor also told the Bells about his identical twin sons, Gilbert and Edwin, who were students at Amherst College. He invited Elsie and Daisy Bell to attend the twins' graduation from Amherst, later in the spring. Mabel responded with an invitation to the boys to visit Beinn Bhreagh in July. They were "certainly fine fellows," she told her mother, "clever and manly with no nonsense about them." She far preferred both of them to some of Elsie's other admirers—particularly one who, she confided to Alec, was said to be "mean and not careful about paying his little card debts."

That autumn, Alec once again managed to spin out the season in Cape Breton. He had the perfect excuse this time: Lord Aberdeen, the governor general of Canada, and his wife, Lady Aberdeen, had announced that they wanted to make a visit to Canada's Maritime provinces and that they would be delighted to call at Baddeck and take lunch with the Bells at Beinn Bhreagh. Baddeck was festooned with bunting for the occasion, and Their Excellencies were met at the wharf by militia bands and a swarm of schoolchildren. The vice-regal couple was satisfyingly glamorous in the eyes both of the village's royalist residents and of Mabel, a good Yankee Democrat. Lord Aberdeen himself was like a tiny chattering wind-up doll, oblivious to everyone around him but his formidable wife. Lady Aberdeen towered over him, chivied her husband through the ceremonies, and radiated grandeur as she sailed through the program. It was raining so hard that the road to Beinn Bhreagh was a sea of mud, but the Aberdeens arrived safely at the Point. There they were served a gargantuan lunch, including local oysters, mutton chops, and duck breast, but Mabel found Lord Aberdeen, she told her mother, "the most fearfully nervous man I ever met. He could not keep quiet a moment, [and] talked incessantly." The diminutive, high-strung governor general took a great fancy to Alec and Arthur McCurdy, with whom he had "a beautiful time . . . talking flying machine and sheep."

Alec had several new enthusiasms in 1897. The first was for glass tubes, which he regarded as a far more efficient way of drinking soup than spoons. (His wife reluctantly told him he could use them when they were alone at Beinn Bhreagh, as long as he made sure they were washed after each use.) Next, Alec returned to the vacuum jacket that he had developed in 1881 after breathing difficulties had caused the death of his infant son, and decided to test it on one of his sheep. Underneath Alec's rather sentimental benevolence, there was a cold, clinical streak when it came to pursuing science: this experiment involved deliberately drowning the sheep, then trying to revive it. The experiment was a success on the second go-round, but not before Alec had lost not only a sheep but also one of his Baddeck workmen, who stomped off the headland, muttering to himself about black magic.

Then Alec became fascinated by the new medical technique of X-rays. A man had arrived at the local doctor's office in great pain because a needle was embedded in his foot. The doctor asked Alec if the telephone probe that he had developed for President Garfield could be used. Alec knew Roentgen's work with X-rays and determined to try the technique. The sufferer, a tall, simple-minded fellow called Donald McDonald, was overwhelmed to discover himself in the famous Dr. Bell's laboratory, with an imposing bearded figure pulling a woolen sock onto his leg and an assistant carefully arranging a camera above the offending foot. Then the party repaired to the Point for dinner. "Donald must have had a lovely time listening to all the doctors' conversations of the pros and cons of cutting him up," Mabel reported to Daisy. Finally, McCurdy produced the photograph, which showed the needle in poor Donald's big toe. It was removed the following day.

X-rays, flying machines, sheep—there was so much, Mabel sighed, to keep her husband at Beinn Bhreagh, and none of these ideas appeared to offer any hope of a quick commercial payoff. But in December, there came news from Washington that had both Mabel and Alec hurrying south. Mabel's father, seventy-five-year-old Gardiner Hubbard, was dying. In recent months, the once-energetic entrepreneur, who had served as regent of the Smithsonian Institution and played a leading role in several organizations, including the Association to Promote the Teaching of Speech to the Deaf, had visibly aged. Suffering from diabetes, he had withdrawn into his family and his collection of Napoleonic prints.

The idea of losing him panicked Mabel. "He is not old as men go nowadays," she wrote tearfully to a friend. "I cannot see why my father should be ill past recovery." Gardiner Hubbard had been a pillar of good sense and equable temperament in his daughter's life, a vivid contrast to her impractical and volatile husband. He was the person she could thank for making sure that she lived in the hearing world and that Alec secured a patent on the telephone. When Gardiner finally passed away on December 11, Mabel was distraught. "I cannot believe that he is really gone, it must be all some dreadful nightmare from which I will awake and go and tell Papa and over which he will laugh in his quiet, humorous, amused way." She refused to accompany Alec on his return to Beinn Bhreagh the following spring, remaining at Twin Oaks to help her mother with Gardiner's papers. "Twin Oaks is horribly sad and lonely without my father," she told a cousin. And when she finally arrived in Cape Breton in the summer, the ache lingered. She wandered around her garden, remembering how he had advised her where to plant hydrangeas, marigolds, and sweet peas and how to construct an arbor. "I never ceased to miss him daily and hourly at Twin Oaks, I miss him scarcely less at Beinn Bhreagh."

Alec's own mother had died the previous year, but he had weathered that loss far better than Mabel was able to absorb her father's death. Eliza Bell had died quietly in her Georgetown home, aged eighty-seven, with her only surviving son by her bedside. Since then, Alec and Mabel had seen even more of Melville Bell, who was now as dependent on Alec as he had once wished Alec to be on him.

Perhaps it was his wife's profound grief that finally persuaded Alec to drag himself away from his Cape Breton aerie. In November 1898, he agreed to abandon sheep and flying machines and to accept the invitation from the emperor of Japan. He, Mabel, their two daughters, and Charles Thompson sailed off across the Pacific. Japan was undergoing a great spurt of modernization under the Meiji dynasty, and the inventor of the telephone was treated with almost religious reverence. "There's nothing like coming to Japan to find out what a big man my husband is," Mabel told her mother. "For his sake the children and I are received with such tremendous attention that I at least am beginning to think myself a very big personage indeed." The emperor himself granted an audience to the famous Dr. Bell, and there were Chamber of Commerce banquets in all the major cities to honor "Bell-san." Soon, as Mabel enthusiastically snapped away with her new Brownie camera, Elsie and Daisy were addressing their father as "Daddy-san."

Although appreciative of the hospitality and respect shown him, Alec was less enthusiastic about the country than was his wife, because he was big in every sense. "He thinks crucifixion couldn't be much worse than having to sit on his knees for two hours at a Japanese dinner that smelt good, but which he couldn't get to his mouth! . . . He had to double up and then bow down over his knees as low as his back would allow and then try to eat off the floor. It didn't help to see all the other Japanese guests just as comfortable as possible and have to chat and laugh while the perspiration was dropping like rain on the floor from pure agony. At present his one idea is to escape from another threatened banquet."

Rickshaw travel was as challenging as chopsticks. Apart from the discomfort of squishing his 245-pound bulk into the flimsy little vehicle, he could not help wondering what would happen if the "little fellow in the shafts" slipped and lost control as he pulled his distinguished passenger down a steep incline. He confided to Charles Thompson, "I cannot get used to a man pulling me about in those terrible things, especially when I see that fellow begin to perspire."

The death of his father-in-law had a second, more onerous impact on Alexander Graham Bell. Without its founder, the National Geographic Society began to wobble. Membership had stalled, its journal was boring, and its debts were growing. Alec had little interest in this dull little Washington club, but both Mabel and her mother were determined that Gardiner Hubbard's creation must survive. Despite his misgivings (and a dread of getting tied down more firmly in Washington), Alec buckled under their pressure and agreed to become the society's second president. In the first months he watched its membership decline to less than a thousand, and glumly acknowledged he had to find a way to give the society a new lease on life. Since the money-losing journal was, in his words, something that "every one put upon his library shelf and few people read," he decided to relaunch it as a much more popular publication. It would be the hook to bring new members into the society. Two of his favorite periodicals were *McClure's* and the *Century* (both had carried articles about him)—he liked their choice of material and their lively writing. The new *National Geographic Magazine* would have as its slogan, "The world and all that is in it." But he wanted to add an additional element: lots of good illustrations and photographs—"pictures of life and action," he explained. "Pictures that tell a story!"

The first requirement for the new magazine was a full-time editor. Alec dug into his own pockets, as he had for *Science* in the 1880s. He announced to the society's board that he would underwrite for the first year a monthly salary of $100 for an editor. Then, urged by Elsie, he persuaded young Gilbert Grosvenor—always known as "Bert"—to quit his teaching job in New Jersey and start work as the National Geographic Society's first full-time employee on April 1, 1899. Once Gilbert started work in the magazine's one-room office at 1330 F Street, Alec was a fount of support. When he heard that the new editor had found the room "a perfect pigsty," with paper all over the floor, dust everywhere, and a spittoon but no desk, Alec sent over his own rolltop desk from Connecticut Avenue. He bombarded Gilbert with ideas for articles; the list included China's influence in the Philippines, polar exploration, Spanish earthquakes, waltzing mice, waves, and auroras. He urged Gilbert to write personal letters to newspapers, suggesting to editors that they review the new publication. Every issue of the magazine included a blank subscription form—the first time such a tactic, dreamed up by Alec, had ever been used.

Bert must have found his proprietor's enthusiasm rather suffocating. But if he did, he kept his cool—particularly since he enjoyed his visits to the Bells' Washington household. Bert had become a fixture there, always happy to make up the numbers at dinner and to escort Elsie to events. One evening, Mabel watched as Elsie flirted with Bert. She held his hand, teased him for being "so sweet," and put her own hand on his knee when she casually leaned over him to speak to her mother. Bert sat silent, smiling but unresponsive. "They are evidently extremely good friends," Mabel reported to Alec. But Gilbert's stolid, throttled-back style was certainly at odds with the passionate courtship Mabel had experienced from Alec. "Is it possible for a young man in love to be so perfectly self-possessed when his lady-love is around and near him?" she mused. "I am sure you weren't." Mabel was not thrilled by Gilbert's prospects, either. She felt that Elsie was "fitted for a more brilliant position than Gilbert can give her," she confided to a friend:

a penniless young man employed by her husband was not her dream son-in-law. "It will never do for her to marry a poor man and have to live in a small house," she wrote, "yet she seems drifting that way."

Mabel had expected that the National Geographic Society would force Alec to spend more time in Washington, where he might mingle with eminent scientists as well as spend more time with his wife and children. He might even, she hoped, pursue money-making inventions in the little laboratory he still maintained behind his father's Georgetown house. But Alec had other ideas: his priorities did not include his daughter's flirtations, the society's day-to-day health, or Mabel's social ambitions. A new note of urgency had crept into his ruminations about flight: "Every bird that flies is a proof of the practicability of mechanical flight by objects heavier than air." He was not alone—even the U.S. War Department was getting interested in man-carrying flying machines. In 1898, when the United States went to war with Spain over Cuba, the American military brass started to put money into Samuel Langley's aerodromes. Alec hungered to keep abreast of his friend and to have one more grand invention in the years ahead. "The more I read of the war news, the more I realize the importance of a flying machine in warfare," he wrote. "Not only for scouting purposes—but for actual offensive work. . . . I am not ambitious to be known as the inventor of a weapon of destruction but I must say that the problem, simply as a problem, fascinates me." He had already decided that, unlike most other American aviation enthusiasts, he was going to explore the potential of kites rather than winged machines. He had established a kite-flying station on Beinn Bhreagh's highest point. Now he ached to head north to the headland's windy slopes, so he could continue his investigation of the airborne characteristics of kites.

Alec didn't have to wait long. Within a year of Bert's arrival, the number of subscribers (always known as "members" within the society) to the *National Geographic Magazine* had almost doubled. Confident that the redesigned magazine was in good hands, Alec returned to Cape Breton. Once back at Beinn Bhreagh, he resumed his correspondence

Alec's ring kite proved surprisingly airworthy.

with Mabel. There were continued mentions of his sheep, and from time to time he left Cape Breton to attend deaf-education conferences and visit schools for the deaf. But the dominant theme was kites. He built kites shaped like stars, large cylindrical kites, big round kites. He built a box kite, resembling a box that had had both ends and a section of the middle removed, that was *huge:* fourteen feet long and ten feet wide—"a monster, a jumbo, a full-fledged white elephant." (His description was correct: he and Ellis had to remove a wall of the laboratory in order to get the kite outside, but they never could get the monster to fly.) His most interesting innovation during this period was the idea of constructing kites out of tetrahedrons: four equilateral triangles joined together to form a pyramid. The cells could be arranged in twos or threes or thousands. Kites constructed out of several of these cells were lighter and less wind-resistant than more conventional kites. The sight of all these weird and wonderful shapes aloft in the Cape Breton skies prompted one boatman, who was rowing a visitor across to Beinn Bhreagh, to describe Mr. Bell's experiments as "the greatest foolishness I have ever seen."

Alec admitted to his wife that building a manned flying machine (he and Langley persisted in calling their machines "aerodromes") "will be an expensive thing to construct, quite apart from the money that must be sunk in abortive experiments." But he tried to persuade Mabel—and himself—that his obsession with kites could well pay off financially. "Suppose a new form of flying toy—or kite—could be put upon the market. . . . There might be money enough made from the toy to build an actual machine." Hoping to convince his wife that these weren't castles in the air, he argued that if only a quarter of America's seventeen million children bought such a kite, he could raise as much as $100,000.

Mabel gently inquired in one letter how he was getting on with drafting his presidential address for the *National Geographic Magazine*. Her question triggered an outburst of frustration. Alec could barely stop his hand shaking as he sat at his desk upstairs in Beinn Bhreagh, and wrote back, "Simply can't do it!!" He considered all his unfinished projects. Ever since they had rushed down to Washington when Gardiner Hubbard was dying, he complained, he had been unable to achieve anything: "I will not give up my work again excepting for matters of *life and death*. I have given up too much of my time already. I am no longer young, and the experiments on which I have been engaged for years should be completed sufficiently for publication. . . . I am sick at heart when I think of the waste of my life and ideas during the last year."

In vain, Mabel argued that Alec's daughters still needed him to accompany them on another trip to Europe to widen their experience and social circle. "They have not had the opportunities that most other girls of their position have. They do suffer from having a deaf Mother, and a Father so absorbed in his work that he won't go out and make friends for them. They have had to do the best they could almost alone." She urged him to rethink his priorities. "Are you willing that Elsie should drift into an engagement with Gilbert without further opportunity of seeing other men? . . . Elsie and Daisy are also works of yours. . . . You started them before you started aeronautics so they ought to be finished first!" Alec was adamant. Convinced he was wasting both time

Charles Thompson holding a
tetrahedral construction.

and inspiration, he begged, "Don't ask me to spend my summer abroad this year. I cannot do it." She, the girls, and "Charles the faithful" were welcome to go abroad; he would not leave Baddeck.

Mabel knew when she was beaten. She could see that Alec would be miserable anywhere but at Beinn Bhreagh because he was so determined to be part of the race to build a flying machine. And she also knew that only on Beinn Bhreagh was Alec capable of relaxing and enjoying his family at the same time as he built bigger and better kites. So she changed tactics and arranged for the whole family to spend the summer of 1900 in Baddeck and enjoy visits from an endless stream of friends and relatives.

Among the visitors was Dr. Simon Newcomb, the irascible director of the American Nautical Almanac with whom Alec often dined in Washington. Nova Scotia–born Newcomb was a brilliant mathematician and astronomer who had never graduated from a university. Since his career path so closely mirrored that of Samuel Pierpont Langley, one might have imagined that the two men were friends. But nothing could be further from the truth: rivalry crackled between them. Mabel was appalled to discover that, somehow, the two Washington scientists were scheduled to stay at Beinn Bhreagh the same week. Alec, however, just chuckled.

It was a Beinn Bhreagh ritual for the family and guests to spend the early evening lounging on the comfortable wicker furniture on the sunporch, gazing at the patterns that breezes made on the lake. During the Langley-Newcomb week, however, the preprandial conversation was anything but amiable. The two men just couldn't stop arguing. "It was very amusing to Mr. Bell," Charles Thompson would recall after Alec's death, "to sit and listen to those two great scientists discuss both sides of a given subject, often very heatedly." Alec often disappeared into the library next door when the two men started squabbling, but Charles would find him still in earshot, "holding his sides with laughter." One day, a fierce argument arose between Langley and Newcomb as to whether a cat would always land on its feet when in free fall, even if it was upside down when it was dropped. Langley insisted it always happened; Newcomb said that was ridiculous. "Here's a good chance for an experiment," said their host. "Let's try it." Charles was dispatched to "round up some cats," as Elsie recalled years later, while she and Daisy rushed around and found mattresses and put them under the porch. Newcomb was ceremoniously presented with a kitten: he held it upside down, squalling and squirming in discomfort, over the balustrade and dropped it. Somehow, it landed on its feet. Newcomb hurrumphed that this was "an accident." Charles was quickly summoned. "I had to recapture [it] a dozen times or more," Charles remembered. "But the cat's landing on its feet every time did not end the discussion, it only added fuel to it."

How Alec loved such exploits at Beinn Bhreagh! Mabel watched her husband laugh as he puffed away on his pipe (he had finally abandoned his cigar habit in 1898, when Mabel pointed out it was costing more than sixty dollars a month). She realized he would never be the intellectual leader within Washington society that her father had been and she had always hoped her husband might become, so she didn't take the girls to Europe. By the end of the summer, Elsie had accepted a proposal of marriage from Bert, and Mabel had welcomed Bert into the family as warmly as her own mother had welcomed Alec. Mabel had accepted that, if Alec would not go off and mingle with men of science in the metropolis, the men of science would have to come to him, in remote Cape Breton.

Chapter 18

FAMILY REMAINS
1900–1906

The new century dawned in a rush of energy and giddy optimism. The English-speaking world had been transformed in the previous century. Britain had built an empire that spanned the globe; the United States had transformed itself from a rugged pioneer society into an industrial power; the new Dominion of Canada reached right across the continent and was starting to assert its independence from Westminster. The momentum seemed unstoppable as the bells rung out at midnight. In London, there were street parties and fireworks in Trafalgar Square. New York City was draped in "colored electric fire, hung from invisible wires," according to the *New York Herald,* and the crowd gave a lusty rendition of "The Star Spangled Banner" beneath a banner that read "Welcome 1900." There was such a fear of riots in San Francisco, reported the *Nation's Business,* that Police Chief Sullivan had planted five policemen at each corner of Market Street to prevent "indiscriminate public kissing on the part of persons who had not been properly presented to each other."

As the *Washington Post* would point out on January 1, 1901, "We have sanitation, surgery, drainage, plumbing, every product of science and accessory of luxury." Cities were becoming pleasanter places to live, as automobiles started to replace horse-drawn vehicles and eliminate the need for street-sweeping and stabling. (In 1896, an enterprising reporter in Rochester, New York, had estimated that the town's 25,000 horses produced an annual pile of manure 175 feet high, covering an acre, which hosted 16 billion flies.) Manned flight seemed just around the corner. What other excitements would the new century bring? Alexander Graham Bell was determined to be part of the ferment of new ideas, new advances. But there were so many damned distractions.

In October 1900, Alec tore himself away from his kites and sheep in Cape Breton to escort his mother-in-law, Gertrude Hubbard, to London. There they joined Mabel, Elsie, and Daisy, who were knee-deep in white satin, point lace, and orange blossoms. Elsie and Gilbert Grosvenor had decided to get married in London because Mabel and her daughters had scheduled a visit there for the fall. The wedding took place at the Eccleston Square Congregational Church, where Elsie, in the opinion of her proud mother, "carried herself splendidly. . . . I could read each slow word as it fell from her lips." It was a grand occasion, as befitted the prestige of the bride's father. Joseph Choate, the newly appointed American ambassador to Great Britain, attended both the ceremony and the reception at the Alexandre Hotel. "Oh it cost a lot!" Mabel confided to her friend Mrs. Kennan. "We began with thirty [invited for breakfast] and ended seating seventy at two dollars and something a head, with champagne extra." To outsiders, the father-of-the-bride's exuberance and pleasure appeared whole-hearted, but the mother-of-the-bride knew his mind was elsewhere. Before he arrived in England, he had made it clear in a letter to Mabel that "I MUST RETURN IMMEDIATELY AFTERWARDS."

Mabel insisted on remaining in England after the wedding. In February 1901, she and Daisy watched the event that would come to symbolize the end of the old century: the state funeral of Queen Victoria, the monarch who had been on the British throne since 1837. Mabel described the event to Elsie: "At the slow, slow tread of the soldiers marching with guns reversed, all stood so still and motionless and presently all heads were bared. Daisy said you could almost have heard a pin drop . . . in all that great multitude. The soldiers passed, the gun carriage rattled on, the King followed grave and anxious-looking with the Kaiser reining in his horse so that the King could ride a neck ahead."

Alec, however, had insisted on returning to the United States before Christmas. On the voyage home, on the S.S. *Ivernia,* he amused himself by developing an apparatus to desalinate seawater through evaporation. Once back in Washington, he got caught up in yet another project: developing a way of tabulating the results of the 1900 U.S. Census. He quickly found himself overwhelmed by business matters— census details, National Geographic Society minutes, the next issue of the *Volta Review* (a magazine he had founded about deaf education), begging letters from would-be inventors and entrepreneurs. "I am mad, whopping mad," he scribbled to Mabel. "My time has all gone." As soon as he could, he extricated himself from his Washington activities and escaped to Beinn Bhreagh.

Alexander Graham Bell was now fifty-four. Each year brought more accolades, and many men his age would have been happy to rest on their laurels. He was president of the National Geographic Society, a regent of the Smithsonian Institution, and a frequent recipient of honorary degrees. (He would collect over a dozen altogether, including from Harvard University in 1896, Edinburgh University in 1906, Oxford University in 1907, and Canada's Queen's University in 1916.) Requests for speeches, articles, and personal appearances by the genius who invented the telephone continued to flood in from all over the world.

But the inventor's enthusiasm for new experiments remained undimmed. He read newspapers avidly, and always remarked when contemporaries such as Thomas Edison and Sam Langley continued to make headlines. Alec dreaded growing old and being left out of the game; he wanted to stay abreast of scientific innovation and play a role in the race to build a manned flying machine. "I am anxious that my WORK shall live after I have gone," he told Mabel—and he was talking about his current kite experiments, not the telephone. "I am not willing *to die* without completing some of the problems I have in hand."

There was an element of the intense egomaniac, accustomed to having his world revolve around him, in such statements. But Alec insisted on pursuing his research interests in Cape Breton, despite its distance from such centers of technological innovation as Boston, Washington, or London, where eager innovators and thrusting capitalists eager to exploit their inventions congregated. Even Alec admitted this was a disadvantage. He decided not to give a prestigious lecture on flying machines in London in 1901 because, as he told Mabel, "I would run a great chance of MAKING A FOOL OF MYSELF on account of my ignorance of what others had done." Despite his talent and ambition, Alexander Graham Bell lacked the killer instinct that would be the hallmark of so many successful twentieth-century innovators. He remained a loner, driven by curiosity and philanthropic motives—the same urge to advance science and improve the world that had motivated him from the earliest days of helping deaf children enter a speaking world.

Mabel's admiration for her husband's energy rarely wavered. "I never saw anybody," she confided to her daughter Daisy in 1906, "who threw his whole body and mind and heart into all that interested him in a hundred different directions." Nevertheless, she was exasperated by the way that Alec's genius bordered on mania, as he recklessly spawned ideas. His lack of intellectual discipline appalled her, even as she created the context in which he could follow his dreams. His brain seemed to work as carelessly as waves beating on the shore, she once remarked, delivering flotsam and jetsam in untidy and unpredictable heaps along the beach. Why couldn't he focus on just one invention that might, God willing, even have a commercial application?

As early as 1899, Mabel had quietly suggested to Alec that he might make faster progress with his kites if he had six or eight thoroughly competent workmen in the laboratory, instead of just Arthur McCurdy plus a local handyman. This system was working well for Thomas Edison, at Menlo Park. "Why not try," she wrote, "to see if by having a lot of men you cannot advance into one month the labors of several."

At this point, Alec brushed the suggestion of a teamwork approach away. "You must remember that this is not a question of Invention but *Discovery*—and discovery is *groping*—a slow laborious systematic groping after knowledge—disheartening, in the number of blind alleys explored." Only when he knew where he was going, he told Mabel, "will be the time when I can follow your suggestion of half a dozen men or more in my laboratory instead of two." So Mabel quietly withdrew, leaving behind in the laboratory a poem she had seen in a newspaper:

> If things seem a little blue,
> Keep on fighting.
> Stay it out and see it through
> Keep on fighting.

The downside of eminence is the obligations it can entail. In December 1903, Alec and Mabel found themselves sitting in the drafty, dimly lit foyer of the Eden Palace Hotel in Genoa, staring at a relentless and chilly rain as it fell on the cobbled streets outside. They had come in haste to Italy on an unexpected mission. The remains of a minor eighteenth-century mineralogist, who was the illegitimate son of a British aristocrat and had never crossed the Atlantic, were buried in Genoa's old British cemetery, on a cliff high over the port. Now the cemetery was scheduled for demolition as the port expanded. It had already been eroded by blasting in the adjacent marble quarry, and coffins had begun to slip and slide down the cliff face, emptying their contents into the Mediterranean as they tumbled. Most of those British bones disappeared unnoticed, but those of the long-dead mineralogist had huge importance for Americans. Seventy-five years earlier, for unknown reasons, this little-known Englishman called James Smithson had bequeathed half a million dollars (the equivalent in today's currency of $50 million) to the United States to fund in the capital an institution "for the increase & diffusion of Knowledge among men." Samuel Langley, secretary of the Smithsonian Institution, had agreed with Alexander Graham Bell, one of the institution's regents, that Smithson's remains should be rescued.

Alec had volunteered for this assignment, but he had not planned to be in Europe, and he developed a severe cold during the transatlantic crossing and the journey by rail down the Italian coast. On his arrival in Genoa, he discovered that Italian authorities were not inclined to release Smithson's skeleton without a lot of formalities. Snuffling and grumbling, his dark eyes heavy-lidded with sleeplessness, he retired to bed in disgust, leaving Mabel (who understood and spoke more Italian than he did) to meet the American consul, Mr. Bishop, and drive up to the gravesite.

Mabel was quite bewitched by the cemetery, despite the endless drizzle and gloomy skies. Bright red lilies and carnations were in bloom. "Everything was of the simplest, and the effect was of thorough good taste," she wrote in her journal. "It was like a beautiful park of cypress trees whose dark foliage was relieved against the white marble paving, the pink and white chapel." The following day she went to the bank and withdrew several thousand lire, to pay for a permit to open the grave, a permit to purchase a coffin, a permit to export the body, and permissions from the national government, the city government, the police, and the health officials. "There seemed to be no end to the red tape necessary before the Government's order to remove the body could be complied with."

The permits and permissions did not do the trick: the Italian authorities erected more roadblocks. Only when Alec rose from his sickbed, announced that he represented President Theodore Roosevelt himself, and distributed several more bribes was the final permission granted.

Then Alec, Mabel, Mr. Bishop, and a troop of gravediggers tramped up to the cemetery. As the gravediggers maneuvered Smithson's decaying coffin out of the damp earth, Mabel, shrouded in a long, dark gabardine raincoat, took out her camera and started snapping. Playing to the camera, Mr. Bishop picked up Smithson's skull and stared into the eye sockets. The disinterment was otherwise conducted as fast as possible, in order to get Smithson's remains, now in a new zinc coffin, aboard the steamer *Princess Irene* and on the high seas, "out of reach of interdict by Italian or French warrant."

When the bones arrived in Washington, D.C., a few weeks later, an elaborate funeral procession was organized to accompany the coffin from the Navy Yard across town to the Smithsonian. But Alec's mind was already elsewhere. The previous night, he had worked in his study in 1331 Connecticut Avenue until 4 a.m. When Charles went to wake him at nine the following morning, so he might participate in the ceremonies, Alec was drowsy and reluctant. "Why am I to get up if I don't want to?" he protested.

"Because you are to be at the Navy Yard at 10 a.m. sharp to escort the remains of James Smithson to their last resting place," Charles told him.

Alec's eyes remained closed. "Nonsense. What are you talking about?" he snapped. "He's been dead for fifty years."

"Can't help it sir," Charles persisted. "He is in Washington now and you brought him here."

After a brief silence, Alec grunted, "What did he come back here to bother me for?"

Smithson's remains had barely arrived at the Smithsonian before the institution's most famous regent was asked to participate in another exercise. This time, it involved the art collection of Charles Lang Freer, a Detroit art lover who had made a fortune manufacturing railroad cars. Freer had a fabulous collection of Asian and American art, including several paintings by his friend James Whistler. In 1904, he offered to leave the Smithsonian his art collection, and half a million dollars to maintain it. Alec found himself, along with Langley and two other regents, heading off to Detroit to look at paintings. None of these elderly scientists had any expertise in art, or any sympathy for the collecting instinct. They yawned their way through five hundred works of art and a thousand pieces of pottery. "I admire and appreciate [Mr. Freer] himself much more than I do the objects of his affection and even adoration," Alec would later write. "He . . . throws his whole heart and soul into the study of things that unfortunately appear to me to be trivial and unimportant in themselves." Luckily, Alec had invited his younger daughter, Daisy, who like her mother had a strong interest in art, to accompany him on the trip. She convinced her father and his colleagues that the Freer collection was not "trivial and unimportant," and the regents of the Smithsonian Institution decided to accept the millionaire's offer. When the collection finally opened in a specially built gallery on the Mall in 1923, it was appraised at $23 million.

Daisy herself was another reason why, during these years, Alec had great difficulty in devoting himself to the workbench. In the spring of 1904, she announced that she wanted to explore her interest in art rather than become a Washington debutante like Elsie. She and a friend took an apartment in New York and began attending drawing lessons in the studio of Gutzon Borglum, an American sculptor who had studied in Paris with Auguste Rodin. Mabel was sympathetic to her daughter's aspirations but shocked by the raffish mores of New York's art world. Alec was no help: it was already lambing season in Cape Breton, and he had fled society life to pursue his sheep-breeding experiments and fly his huge kites at Beinn Bhreagh. Mabel felt ill equipped to venture alone into the art-student scene. After years of taking most of the responsibility for their daughters' upbringing, Mabel felt that headstrong Daisy's bohemian ambitions were more than she could handle. She wrote to Alec outlining Daisy's plans, which she knew would startle him.

Whatever Daisy's spoken reasons for her dive into New York, her unspoken motivation was clear: she wanted to catch her father's notice. And she more than succeeded. Alec was so shocked that he downed his tools in his laboratory. "I do not approve of this at all," he announced. He complained that Daisy had not consulted him about her new direction. He worried that Borglum "is a man I never heard of before, and of whose moral character I know nothing." He insisted on making a special trip to New York to "call upon Mr. Borglum and satisfy myself concerning him." Gutzon Borglum had a way with wealthy men, whom he cultivated assiduously as patrons. He had recently completed a sculpture of the apostles for New York's St. John the Divine Cathedral, and would later be commissioned to carve the monumental sculptures of presidents George Washington, Abraham Lincoln, Theodore Roosevelt, and Thomas Jefferson into Mount Rushmore, in South Dakota. He seems to have worked his magic on Alec. Alec left New York reassured that his daughter was in safe hands: his "darling Daisums" was allowed to stay.

One day, however, Daisy airily mentioned to her mother that she had accompanied her friend Alice Hill on a visit to a gentleman's boardinghouse. Such behavior was simply "not on" for the daughters of America's ruling class. Visions of her daughter getting a reputation as fast and loose gripped Mabel. "I cannot see any need of your breaking the conventions so far in order to have a good time," she scolded Daisy. "I am liberal enough and broadminded enough to go against every convention in the land, if enough is to be gained by doing so, but I cannot see that you are gaining enough to warrant running counter to your mother's and father's feelings of propriety. . . . [I]t is not customary for young ladies to go to a gentleman's boarding house and club and I do not like you doing it." Mabel knew that if Alec heard about this, he would hit the roof. This time, Mabel swallowed hard and dealt with the situation by herself: "I have not said anything to Papa about this or shown him your letters."

Mabel needn't have worried. In November 1904, Daisy's brother-in-law, Gilbert, introduced her to a distinguished botanist, David Grandison Fairchild, who had recently lectured to the National Geographic Society. A tall, thin thirty-four-year-old with sandy hair, a pale mustache, and a pair of dark-rimmed pince-nez, David worked for the U.S. Agriculture Department and had just returned from Baghdad for the Office of Seed and Plant Introduction. He was far more interested in plants than paints. But the affable, adventurous scientist was dazzled by this self-assured young woman eleven years his junior, who drove through the Washington countryside in her own electric automobile "at incredible speed, twelve miles an hour." And for all her bohemian escapades, at heart Daisy was as conventional as her parents. Romance quickly developed, and David and Daisy were married in April 1905. Both Bell girls had met their husbands through the aegis of the National Geographic Society. Mabel had been right to bemoan Alec's detachment from his daughters' prospects: it was thanks to the society founded by her father rather than to the fame of her husband that her daughters had found suitable partners within Washington's status-conscious society.

Alec's mind, as usual, was on other things—specifically, on tetrahedral cells, the four-sided cells in which each side is an equilateral triangle. He could see that tetrahedral construction might have all kinds of applications in engineering projects, owing to its three-dimensional strength. As he wrote in an article for the *National Geographic Magazine* in 1903, the tetrahedral cell combined the qualities of strength and lightness: "Just as we can build houses of all kinds out of bricks, so we can build structures of all sorts out of tetrahedral frames. . . . I have already built a [sheep] house, a framework for a giant wind-break, three or four boats, as well as several forms of kites, out of these elements." Mabel had even persuaded him to complete all the paperwork required to patent his work, which she was convinced had potential for railroad-bridge trusses. ("Alec would never have done anything more than talk about it, I am pretty sure," she wrote later.) On September 20, 1904, Patent No. 770,626, for "aerial vehicle or other structure," was granted to Alexander Graham Bell— the first patent he had received since 1886, when he and his associates patented the graphophone, an early form of phonograph. He was also granted a joint patent with his Baddeck assistant Hector P. McNeil for a tetrahedral connector.

It was the potential of tetrahedral cells in kite construction that intrigued Alec. He had started making tetrahedral cells, each side measuring ten inches, out of aluminum tubing rather than the black spruce of his earlier experiments. Teams of girls from Baddeck were hired to sew red silk onto two of the four triangular faces of each cell. Alec had selected silk because it was light and strong; he preferred the color red because it offered more contrast in the black-and-white photographs that he or his assistant Arthur McCurdy now took of all his experiments.

Kites appealed to Alec because they were inherently safer and more stable than biplanes and presented less risk to their pilots in test flights. He was consumed with the challenge of building the biggest kite he could—a kite that, eventually, might both carry a man and be powered by an engine. He watched his friend Dr. Langley come to terrible grief in December 1903 with his first two attempts at manned biplanes, powered by gasoline engines. The first of Langley's manned aerodromes, launched from the houseboat on the Potomac River, plunged into the river like "a handful of mortar," according to the *New York Sun*. The reporters present, who had always found Langley arrogant and rude, guffawed. The second biplane (nicknamed "Buzzard" by the press) collapsed in midair. Langley's chief engineer and designated pilot, Charles Manly, scrambled to safety, but the newspapers had a field day mocking the director of the Smithsonian Institution. Langley himself never recovered. Alec believed that the ridicule "broke his heart" and contributed to his death from a series of strokes only three years later.

But Alec was swimming against the tide by concentrating on kites rather than on winged planes. Many of Alec's friends thought his obsession with kites was leading him up a blind alley. His former patron Sir William Thomson, who had been so enthusiastic about the telephone at the 1876 Philadelphia Exhibition and who had become Lord Kelvin in 1892, expressed extreme skepticism. Now elderly but still a highly respected physicist, Kelvin met Alec in Halifax, Nova Scotia, in 1897 and suggested to him that he was "wasting his valuable time and resources." In 1901, Alec's protégée Helen Keller visited the Bells in Cape Breton. She had a fine time holding the guide ropes on one of her mentor's constructions and feeling the powerful tug of the airborne contraption. Nonetheless, she privately told a friend, "Mr. Bell has nothing but kites and flying machines on his tongue's end. Poor dear man, how I wish he would stop wearing himself out in this unprofitable way."

"The word 'Kite' unfortunately is suggestive to most minds of a toy," admitted Alec, "just as the telephone at first was thought to be a toy." But he had no time for any doubting Thomas who didn't realize that *his* kites were "enormous flying structures." "Keep on fighting" had now become Alec's motto, and his kites just grew and grew in size. In 1904, he built a kite with a single wingspan composed of dozens of tetrahedral cells, plus a short horizontal tail. "We have everything ready for experiment with the big kite," he told Mabel. "So we are whistling for wind—not too much, and not too little, and not in the wrong direction." That kite took wing and looked so like a great soaring bird that he named it the *Oionos*—the Greek word for a bird of omen.

That summer, Melville Bell came to Beinn Bhreagh for what would be a last visit. He had not slowed down after the death of Eliza Bell in 1897: he had always had an eye for a pretty girl, and he soon remarried. His bride was a fifty-four-year-old Canadian widow named Harriet Shipley—"as sweet and good as she can be," according to Alec, who also noticed his father "stepping out briskly like a young man." But as Melville entered his eighties, his health began to fail. He now tried to arrange his life so that he was always close to his son, in either Cape Breton or Washington. Mabel told her mother that her father-in-law's first question in the morning, "constantly repeated throughout the day, [is] 'Where is Alec? Is he well? When will he come?' And Alec reads to him for hours every evening, or he likes to listen to Alec playing and singing."

In November 1904, Alec wrote to Mabel, who had already returned to Washington to be with her mother:

> I am much troubled by my father's weakness. He is not in
> a condition to stand a long journey to Washington at the
> present time, and I am perplexed to know what to do. He
> seems to be happy here, but hardly says a word. He likes
> to have me sit beside him and looks forward *pathetically* to
> the evening, when he knows I will give him all my time. . . .
> Although claiming to be perfectly well, his physical weakness
> is so great, and his somnolency, as to make me fear that he
> may not be long with us. It is a great comfort to me to be able
> to be with him now, and a great happiness that he wants me
> and craves for my society. . . . Sad changes will come soon
> enough. Let us be happy while we can. . . . I can't write more
> now, my sweet little wife, so good night for the present.

A few days later, Alec organized a private sleeping car in which he and his father could travel by rail undisturbed from Iona, near Grand Narrows on Bras d'Or Lake, all the way to Boston. On a sunny December day, Melville was carefully carried onto a steamer at the Beinn Bhreagh dock, then equally carefully transferred to the sleeping car at the other end. Alec settled his father onto the red plush upholstery of the bunk bed, tucked a couple of woolen blankets over him, and switched on the gleaming brass lamps on the opposite side of the compartment. But the light was too dim for his eyes, so he could not read. He could only stare out at the dark woods and the first snow powdering the hills, and reflect on his father's life that was visibly seeping away.

It must have been a poignant journey for both Melville and Alec. Melville was now so dependent on the son from whom he had demanded dependence all those years ago. While the sleeping car swayed and rumbled over the rails, he held tight to Alec's big-knuckled hand and drifted in and out of sleep. For his part, Alec gazed silently at his father's face—a face that, with its dark eyes, big nose, gruff expression, and bushy white beard, was so similar to his own. Did he recall his resentment of Melville's attempts to mold him in his own image as a proponent of Visible Speech, and his hurt when Melville scorned his experiments with wires and dynamos? Did he remember his hunger for his father's approval, even as he struggled to escape Melville's overbearing presence? Only when Mabel had become the emotional center of Alec's life had Melville lost his hold on his son. But now Alec could recognize what he owed his father. If his father had not insisted that his family leave London, after the deaths of Melly and Edward, he himself would likely not have survived. Without everything his father had taught him about speech and hearing, he might not have invented the telephone. Without the example of his parents' happy marriage and of Melville's obliviousness to his wife's deafness, he himself might have thought twice about marrying a profoundly deaf woman.

When Alexander Melville Bell died, Alec would be the last remaining member of his own family, and the last of the Alexander Bells trained to be a "professor of elocution." The journey to Washington, with a change of trains at Boston, took more than two days. Melville felt entirely cocooned in his son's care. Alec felt entirely alone.

The end came in August 1905. Alec was not present, but Mabel was with Melville in his Georgetown house as he struggled to take his last breaths. He was as theatrical in death, she wrote to her husband, as he had been so often in life. He "held up his hand and then spread it down as he does when a thing is finished, the characteristic elocutionist's gesture, marking the last fluttering breath."

Alec left no written record of his feelings as he contemplated the passing of his father. It was one of the many occasions in his life when he retreated to his laboratory and refused to examine his emotions because the experience might have been too painful. Instead, in Cape Breton, he plunged into his newest project: the construction of a huge kite that consisted of two banks of tetrahedral cells and was so massive

*Mabel (left) encouraged
her husband to patent
the tetrahedral cell
system of construction.*

that it required a howling gale to lift it from the ground.

In mid-November, an Atlantic gale came shrieking up the Bras d'Or, whipping the water into mountainous waves and tearing the sloops in Baddeck harbor from their moorings. This was what Alec had been waiting for. Early the following morning, Mabel watched her husband roar out of the house with excitement and stride toward the newly built kite house with the vigor of a man half his age. But the Baddeck workmen had looked at the raging bay and decided not to row across a mile of choppy water to Beinn Bhreagh that day.

Alec was devastated by what he regarded as his employees' treachery. Perhaps his father's death had affected his emotional stability after all, although he had appeared to survive it without a problem. Mabel watched her husband slowly walk back to the Point, his shoulders slumped in dejection as he muttered "Sheesh, sheesh" between clenched teeth. "The shock was terrible for Father," she told her son-in-law Bert. "He looked gray when he came home, wrote a short note dismissing the staff and closing the laboratory, turned his face to the wall and never spoke again that day or night."

Mabel sat in her study for a few hours, watching the whitecaps on the lake water and remembering all the times in their marriage when Alec had overreacted to setbacks or plunged into despair. Then, as evening approached, she took matters in hand. "While he slept," she admitted to Elsie, "I called for volunteers as I knew he would never ask himself." The following morning, as the gale continued to blow, she rounded up all the men on the estate—Charles Thompson, Arthur McCurdy, the shepherd, and various boys hired to help with the garden. Mabel roused Alec, and the small party made its way down to the kite house. Braving a wind that whipped the caps off their heads and brought tears to their eyes, they lugged the giant kite up to the kite field and maneuvered it into position. The red silk in each tetrahedral cell bellied out, and as a particularly strong gust tore across the headland, the kite rose. While the men hung on to the guide ropes with all their might, the inventor almost took wing himself with glee. "The experiment was so satisfactory," reported Mabel, "that it demonstrates that this form of kite could sustain a much greater weight than he had dared hope."

As usual, success sent Alec's spirits soaring again. He quickly hired back all his Baddeck workmen and sketched for them a new design: a kite composed of 1,300 cells, providing 440 square feet of lifting surface, packed together into a wedge shape. He named his enormous new creation *Frost King*, because Arthur McCurdy's daughter Susie had recently married a man called Jack Frost. Within a month, *Frost King* was not only airborne but also had proved one of its inventor's key hypotheses: that a kite could have sufficient lift to bear the weight of a man. Neil McDermid, brother of the Bells' coachman, forgot to let go of the tether rope he was holding and was wafted thirty feet into the air. The kite's designer was torn between euphoria and fear—euphoria that *Frost King* supported McDermid's weight, and fear that McDermid might let go of the rope and be killed. Alec did, however, have the presence of mind to capture the moment on film. Once McDermid was safely back on terra firma, Alec rushed back to the Point and burst into his wife's study, waving the Kodak camera.

When the enormous Frost King *caught the wind, it appeared to vindicate Alec's faith in kites.*

"Develop, develop," Alec shouted at his startled wife. "We've got him!" Mabel took the film down to her basement darkroom and developed the film "slowly and calmly," as she informed Bert, "in spite of Mr. Bell popping in and out the door, jumping about like a boy."

"Mr. Bell is so happy and so excited underneath a very quiet demeanor—it means so terribly much to him," Mabel confided in Bert Grosvenor. "We've just lived for this moment." She hadn't seen Alec so elated since he had invented the photophone in 1880, while she was pregnant with Daisy; his sense of accomplishment infected the whole household. For over twenty years, until the patent for tetrahedral cells had been issued, Alec had been unsettled and struggling—flinging "seaweed on the sand," as she had put it to Daisy, then retreating "to fling more seaweed in some other wildly separated place." Now he had an invention that he could concentrate on.

At the same time, Mabel had watched him enjoy a new experience: the company of his sons-in-law. "He and Bert went off yesterday . . . in Bert's boat," she told her mother in 1904. "They were in such good spirits and pleased with each other. I can't remember Alec's being so much interested in a spree before for years." The sight of her husband bonding with men young enough to be his sons triggered wistful thoughts in Mabel of the might-have-beens of their marriage—if only those two little baby boys had lived. But Mabel was never one to dwell on lost opportunities. She decided to act on her own suggestion of years past and bring in some competent men to help her husband and fill the space those long-lost sons might have occupied. "You want heirs to your ideas," she told her husband firmly. She began to put together a team that would, within a few years, have a significant impact on the development of flying machines.

The first young man Mabel recruited to help her husband was a Baddeck native: Douglas McCurdy, son of Alec's longtime secretary Arthur McCurdy. Douglas was a dark-haired, long-limbed, bright fellow who had enrolled as an engineering student at the University of Toronto. Since childhood he had been like an adoptive son to the Bells, and now he spent his summers back in Baddeck, often helping on the Beinn Bhreagh estate. In 1906, Mabel wrote to him and asked him to see if he could recruit any of his fellow engineering students to come to Baddeck and help Alec with tetrahedral constructions. Mabel knew that her husband needed the assistance of someone who was on top of the advances in mechanical and electrical engineering in recent years. Alec could operate brilliantly on intuition—his notebook shows a sketch of a trestle bridge constructed of tetrahedral forms, with a train chuffing over it, a scribble of smoke rising from the engine—but he always needed a skilled craftsman for scale drawings, models, and execution. Who would be his new Tom Watson? In particular, which of Douglas's friends was trained in the kind of calculations required for a construction that Alec had had in mind for some years: a tower on Beinn Bhreagh's highest point? Mabel was eager to see the tower built, to demonstrate the potential of tetrahedral construction.

Douglas talked to a Toronto crony, Frederick Baldwin—known to all his friends, on the strength of his baseball abilities, as "Casey,"

*Flags waved when
the tetrahedral
tower was opened in
August 1907.*

after the poem "Casey at the Bat." The grandson of Robert Baldwin, a famous statesman in pre-Confederation Canada, Casey combined the easy congeniality of an upper-crust Torontonian with a sportsman's enthusiasm for outdoor life. He arrived in Baddeck in the summer of 1906 for a two-week visit, met the famous Alexander Graham Bell, and promised to come back in the fall. On his return, he began hammering out of half-inch iron pipe the four-foot tetrahedral cells from which the tower would be assembled. Mabel wrote to Daisy, "Mr. Baldwin begins tomorrow on the construction of a steel tetrahedral tower. He expects to get the whole thing up with just a jackscrew instead of the expensive and complicated machinery usually necessary, so perfectly are the cells fitted one into another."

By November that year, construction of the huge tower, looking like a giant camera tripod with its three seventy-two-foot legs, was well underway. Mabel's spirits rose alongside the construction: as she confided to her mother, "It is such a lovely thing to see my husband at last, before it is too late, working in company with a capable young man who so thoroughly believes in him and his latest invention that he is staking his whole future on it." The following August, Alec organized an elaborate opening ceremony for the tower, with bunting, speeches, and a commemorative article in the *National Geographic Magazine*.

Tetrahedral cell construction had obvious advantages: Buckminster Fuller's patented octet truss uses a similar concept, and some large geodesic domes are reinforced with struts that form tetrahedrons. But Alec's invention received no attention in 1907 because no structural engineers were ready to make the three-day journey to check out this amazing tower in the backwoods of Cape Breton. In any case, Alec's mind was on flying machines, not three-legged towers. He had just learned that aviation science had taken a great leap forward—and he wanted to be part of it.

Chapter 19

BELL'S BOYS
1906–1909

On December 17, 1903, a young bicycle mechanic called Wilbur Wright had made the first-ever flight in a powered biplane, named the *Flyer*, at Kitty Hawk, North Carolina. The plane had been built by Wilbur and his younger brother, Orville, who together ran a bicycle shop in Dayton, Ohio.

Like Alexander Graham Bell, these two young men, sons of a bishop in the Church of the United Brethren, had followed with keen interest the exploits of pioneer aviators such as Otto Lilienthal and Samuel Langley. In 1899, they had written to the Smithsonian Institution requesting everything available on aerodynamics, gliders, and planes, and then they had started building gliders in their bicycle shop. But to fly their gliders, they needed big winds. The U.S. Weather Bureau recommended to them Kitty Hawk, a long sandy beach on the one-hundred-mile Outer Banks of North Carolina, and one of the windiest stretches of ground in the country. It was ideal for the grimly brilliant Wrights, since only a handful of people lived there and the brothers were obsessed with secrecy. At Kitty Hawk, from 1900 onward, the brothers graduated from unmanned gliders to manned gliders to powered, manned biplanes. Through painstakingly methodical research, they managed to develop what had eluded their competitors in the race to the skies: a system of flight control for their creations. While other aviation pioneers, including Langley and Alec Bell, directed all their attention to getting a machine into the air, Orville and Wilbur Wright had worked out how to control the movements—the pitch, roll, and yaw—of a flying machine once it was airborne.

Although the *Flyer's* successful flight took place only six days after Langley's aerodrome made its disastrous plunge into the Potomac, the Wrights were not keen to publicize their achievement. They wanted to patent their innovations before rivals could steal them, so they refused to release photographs of the *Flyer* and they discouraged reporters. When their first attempts to fly an airplane in a cow pasture outside Dayton failed, newspapers lost interest. There were so very many avid aviators claiming supremacy in the air for their colorful assortment of balloons, dirigibles, and heavier-than-air flying machines—who knew which ones had really made a breakthrough?

However, in 1906, Alec had held one of his Wednesday-evening get-togethers of scientists in Washington. Among those present was Octave Chanute, an elderly civil engineer who had published a volume entitled *Progress in Flying Machines* in New York in 1894 and who had been a mentor to the Wrights. With great aplomb, Chanute stuck his thumbs in his vest pockets and stood up to announce that the Wrights had built a flying machine that really flew. According to David Fairchild, who was attending the distinguished gathering in his new status as the host's son-in-law, Alexander Graham Bell asked, "What evidence have we, Professor Chanute, that the Wrights have flown?" Chanute, who had visited the Wright brothers at Kitty Hawk and loved the drama of this moment, paused until he had everybody's attention. Then, with great solemnity, he replied, "I have seen them do it."

Alec could hardly wait to get back to Beinn Bhreagh and share the news with Douglas and Casey. When he discovered an account of the

Wright brothers' flying machine in the French journal *L'Aérophile*, he expressed his exasperation to Mabel: "It seems strange that our enterprising American newspapers have failed to keep track of the experiments in Dayton, Ohio, for the machine is so large that it must be visible over a considerable extent of country. . . . This seems to be due to the desire for secrecy."

Alec was convinced that experiments with an airborne structure made of tetrahedral cells might move the science of flight along, because they might yield valuable information about aerodynamic stability. And he now had an additional assistant. Lieutenant Thomas E. Selfridge, a twenty-five-year-old graduate of West Point, was an aviation enthusiast who had introduced himself to Alec in Washington and wangled an invitation to Cape Breton to see the kites. Mabel was charmed by this tall, soft-spoken Californian who always held the door open for her, helped her carry parcels, and quickly learned to speak directly to her so she could read his lips. At her urging, in September 1907 Alec persuaded the U.S. Army to dispatch Lieutenant Selfridge to Beinn Bhreagh as an observer. As Robert V. Bruce describes in *Bell: Alexander Graham Bell and the Conquest of Solitude*, "the lieutenant and the two engineers introduced a new precision to Bell's meticulously recorded but poetically calibrated measurements. Now wind velocity was read from an anemometer, not the look of the waves; altitude from a clinometer rather than the snapping of a manila rope or the pulling loose of a nailed cleat."

At this stage, the Bell team was still trying to develop a powered kite, and for this they needed a lightweight engine. Alec ordered one from G. H. Curtiss Manufacturing Company in Hammondsport, a small town fifty miles south of Rochester in northern New York State; the company had already provided motors for several makers of dirigible balloons. Alec had met Curtiss in January 1906 at the New York City Auto Show, at which aviation pioneers had also been invited to participate. The Wrights had treated the show with their customary suspicion, telling the organizing committee that they would display only

1907.July.12.

As Alec's obsession with manned flight grew, his Baddeck workforce expanded.

the crankshaft and flywheel of their 1903 engine, because "[i]t would interfere with our plans if we should make public at once a description of our machine and methods." Alec had not only exhibited a large tetrahedral kite but had also taken a great interest in other exhibitors, especially Curtiss's engines. He started referring to the young engine-builder as "the greatest motor expert in the country," and when the Beinn Bhreagh group decided it needed an engine, Curtiss was the obvious source. But first Glenn Curtiss took a long time to fill the order, and then the model he supplied proved unsatisfactory. Alec ordered a larger one and offered Mr. Curtiss twenty-five dollars a day to deliver and demonstrate it in person.

When Glenn Curtiss arrived in Cape Breton, he struck Alec and his young colleagues as a rather glum young man who rarely joined the college-boy pranks enjoyed by the others. But Casey, Douglas, and Tom had to admire the twenty-nine-year-old's technical skills. Curtiss had started a bicycle shop from scratch and then had begun building motorcycles. He was creative and ingenious (tomato cans had served as both carburetor and gas tank on his first makeshift engine), and in

1903 he had won the U.S. national motorcycle championship. A few weeks before he traveled to Canada, he had broken the world record for speed, averaging 136 miles an hour on his motorcycle over the course of a mile. And although he was a fish out of water in rural Cape Breton, Curtiss was intrigued by the Bell setup and by the Grand Old Man's ambitions. He also felt a special bond with his deaf hostess: since his own sister Rutha had lost her hearing to meningitis as a child, he was used to enunciating his words clearly so his lips could be read. He decided to stay. He was the fourth and final member of the team.

Alec now had four enthusiasts at his side. He "enjoys his boys immensely," Mabel told her mother, "and they all seem pleased with him." After dinner each evening, once Mabel had retired to bed, "the four make straight tracks for Alec's study." While their benevolent patron puffed away at his pipe, the young men sketched out plans for flying machines. Alec listened carefully, pointing out different ways of coming at a problem or suggesting refinements to their ideas. After the younger men had gone to bed, Alec would "put on his dressing gown," according to Charles Thompson, and "get his note-book and pen and settle down to work." He would still be working when Charles bought him coffee at 6 a.m., but then he would retire, leaving his note-book open on his table. While their patron slept, the younger men would translate his drawings, sketches, and notes into prototypes and experiments.

This was just the kind of collaboration that Mabel had wanted for Alec: a group who would provide the structure, drive, and skills to keep his experiments and spirits on the rails. The creative atmosphere that now suffused Beinn Bhreagh was, in the words of Seth Shulman, biographer of Glenn Curtiss, a cross between "the rigors of a fast-track engineering laboratory and the playful pleasures of a child's summer camp." That September, after an afternoon of messing about with boats and kites, Alec and his four protégés returned to the Point to dry themselves in front of the fireplace. Mabel poured them all tea, handed around buttered toast, and made her proposal.

She had recently received $20,000 ($415,000 in today's currency) from the sale of a piece of property in Washington that she had inherited from her father, she announced. She wanted to put the group on a formal, legal footing by underwriting their expenses, construction materials, and salaries. Douglas McCurdy and Casey Baldwin would each receive a salary of $1,000 a year. Curtiss, already an established businessman, would receive $5,000 a year while in Cape Breton and half that amount when away. Tom Selfridge would not receive a salary since he was already on full pay as an army officer; Alec would serve as chairman without salary. "My special function, I think," Bell decreed, "is the co-ordination of the whole—the appreciation of the importance of steps of progress—and the encouragement of efforts in what seem to me to be advancing directions."

The Bells and the four young men traveled to Halifax to sign the Aerial Experiment Association agreement on October 1, 1907, in front of a notary and the U.S. consul general. The AEA had a life expectancy of at least one year, and a goal (as Tom Selfridge put it) of getting "into the air." Bell's boys took up residence at the Point. "I have a perfect forest of them, 30 feet of Americans and about fifteen of Canadian," Mabel told her mother. She loved the vigor and gusto of the young men. They had transformed her and Alec's life at Beinn Bhreagh from that of an aging and now childless couple exchanging desultory comments on the day each evening into a lively extended family eagerly planning their next adventure. Her guests were "as nice as they can possibly be and a hundred times less trouble than girls to entertain." And for Alec, this was one of the happiest periods in his life, as he and his new colleagues lived and breathed aviation.

The AEA's first manned craft was a huge tetrahedral kite, composed of 3,400 pyramidal cells, each covered in hand-sewn bright red silk. Alec had named this extraordinary creation, which looked like a giant slab of scarlet honeycomb, the *Cygnet*. "Friday, Dec. 6, 1907," wrote Alec, "is a day ever to be remembered." This was the day that Tom Selfridge crawled into a small space in the center of the *Cygnet*. The

The Aerial Experiment Association: (left to right) Glenn Curtiss, Douglas McCurdy, Alec Bell, Casey Baldwin, and Lieutenant Thomas Selfridge.

kite was then placed on a small boat, and the *Blue Hill* (a local steamer that had a regular run between Baddeck and Grand Rapids) towed the whole assemblage out of Baddeck Bay. Tom "lay on his face on the ladder floor provided," recalled Alec, "covered up with rugs to keep him warm for he was lightly clad in oilskins and long, woollen overstockings without boots." Once the *Blue Hill* had steamed well out into the chilly, wind-whipped waters of Bras d'Or Lake, the men on the *Cygnet*'s launch boat cut the ropes that held the kite down and gave the signal to the *Blue Hill* to go full steam ahead. Alec, Mabel, Douglas, and Casey all held their breath as they watched the *Cygnet* give a slight tremor, then rise gracefully at the end of its towline and fly steadily about 150 feet above the lake, where it hovered for seven minutes until the wind dropped. The spectators cheered like mad, but Selfridge was now in difficulties. His forward vision was obscured by red silk, and the kite's descent was so slow that he did not realize he should cut the towline until he was already in the water. After alighting gently and safely upon the water, Lieut. Selfridge found the kite being towed through the water at the full speed of the steamer *Blue Hill*.

The spectators were terrified. Where was Tom? They couldn't see him in the tangle of ropes and spars, and they knew that nobody could survive long in the icy water. If he had been knocked unconscious, he would drown.

Within minutes, though, Tom appeared—he had managed to swim clear of the fragile contraption. But the *Cygnet* was dragged to pieces. The next day, little girls from Baddeck ran along the lakeshore gathering up the red silk remnants of the tetrahedral cells from which to make dolls' dresses.

Alec's immediate reaction was that the kite's destruction was a "catastrophe." Yet he proved unusually resilient on this occasion— within a few days, Mabel told her mother that he was "jolly again." He felt vindicated: his kite experiment had not resulted in any injury to the pilot, and the *Cygnet* had proved, in Alec's words, that "the tetrahedral system can be utilized in structures intended for aerial locomotion." The kite had proven itself incredibly stable in the gusts of wind over the lake. With a bit more research, he speculated, he might be able to construct a kite so that a man could climb up a rope ladder, start the motor, and fly off to Halifax. (Tom Selfridge, however, challenged Alec's calculations by pointing out that the *Cygnet* was seriously overbuilt: only 41 percent of the tetrahedral cells contributed any lift.)

At the same time, the Aerial Experiment Association was performing just as Mabel had hoped: Alec's young colleagues were keeping him abreast of scientific developments elsewhere. Bell's Boys, as they were widely nicknamed, now took their boss in a different direction. They convinced Alec that it was time to look at issues of control in the air, as well as stability. This meant moving on to manned gliders and, for the sake of milder winter weather, shifting the AEA's seat of operations to Hammondsport, where Glenn Curtiss had his machine shop and where prototype biplanes could take off from the frozen surface of Lake Keuka. By January 1908, the young men were hard at work on a rigid biplane design similar to an early Wright model. Alec had too many commitments in Washington to let him join his boys, so he and Mabel remained in Washington, four hundred miles away. He was kept up to date on their work through a weekly newsletter with the splendidly important title the *AEA Bulletin.*

The AEA members agreed that each of the five of them should design his own aircraft but would be on call to help any of the others. The group's major achievements, however, all came from the younger men. Tom Selfridge was first out of the gate, with *Red Wing,* a biplane named for the red silk that covered its wings—the same red silk that they had used on the Beinn Bhreagh kites. "It is a beautiful machine," Alec, who could keep away no longer, reported to Mabel in March. Its propeller, mounted at the rear, was powered by a forty-horsepower Curtiss engine, and the pilot had two controls: an elevating device at the front and a rudder at the rear, for direction. On March 12, Tom was absent, so Casey was the pilot as Douglas and Glenn pushed it out onto the frozen lake. Its forty-three-foot span of scarlet wings glowed against the blue winter sky and the brilliant white ice. On its first flight, in front of a crowd of excited spectators, the little plane managed to climb for 10 feet and fly 319 feet in a straight line (the rudder's only purpose seems to have been to prevent it from veering to right or left). Since the Wright brothers had done almost nothing to publicize their test flights, this was the first public demonstration of a powered, manned flying machine in North America. Casey had become the first British subject and the seventh human being to fly. Unfortunately, on its next flight, a gust of wind from the wrong direction caught the *Red Wing* and it tipped sideways and crashed into the ground. Casey walked away.

"In scientific experiments," Alec believed, "there are no unsuccessful experiments. Every experiment contains a lesson. If we stop right here, it is the man that is unsuccessful, not the experiment." The lesson from the *Red Wing* was that a biplane needed a better system to ensure horizontal stability and prevent roll than simply a shift of weight by the pilot. "Movable wing tips," suggested Alec, who had noted years earlier that some birds altered the angle of their wing feathers in flight. Casey Baldwin incorporated this suggestion in his biplane, the *White Wing* (white cotton muslin had replaced the red silk). The *White Wing*'s wings had hinged tips, attached by wires to the pilot's seat. Casey controlled these devices, called "ailerons," by leaning in the direction he wished the plane to go. The plane also boasted another innovation: a tricycle undercarriage for takeoff and landing on solid ground. By now Lake Keuka was no longer frozen, so this plane needed to take off from an old abandoned racetrack outside town. The track was partly covered in grass and uncomfortably close to a potato patch and vineyard.

Obsessed as always by safety, Alec was extremely nervous: "[T]he machine is distinctly of *the dangerous kind*. . . . [T]he young men here are prepared to take risks." He longed for Mabel's calm presence at his side ("I feel *awfully lonely*") and confided to her, "[F]or my own part, I should prefer to take my chances in a tetrahedral aerodrome, going more slowly over water. However, I have not the heart to throw a damper over the ambitious attempts of the young men associated with me." Thankfully, there were no mishaps. Between May 18 and 23, when Douglas McCurdy crashed the *White Wing*, all four young men successfully flew it.

The race to fly was now the biggest story in North America. In the fall of 1907, *Scientific American* had offered a trophy for the first public flight over a measured course of one kilometer. Charles Munn, the magazine's publisher, hoped to lure the Wright brothers into a public demonstration of their aircraft—they were being widely challenged to prove they were either "flyers or liars." The Wrights were way ahead of their competitors: they had already developed a way to make their Flyers turn to left and right, as well as fly in straight lines. But they had not yet developed a wheeled plane that took off under its own power. Their machines were launched from a track, propelled forward by a cable and pulleys attached to a half-ton weight that was dropped from a tall derrick. And despite Munn's nudging, they spurned the opportunity to participate in the *Scientific American* race because of their fear of copycats. This left the field clear for every other aviation pioneer on the continent to dream of winning the coveted trophy—an elaborate silver sculpture of an airplane circling the globe, on a pedestal ringed by flying horses.

Bell's Boys were eager to go for it. The AEA's obvious candidate was the fourth biplane it had developed: Glenn Curtiss's *June Bug*—named, Mabel told her mother, "because Alec said its wings resemble those of an insect in flight and it is to fly in June." Both Alec and Mabel were in Hammondsport to watch Curtiss take the little plane through its early test flights, but they did not stay long; as usual, Alec was determined to return to Cape Breton. He agreed, though, that his boys should continue their aviation experiments in Hammondsport, and that the *June Bug* should be entered for the *Scientific American* trophy.

That summer, Hammondsport racetrack had a carnival atmosphere, as flying-machine buffs, schoolboys playing truant, and cranky inventors milled about. The inventors all learned technical tricks from each other. Although the Wright brothers had patented a version of ailerons two years earlier (they called their system "wing warping"), most people saw them for the first time on the *White Wing*. For his part, Alec had been fascinated by a Mr. Myers, "the inventor of an Orthopter (or Ornithoper)—a machine akin to the beating-wing type, but which does not operate by beating wings. . . . 'The boys' call it the 'Wind-grabber.'" Each day, various flying machines were trundled out to wait for either a good wind (for kites) or no wind (for powered biplanes). A pack of reporters did not dare visit the local town, "even to eat, for fear they might be absent at the critical time."

After days of waiting for the right conditions, Curtiss took the *June Bug* up on the Fourth of July 1908. David and Daisy Fairchild were among the crowd that clustered around the aircraft before takeoff. It looked such a fragile contraption, with its vast yellow wings of fabric and bamboo atop three rickety bicycle wheels, and the forty-horsepower motorcycle engine behind the pilot's seat. The top set of wings arched gently down, while the lower set arched upward. A new feature of this machine was that the wing fabric had been soaked in a mixture

At Hammondsport, New York, Daisy Fairchild posed in the June Bug *before it was trundled out of its hangar.*

of paraffin, gasoline, turpentine, and yellow ocher, to reduce air resistance. (This "doping" dramatically increased the little plane's lift and was incorporated into all subsequent machines that employed cloth wings.) A frowning Curtiss was perched on a seat in the middle, with the whirring propeller in front of him, as the biplane trundled down the runway. Liftoff seemed at first impossible, as the tricycle wheels rumbled over the stony ground; then, as Curtiss pulled on the front "lift" control, it happened almost effortlessly.

Against the pale gray of the evening sky, the *June Bug* rose thirty feet above the ground. Curtiss's hair streamed out behind him as the flimsy little plane hurtled forward, straight toward the finish line. Daisy was ecstatic at her first glimpse of "a man flying through the air." She wrote to her father, "As Mr. Curtiss flew over the red flag that marked the finish and way on toward the trees, I don't think any of us quite knew what we were doing. One lady was so absorbed as not to hear a coming train and was struck by the engine and had two ribs broke. . . . We all lost our heads and David shouted and I cried and everyone cheered and clapped and engines tooted."

Victory! The June Bug, *piloted by Glenn Curtiss, wins the* Scientific American *trophy in 1908 for being the first flying machine to fly one kilometer in a public demonstration.*

Curtiss had flown 5,090 feet—just short of a mile, or 1,810 feet more than the *Scientific American's* specified distance of a kilometer. His flight took the coveted trophy. The Wright brothers might have achieved the first manned flight, but Alexander Graham Bell's AEA had won the first *Scientific American* trophy. It was a huge achievement. Bell's Boys did what the Wrights had failed to do: they finally convinced the world that human flight—the dream of centuries—was real.

Bell's Boys were on a roll. Casey Baldwin agreed to return to Baddeck, to help Alec with the next generation of tetrahedral kite. Douglas McCurdy stayed in Hammondsport to work on his flying machine, which, since its wings were covered in a silvery waterproofing material, was already nicknamed the *Silver Dart.* "All the A. E. A. have learned a lot from the Hammondsport machines," Mabel told her mother, and they were all eager to challenge the Wrights' machines in the next competition.

But then tragedy struck. Tom Selfridge had been appointed to the U.S. military board that was conducting trials of the Wrights' planes, with a view to offering them what they had sought for years: a lucrative army contract. On September 17, 1908, Tom volunteered to be Orville Wright's passenger on a required two-man flight at Fort Myer, an army base in Virginia. Everything appeared fine at first—"Wright could be seen, hands on levers, looking straight ahead, and Lieutenant Selfridge to his right, arms folded and as cool as the daring aviator beside him," reported one observer. Suddenly, after four minutes aloft, during which the plane circled the airfield, there was a loud crack, and a piece of the propeller blade flew off. The plane shivered, then nosedived into the parade ground. The engine tore loose and thudded into the earth. Dust exploded like the burst of a mortar shell. Orville Wright was badly injured; Thomas Selfridge never regained consciousness and died later that day. He had the melancholy distinction of becoming the first person to die in an airplane accident.

The Bells were devastated. "I can't get over Tom being taken," Mabel wrote to her mother. "Isn't it heart-breaking?" She had always cared about all four young men in the AEA, but Tom was the one who went out of his way to care for *her,* looking after her "in a hundred little ways I have never been looked after before." The loss of Tom deeply depressed Alec, who had always been so conscious of the risks of flying. "I am still quite stunned by the news from Washington," he confided to Mabel.

There was one more success ahead for the Aerial Experiment Association. Mabel Bell had extended the life of the association's original agreement by a further six months and by an investment of a further ten thousand dollars. Douglas McCurdy completed his *Silver Dart* and shipped it to Beinn Bhreagh. On February 23, 1909, the little biplane was wheeled onto the ice of Baddeck Bay. Baddeck townspeople streamed onto the ice to watch Dr. Bell's latest madness; they stroked the little plane's silver wings with mittened hands, watched Douglas as he checked the machine, and pulled their woolen coats tighter against the winter wind that was tearing down St. George's Channel. Then John McDermid, the Bell's coachman, cranked the

Piloted by Douglas McCurdy, the Silver Dart *soars over Bras d'Or Lake in February 1909.*

propeller and Douglas revved the engine. When it had reached top speed, he put the engine in gear and tore down the lake, followed by a crowd of schoolboys on skates. Once the *Silver Dart* took to the air, the schoolboys were left far behind. In the crisp winter air, the machine twinkled like a brilliant gem in the sunlight as it roared off, at forty miles an hour, then turned and completed a flight of more than half a mile.

A cheer had arisen from the crowd below, but nobody's grin was wider than Alexander Graham Bell's. Knowing that the Wrights had already achieved this distance and speed in the United States, Alec reverted to his British origins as he declared that this was the first heavier-than-air flight by a British subject in the British Empire. Then, flinging his arms wide open with characteristic generosity, he invited everybody present back to Beinn Bhreagh for sandwiches, coffee, and raspberry juice made from berries that grew wild on the headland.

The *Silver Dart* incorporated every piece of knowledge then available about aviation. By now the Bell team had developed the technology required to fly the plane in circles. The next day, Douglas covered more than four miles in a flight around the bay. In this rugged, ravishing corner of Canada, aviation history was made. Over the next month, Douglas would make thirty more flights across Bras d'Or Lake. "Silver Dart Flies for Eleven Minutes," reported the *New York Times* on March 9, 1909. "Dr. Bell's Flying Machine Covers Twelve Miles in Circular Course. Hopes to Beat Record."

But this was, effectively, the end of the Aerial Experiment Association. Tom was dead, and Glenn had returned to his Hammondsport machine shop to set up his own aircraft company. On the evening of March 31, 1909, Alec, Douglas, and Casey assembled in front of the fireplace in the Point's great drawing room to watch the clock tick away the last minutes of the AEA. It was, in Alec's own words, "a pathetic little group" that assembled under the Beinn Bhreagh clock. "Casey moved the final adjournment," Alec wrote later, "Douglas seconded it, and I put it formally to the vote. We hardly received the response 'aye' when the first stroke of midnight began. I do not know how the others felt, but to me it was a really dramatic moment."

The men of the Aerial Experiment Association had achieved so much. They had put five craft aloft, four of them self-powered, heavier-than-air machines, and they had done so in less than eighteen months. Although no valuable patents had emerged from the association's endeavors, its experiments had edged the science of aeronautics forward by refining inventions, such as ailerons, made by others. Two of the AEA's innovations—the steerable tricycle undercarriage and wing doping—were genuine world firsts that all subsequent experimental craft in these early years copied. The AEA's activities, and particularly the speeches and interviews about aviation that Alec had given throughout North America, contributed to the fizz and thrill of the flying-machine craze, and to public confidence that manned flight was possible. One newspaper described the group as "a brilliant coterie." The association's members had formed a "band of comrades on a grand adventure," as Mabel put it. Her initiative had also given her husband a taste of the collective effort that increasingly characterized the field of scientific invention. And the surviving members all benefited from the training it had given them.

Only Casey Baldwin and Alec Bell himself now remained in Baddeck. The various AEA flying machines were not the last inventions that Alec would have a hand in—he was already exploring a new idea. Nevertheless, that chilly Cape Breton night in 1909 was a melancholy moment for the sixty-two-year-old inventor, who would never again enjoy anything as stimulating as Bell's Boys.

Chapter 20

THE AULD CHIEF
1909–1915

By 1909, Alec was almost indistinguishable from illustrations of Santa Claus. An aureole of silvery curls topped his wide forehead, and his bushy white beard and mustache broadened his face into a permanent expression of goodwill. His dark eyes were as intense and observant as ever, but his cheeks usually had a ruddy glow and, despite his wife's efforts to control his diet, his weight hovered around 250 pounds and he boasted a well-padded paunch. ("You can have your dear bacon and Rochefort cheese," Mabel instructed him in 1906, "if you will give up sugar and bread." He didn't.) To complete the Jolly Saint Nick image, he was often followed by a gaggle of small grandchildren as, clad in his distinctive long wool stockings and grey tweed knickerbockers, the sixty-two-year-old inventor strolled over his Cape Breton estate.

The Bells' first grandchild had arrived in 1901. He was named after his great-grandfather Melville Bell, who died four years later. (Alec recorded Melville's birth in his notebook, sandwiched between notes for a speech on heredity and studies for tetrahedral kites.) The Grosvenors would eventually have seven children. Their family included five daughters (Gertrude, born in 1903; Mabel, 1905; Lillian, 1907; Carol, 1911; and Gloria, 1918) and two boys (Melville and Alexander Graham Bell Grosvenor, who was born in 1909 and died—likely of appendicitis—when he was only five). The first Fairchild grandchild, Alexander Graham Bell Fairchild, was born in 1906, and was followed by Barbara in 1909 and Nancy Bell in 1913. The permanent residences of both the Grosvenor and the Fairchild families were in Washington, where Gilbert Grosvenor continued to edit *National Geographic Magazine* and David Fairchild pursued his plant studies in the Department of Agriculture. But each summer, the breezes over Beinn Bhreagh carried the sounds of children's chatter and laughter.

The Grosvenor clan had taken over the the Lodge, the Bells' original house on the property, while the "Fairchildren," as Mabel liked to call them, stayed at the Point, where the whole clan would meet for meals. At dinner, Alec, in a black velvet jacket, would preside at one end of the long, linen-covered table, Mabel, with narrow gold bracelets on her smooth white arms, at the other end. Each grandchild was expected to have a story or some interesting item to relate at the dinner table, the little speech "always beamed at Grandmamma," recalled Daisy, "who had to see the lips to understand." Now in her fifties, Mabel was a benevolent presence at family meals. She wore her thick brown hair, still untouched by gray, in a loose knot at the back of her head. With her rimless spectacles (essential for seeing exactly what people were saying), loose linen gowns, and tweed cloaks, there was an air of the brainy Cambridge intellectual woman about her—lean, stylish, but a little disheveled.

After dinner, a scientific experiment might be performed. Alec would ask Mary, the kitchen maid, to bring an empty milk bottle, then he would insert a lighted piece of paper and balance a shelled hard-boiled egg in the mouth of the bottle. The children stood on tiptoe to watch the egg being sucked into the bottle. Then the inevitable question, "Why?"—which the children were expected to reason out.

Later, Alec might sit at the piano in the big hall and start pounding out Scottish ballads, spirituals, or Gilbert and Sullivan favorites such as "Tit Willow" and "I'm Going to Marry Yum-Yum." On other occasions, Mabel would organize charades or amateur theatricals, in which VIPs from Washington, guests from Baddeck, and the grandchildren were all encouraged to join.

Mabel Bell's devotion to her children and grandchildren was even stronger after the tragic death of her mother in October 1909. Eighty-two-year-old Gertrude Hubbard had been riding along Connecticut Avenue in Washington, in the back of her chauffeur-driven car, when a streetcar crashed into the rear of the vehicle. She died two hours later, in the hospital. Alec and Mabel were in Cape Breton at the time, preparing for the visit of yet another Canadian governor general eager to visit the famous inventor in his Cape Breton lair. Earl Grey had expressed a particular interest in the flying machines that he had often read about in the newspapers. On hearing the ghastly news from Washington, however, the Bells immediately traveled south, leaving Casey, and his new wife, Kathleen Baldwin, to receive His Excellency. After the funeral, Alec returned to Beinn Bhreagh, but Mabel stayed in Washington to help her last surviving sister, Grace, pack up her parents' possessions and move her own family into Twin Oaks.

Losing her mother was a severe blow to Mabel. Gertrude Hubbard had been a champion of and advocate for Mabel throughout her life, a constant source of the kind of encouragement that enabled her daughter to dismiss her deafness as no great handicap. One afternoon she sat down at her mother's desk and gazed out at the marvelous garden, filled with fruit trees and exotic perennials, that Gertrude Hubbard had created. Then she wrote sadly to Alec, "Nobody was as proud of me as she was, nobody else ever made the most of any bright little thing I ever did or said. It is growing more and more strange to have to do without that underlying sense of her love which has lain deep down at the bottom of my heart and comforted me in my moments of deepest depression. 'Mama would see, mama would care,' and it

Alec with three of his ten grand-
children: Lilian (in his arms),
Gertrude (left), and Mabel
Grosvenor.

didn't matter that she really neither knew nor cared, I had but to tell her to get all the sympathy and understanding I wanted." The woman on whom she had relied so heavily—who had kept her in the speaking mainstream, encouraged her to marry Alec, and helped her through personal crises such as the deaths of her sons—was gone. Mabel had needed her mother's support, and now she was much more alone than she had ever felt before.

As usual, Mabel weathered her personal crisis by turning her attention to those who needed her. Alec left all domestic arrangements to her, including the job he hated most: overseeing the finances. After a visit to Cape Breton in 1911, Mabel's cousin Mary Blatchford reported, "Everything here begins with her and works around again the same. . . . Mabel is a determined person, and what she demands is generally accomplished." The estate's considerable running expenses often troubled Mabel, who struggled to balance the books. With around forty employees, including domestic and grounds staff, coachmen, shepherds, laboratory assistants, and boatmen, plus maintenance of roads and an expanding settlement of outbuildings, Beinn Bhreagh swallowed up the Bells' considerable investment income. Mabel was constantly borrowing against future dividends.

However, the Bells still enjoyed a sizable fortune, and Alec was well protected from financial pressures. Mabel never stinted on her husband's laboratory expenses, and in addition gave him an allowance of $10,000 a year (around US$200,000 in today's values) throughout their marriage. But she never abandoned her campaign to make Alec overcome his reluctance to patent and commercialize his ideas. Sometimes she would go over the laboratory accounts and ascertain the approximate cost of a particular invention. According to Charles Thompson, she would present this to her husband and ask him, "Why can't this thing be put on the market?" Alec would reply, "I cannot waste my time doing that sort of thing, if any one wants to, let them go right ahead." If Mabel suggested he was wasting money, he reacted with indignation: "I have used the money to accomplish my purpose. The end justifies the means." Mabel would just sigh. "Our expenses are out of all proportion to our income," she confided to her son-in-law Bert Grosvenor. "But . . . one lives only once and [Mr. Bell] has the right to do what he will with the products of his own intellect."

Mabel's happiest moments in her later years came when a grandchild was left in her care at Beinn Bhreagh. She was always content to hold a sleeping infant or to sit at the kitchen table with a toddler who was playing with candle wax. "All the plans, the hopes and the ambitions that have lain buried in the graves of my own little sons," she wrote, "sprang to life with the coming of each one of my three grandsons." The granddaughters too got lots of attention, although "Gammie" could be a little severe. "She and I sometimes washed the breakfast dishes together because she thought I wasn't getting enough household training," Mabel Grosvenor recalled later. "She took our education seriously. . . . [W]hen she didn't approve of my reading matter, she gave me the most boring book to read about the lives of the painters."

For the older children, however, "Grampie" was the draw. He was such a fund of fascinating experiments and knowledge—how to make

*Mabel with her granddaughter
Nancy Bell Fairchild.*

a whirligig fly through the air, where to find the biggest frogs, what were the best materials from which to construct a model boat. He kept two big jars of hard candy in his study for young visitors. "When we knocked at his door," Melville later recalled, "no matter how busy he was, he'd welcome us with a 'Hoy! Hoy!' and a cheery 'Come in.' He'd always want to know what we'd been doing and would draw us out with questions. Then with great ceremony like a circus master, he'd reach up to the shelf and hold a candy jar out to us. But Lord forbid if we asked for candy. He would always find some excuse to change the subject with a fairy tale or amusing incident." Alec's indulgence to his grandchildren included ordering a Shetland pony for them by mail from Harrods department store in London.

Yet grandchildren were not enough to divert the inventor's attention from his work for long. The urge to make another important contribution to science never left him. Alongside sketches and notes on his abiding preoccupations (saltwater distillation, improved wing design for airplanes, speedboats) there were new ideas in his notebooks—for solar heating, for example, or composting at Beinn Bhreagh. "Above everything," his son-in-law David Fairchild would write later, "he loved to meditate, to think, to dream in the inventor's sense, and to be free from interruptions. As he expressed it, he had 'a yearning for something deeper than the bare facts.'"

Alec would happily play with his grandchildren, as long as he knew he could retreat when he wanted. And while he left management of the estate to his wife, he insisted on complete control of his own work routines. Every year seemed to reinforce his obsessive-compulsive tendencies. Convinced that his physical surroundings induced specific trains of thoughts, he established particular workspaces for particular purposes. During the summers, he had three. In the little office near the laboratory on Beinn Bhreagh, recorded Daisy, "he occupied his mind with problems connected with the experiments; in his study in the house, he thought and worked over his theories of [flight]; while the *Mabel of Beinn Bhreagh* was the place to think of genetics and heredity." By now, the houseboat in which he and friends had once cruised around Bras d'Or Lake had become unseaworthy, so it had been hauled onto the shore below the Point. Alec regularly retreated to its solitude from Saturday afternoon until Monday morning; there he would think, scribble down ideas, smoke his pipe, and subsist on baked beans and hardtack. On hot summer days, he would strip off and wade naked—"in *puris naturalibus*," as he liked to say—into the lake. (On one occasion, he was sitting in his favorite state of nature on the houseboat roof when he realized he was in full view of a tour boat that was steaming up the lake, with a guide pointing out "the summer home of a great inventor.") "Theoretically," noted Daisy, "no one went near him when he was in retreat but actually John [McDermid, the coachman] always went down on Sunday morning to clear up, take provisions, lay a fire in the stove and bring back a report to Mother that Father was alright."

Alec also clung to his nocturnal work habits. If seized by a brilliant inspiration in the middle of the night, he would frequently wake Mabel to tell her about it and ask her to take dictation so he could

Mabel of Beinn Bhreagh: *the beached houseboat where nobody was allowed to disturb Alec.*

work on it the following day. Everything was grist for invention. One day, a beautiful and expensive set of venetian blinds, specially ordered from Italy, arrived at Beinn Bhreagh. But before Mabel had a chance to hang them, Alec had spirited them off to his laboratory to be transformed, with strong glue, into a spiral-shaped propeller. A few months later he couldn't find any insulating material, so he suggested to his laboratory assistant John MacNeil that they rip up some carpets from the Point. Remembering Mrs. Bell's face when she discovered the fate of her venetian blinds, MacNeil secured another source of insulation.

A young Baddeck woman named Catherine Mackenzie became Alec's secretary on Beinn Bhreagh in 1914, and she soon learned that her boss was more than set in his ways:

> In the office [near the laboratory] Mr. Bell had a small wooden table, cherry-stained, made by a local carpenter, with a drawer for his pipes and tobacco, a spare pair of spectacles and his reading glass. . . . The arrangement of this table was the same as that of his study table, and any

change in its accustomed order annoyed him. There was
a receptacle filled with birdshot, something like a wooden
pyramid cut off 2/3rds of the way up, for pens and pencils.
His lead pencils in the nearest corner, pen and red and blue
marking pencils always in the same places and a piece of
wire for a pipe cleaner stuck down the middle. Occasionally
someone tidying up in my absence would fill up this holder
with all the pens and pencils on the table, and when this
happened it never failed to draw a "what are all these things
doing here?" from Mr. Bell, and a rearrangement before he
settled down to work.

Everything on Alec's desk had its place, including the black inkwell,
the red inkwell, the penknife, the tin of Dill's Best Tobacco, the pipe
cleaners, the bowl for spent matches, and the two pipes he always kept
on hand (one to smoke and a cool one as a relay). The blotter hung on
a particular hook, and the only ashtray he would use was one he had
made himself from an old tin box fitted crossways with two metal bars.
He would ritualistically bang out his pipe on the bars, "and he could
bang out his mood in those raps more fluently than in words," according to Catherine Mackenzie.

Bell employees quickly learned when they could and couldn't disturb
the boss. To break his train of thought was a cardinal sin: nobody was
allowed to knock on the door. "If he was busy he paid no attention to
the visitor, but if he had to interrupt his work to say 'come in,' or worse,
if he had to rise to open the door, it annoyed him. The lab employees
and the maids at the house were schooled in this, and luncheon trays
were brought in and fires replenished, and so on, and unless spoken to
Mr. Bell worked on undisturbed. But if there were many interruptions
a big NO was printed on a sheet of paper and pinned on the study door,
and then even Mrs. Bell dared not penetrate."

Punctuation had its own rules in Alec's papers. "Home Notes [a
daily record of activity, separate from the Lab Notes] had to be taken

always with the date, day and place at the top of the page. Mr. Bell was particular about the margin, the use of the colon, and underlining. He . . . had an antipathy to the use of the period after an abbreviation, it was ugly and a waste of space."

Try as he might, however, Alec was far too impulsive a person to stick to the routines he prescribed for himself. He would announce that he was going to have breakfast at ten, dictate his Home Notes from eleven to twelve, drive to the office for one, be home at five. But within days, the schedule would collapse for any number of reasons—because his dictation had gone on too long, or he had become absorbed in the *New York Times,* or he had worked until dawn on a new idea, or there had been a midnight storm and he had been unable to resist rushing out in a swimsuit and rubber boots to enjoy the raging elements. "But he liked to have other people live on schedule," according to Catherine Mackenzie. "He liked to have the regularity of life there to count upon and be independent of when he chose."

Alec's daily routines were as rigid during the winter months in Washington. There, too, he had three different workspaces: his study at 1331 Connecticut Avenue in which to do correspondence; the Volta Bureau in Georgetown, where he could focus on his work with the deaf; and a small hut in the Fairchilds' backyard, overlooking Rock Creek, for more abstract thinking. "A sofa, a fireplace or stove, a box of shot in which he could stick his pens and pencils, a few pipe cleaners and a rug of some kind to throw over him were all the material comforts which Mr. Bell ever required," recalled David Fairchild. "But the most important thing of all was quiet."

When Mabel was absent, Alec's eccentricities went uncurbed. During a heatwave early one year in Washington, a reporter turned up to interview the inventor of the telephone. He was asked to return later, but when the reporter suggested six or seven that evening, the inventor replied, "Oh no, nothing before 11. You could come at 11, or 12, or 1, or 2 at night and we'll be quiet and I'll be very glad to see you." When the reporter returned at midnight, Alec answered the door clad in a long dressing gown fastened around his ample girth with a length of cord. He ushered the reporter into the pantry with the words, "Mrs. Bell is always very careful about what I eat, so we'll just raid the icebox and get something good to eat." After he had gorged on such forbidden delights as ham and buttered toast, he informed the reporter that they would move on to the swimming pool.

The reporter protested that he did not have a swimsuit. "Oh, you don't need a bathing suit," Alec replied. "There is no water in the tank. It is just an experiment that I am making in cooling the house, as it seems to me it is just as sensible to cool a house in summer as it is to heat it in winter." The two men proceeded down the back stairs of the house into a concrete tank, at the bottom of which, according to Daisy Fairchild's account of the incident, "[m]y Father had a rug, a writing table with a student lamp on it, and two comfortable chairs in it." In an era before either refrigerators or air-conditioning had been invented, Alexander Graham Bell had put an icebox in a third-floor bathroom and then installed a fan that blew the cold air from the ice down through a canvas firehose into the swimming pool. "Of course," added Daisy, "he didn't mention the big expense involved."

The demise of the Aerial Experiment Association left a vacuum at Beinn Bhreagh. Casey Baldwin, along with his wife, Kathleen, remained in Baddeck, and he continued to work on flying projects with Alec. At one point, it had seemed possible that the Canadian government was interested in the aviation experiments of the famous Dr. Alexander Graham Bell. He was invited to travel to Ottawa in March 1909 and speak to the Canadian Club at the Chateau Laurier Hotel. Mabel encouraged him to accept the invitation: "The Govt. may take over the whole expense of further experimenting with the *Silver Dart* and the engine and we may evolve something for the *Cygnet*." The following August, Casey and Douglas McCurdy took their latest biplane prototypes (*Baddeck No. 1* and *Baddeck No. 2*) to Ottawa for military trials. But the trials were held on a bumpy cavalry field, both planes were wrecked, and Mabel's hopes died with them. Alec returned to his lonely (and doomed) crusade to prove that kites constructed from tetrahedral cells had as much potential as biplanes for powered, manned flight.

Yet Alec's fertile mind was already going in a new direction. "Why should we not have heavier than water machines as well as lighter than water?" he had asked himself in 1906, after an article about hydrofoils had appeared in *Scientific American*. An American hydrofoil pioneer explained in the article the basic principle of hydrofoils: plates or blades that would act in water the same way that airplane wings act in air. As a craft fitted with hydrofoils gathered speed in water, the submerged hydrofoils would lift the hull until it left the water entirely, and the vessel would then skim along the surface. In 1909, soon after the Cape Breton flights of the *Silver Dart*, Alec and Casey started building hydrofoils, or "hydrodromes" as Alec insisted on calling them. But disaster dogged attempts to launch their early prototypes, which looked like clumsy rafts with large wooden propellers mounted above the stern. As soon as the vessels achieved any speed on Bras d'Or Lake, they collapsed into a mangled mess of wooden supports and jerry-built propellers.

It was time for a change, and Mabel suggested her favorite remedial activity: international travel. In 1910, the Bells and the Baldwins set off for a year-long world tour—a tour that did wonders for Alec's morale. "One of the nice things," Mabel wrote to her daughters from New Zealand, "is the way so many perfect strangers have come to say how glad or honored they are that Papa should come to their country." The inventor of the telephone was acclaimed everywhere. An Australian reporter described Alec as "one of the two most interesting visitors Australia has ever had from the United States" (Mark Twain was the other). In China, the Bells and Baldwins were invited to an official lunch in the new Foreign Office in Beijing, and feted with fried silver fish, shark's fin soup, Yunnan ham, bamboo sprouts with caviar, pheasant, roast duck, and cakes with almond sauce. (The shark's fin soup left Alec cold: he was much more interested to see the popularity of the telephone in China, where the profusion of Chinese characters had rendered the telegraph too unwieldy.)

Alec never really enjoyed meeting strangers; a few years later, Mabel recalled the "determination he showed to escape any [invitations] while we were in Australia." But he did agree to visit the Melbourne home of Marie McBurney, née Eccleston, to whom the young Alexander Graham Bell had proposed in London forty years earlier. Recently widowed and struggling to support herself by giving music lessons, Marie McBurney impressed Mabel as a "bright, plucky, energetic woman." In India, the party sent off their first airmail letters in a tiny, rickety airplane at Allahabad; the plane hopped bravely across the Ganges and landed the mailbag on the opposite bank.

But Alec's mind still raced with new ideas and new approaches to old ideas. All the way across the Pacific, he experimented with composite photography and made schematic representations of his sheep studies. He studied the soaring flight of albatrosses as the party steamed from Australia to New Zealand, pointing out to Casey how their heavy heads and short tails contradicted all the latest theories about manned flight. He noted plants and insects that he saw for the first time (the stinging tree, the lawyer vine). In the South Pacific, he was fascinated by the speed and efficiency of outrigger canoes and immediately began making sketches of catamaran hulls he intended to build in Cape Breton. Everywhere he went, he visited institutions for the deaf and collected government publications about educational methods, infant mortality, natural resources, rainfall. And he continued to ruminate about flying machines.

Alec's interest in such machines finally found an outlet when the party reached Europe. In Monte Carlo, he and Casey were intrigued to see flat-bottomed hydroplane boats roar noisily across the blue water of the Mediterranean. They were even more excited to see, on Italy's Lake Maggiore, similar vessels designed by the Italian engineer Enrico Forlanini speed across the water with much less commotion than the Monte Carlo boats. The enthusiastic Signor Forlanini insisted that the great American inventor and his assistant take a demonstration ride. Alec, who never rode in any of the contraptions produced by his own laboratories, was nervous, but too polite to decline. "Both Father and Casey had rides in the boat over Lake Maggiore at express train speed," Mabel wrote to Elsie. "They described the sensation as most wonderful and delightful. Casey said it was as smooth as flying through the air. The boat . . . at about 45 miles an hour glides above the water . . . being supported on slender hydro-planes which leave hardly any ripple."

The Forlanini craft fired the imaginations of both Alec and Casey. They were convinced they could do better. By the time the Bell party had reached France, the two men had reverted to inventor mode. In Paris, they saw a seventy-five-horsepower motor, built by the Gnome Company, that Alec decided was exactly what he needed for his newest man-carrying kite: the *Cygnet II.* When the Gnome Company refused to sell the engine unless one of their own engineers was retained to install it, Casey enrolled at the Gnome plant as an apprentice mechanic for six weeks. Then he and the motor steamed back to Cape Breton, and Alec began to focus on building the fastest boat in the world. As usual, however, there were many distractions—sheep, the National Geographic Society, the American Association to Promote the Teaching of Speech to the Deaf. He wrote to Mabel, who was in Washington, "There seems to be always something going on where I am. Nothing, perhaps, that would interest other people, but it keeps me busy and interested all the time. First, I have been working very hard at the sheep records, and now I am off [to help grandson Melville with] experiments concerning ice and water, with other ideas crawling around not yet expressed, relating to a reefing displacement for hydro-aerodromes. . . ."

Soon after the Bells' return from their world tour, yet another preoccupation piqued their interest: the Montessori method of education. A new educational philosophy was taking root in North America in the early twentieth century. In retrospect, it is not surprising that, in a world where scientific inquiry was proving as valuable as classical studies, rote learning and strict discipline were losing ground. Reformers who took a more child-centered approach to education challenged the traditional view of a child's mind as an empty vessel to be filled with knowledge by an authoritarian teacher. Among the most impassioned critics of old-fashioned pedagogy were the followers of the Italian physician Dr. Maria Montessori. "Children teach themselves" was the Montessori slogan. Montessori teachers would place carefully prepared materials in front of their young students, then let them explore and discuss those materials at their own pace. A young Chicago teacher called Anne George studied with Dr. Montessori in Rome and then returned to America to establish a primary school. In Tarrytown, New York, George and her friend Roberta Fletcher pioneered Montessori methods. In 1911, *McClure's Magazine* carried an article about their Tarrytown school.

Alexander Graham Bell never had any patience with traditional teaching methods. He had given his opinion of them to a Chicago newspaper reporter in the 1890s: "The system of giving out a certain amount of work which must be carried through in a given space of time, and putting the children into orderly rows of desks and compelling them to absorb just so much intellectual nourishment, whether they are ready for it or not, reminds me of the way they prepare pâté de foie gras in the living geese." When his own daughters were young, Alec was too caught up in his own pursuits to spend much time on their education. However, show-and-tell sessions at Beinn Bhreagh with his grandchildren inspired him to design—and publicize in the *Volta Review* (published by his Volta Institute)—simple experiments for children. "If their curiosity and interest can be aroused," he wrote, "they will speculate for themselves as to the causes of the phenomena observed. This exercise of the mind is just what children need. It develops their reasoning powers and arouses their interest."

Alec probably heard about the Montessori Method in 1912 at one of his Wednesday evenings, at which S. S. McClure, founder of the magazine, was a regular guest. The new educational philosophy had instant appeal for him. Within weeks, his wife and his daughter Daisy had visited the Tarrytown school. They persuaded Roberta Fletcher to open a Montessori school in Washington. That summer, Miss Fletcher joined the Bells in Baddeck to establish the first Montessori school in Canada.

The Montessori school in Baddeck had a serious educational purpose, but for the Bell grandchildren it was just another wonderful activity sponsored by "Gammie and Grampie." The loft of a Beinn Bhreagh warehouse was given a new coat of whitewash and decorated with potted trees and prints of Norway and Egypt acquired by Mabel on her travels. When the school opened on July 18, 1912, there were twelve pupils: five Grosvenors, two Fairchilds, and five small Nova Scotians. "Little ones from three years of age upwards," Mabel noted with delight, "can experiment with all sorts of things to their hearts' content, and at their own sweet will." Alec was fascinated by the school and had regular afternoon conferences with Miss Fletcher to review progress.

When the Bell household returned to Washington in October, Miss Fletcher discovered the momentum bestowed on any project by the Bell name. A larger school was opened in the annex of the Bells' Connecticut Avenue residence, for 23 children, including the Grosvenors and Fairchilds. Press coverage stimulated more inquiries, and in April 1913, over 250 interested people gathered in the Bells' home to discuss a permanent Montessori school in the U.S. capital. The Montessori Educational Association was formed, and Mabel was unanimously elected president. Within months, she had purchased a large old house and garden at 1840 Kalorama Road, overseen the renovations, furnished it with low tables and chairs and a piano, and rented it to the association. "I want my grandchildren," she confided to a friend, "to be workers and if the Montessori System does not make them efficient men and women, looking on work as the noblest thing of all, I shall be disappointed."

The Montessori bandwagon rolled on for a few more years. The founder herself arrived in Washington in December 1913 to give a lecture at the Masonic Temple. "Mr. and Mrs. Alexander Graham Bell entertained at a reception this evening, complimentary to Dr. Maria Montessori, when 400 members of Washington society met the noted Italian educator," reported the *New York Times*. By 1915, the Montessori Educational Association was publishing a magazine and Alec had replaced Mabel as president. But the Montessori movement began to decline in North America as other educational reform movements, particularly the homegrown "learning-by-doing" approach promoted by Chicago's John Dewey, took off. In 1919, the school on Kalorama Road closed its doors.

But Alexander Graham Bell remained constantly in the news. In 1912, he appeared before a congressional committee on foreign affairs, urging the adoption of an international agreement on standard pronunciation. "You have no idea of the absurdities of our speech," the proponent of Visible Speech explained. "For instance, e-n-o-u-g-h spells 'enuff,' whereas p-l-o-u-g-h spells 'plow.' A foreigner might think that c-o-u-g-h spelled 'cow,' but it doesn't." Alec was regularly invited to speak at school graduation ceremonies. "What a glorious thing it is to be young and have a future before you," he told one such class. "But it is also a glorious thing to be old and look back upon the progress of the world during one's own lifetime." His Wednesday-evening get-togethers continued to be among the most prestigious private events in the city. Between January and April 1912, participants discussed subjects ranging from wireless telephony and the temperature of volcanoes to hookworm and the Scott Expedition to Antarctica.

Alec's world renown, combined with Mabel's achievements despite her deafness, made them celebrities within Washington before the word "celebrity" was even fashionable. And they were such an endearing couple. At Washington dinners, if a speaker was out of Mabel's sight, Alec would position himself so that he could silently repeat the speaker's words to his wife. At the theater, fellow members of the audience were intrigued to watch Alec turn his face toward Mabel and mouth the dialog so she could follow the plot. Once silent movies

The Bell clan in 1918. Top row (left to right): Dr. David Fairchild, Mabel Grosvenor, Daisy Bell Fairchild, Elsie Bell Grosvenor, Melville Grosvenor, Gilbert Grosvenor, Gertrude Grosvenor. Seated: Alexander Fairchild, Mabel Bell holding Gloria Grosvenor, Lilian Grosvenor, Carol Grosvenor, Alexander Graham Bell, Nancy Bell Fairchild. At Alec's feet: Barbara Fairchild.

arrived, the Bells became regular moviegoers. This time, Mabel would whisper to Alec what the actors were actually saying, as opposed to the dialog depicted in the subtitles. During a love scene in one Mary Pickford film, both Bells suddenly burst into laughter. Although Pickford, according to the caption, was assuring Douglas Fairbanks that she loved him dearly, Mabel had watched Pickford say, "Get off my foot, stupid, you're hurting me."

It was now years since Alec had had anything to do with telephones. Since 1894, when his patents and the Bell Company's monopoly had expired, telephone development in the United States had surged. By 1902, more than a thousand new independent telephone companies had emerged, usually to service areas ignored by the Bell system. In 1894, there had been 285,000 telephones in the country, primarily serving businesses and news organizations. By January 1911, there were 7.6 million, with fewer than half in the Bell system. Alec's invention had truly changed the world. Telephone wires crisscrossed North America, linking families and friends. (And, often, strangers. Many of the non-business lines were party lines, shared between several households, each of which had a particular ring. This offered irresistible opportunities to eavesdrop on other people's conversations.) Since Alec's voice had traveled the eight miles from Brantford to Paris, Ontario, in the first-ever long-distance telephone call in 1877, more powerful voice amplifiers had steadily increased the distance possible between callers. In May 1911, Damon Runyon, then a reporter with the Hearst daily the *New York American,* had placed a historic 2,066-mile telephone call to a fellow reporter in Denver, Colorado. The next goal that telephone engineers scrambled to achieve was coast-to-coast transmission.

Finally, in 1915, new telephone technology made a transcontinental call possible. Alec was generally reluctant to appear at staged press events—his telephone years were behind him, he insisted. Anyway, he hated dressing up. He always preferred the simplicity of elastic-sided pull-on boots to shoes that demanded lacing, and he refused to wear any necktie that he had to tie himself. But on this occasion, the Grand Old Man of telephony agreed to officiate at the inaugural transcontinental call. Resplendent in a frock coat and white waistcoat, at 4:30 p.m. on January 25, 1915, he took his seat at a long table at the New York City headquarters of AT&T—the American Telephone and Telegraph company—which had become, by now, a mighty monopoly. He was not in the best of moods: the event organizer had suggested a scripted conversation and had given him some draft dialog. Alec threw aside the script and declared that if the telephone didn't work, he wasn't going to pretend. So nobody knew what Alexander Graham Bell, the founding father of telephony, was going to say when, flanked by company directors, he picked up the telephone.

There was no reason for concern. Alec knew exactly what to say to the man specially selected to receive the call at the other end. With a smile, Alec inquired, "Hoy! Hoy! Mr. Watson, are you there? Do you hear me?"

The historic words sped along copper wires via Pittsburgh, Buffalo, Chicago, Omaha, Denver, and Salt Lake City to the Pacific coast—a distance of over 3,000 miles, straddled by more than 130,000 telephone poles. This time, unlike the very first time these same men had been linked by a telephone wire, the call was two-way. In San Francisco, Thomas Watson, Alec's assistant in the early Boston experiments, lifted the telephone in front of a crowd of reporters and replied, "Yes, Mr. Bell, I hear you perfectly. Do you hear me well?"

"Yes, your voice is perfectly distinct," responded Alec. "It is as clear as if you were here in New York instead of being more than three thousand miles away. You remember, Mr. Watson, that evening thirty-eight years ago when we conversed through a telephone on a real line for the first time?"

"Yes, indeed," said Thomas. "That line was two miles long, running from Boston to Cambridge. You were overjoyed at the success of the experiment."

The men chatted for a few minutes. If the nation's press corps hadn't been watching, each might have risen to do a few steps of the old Mohawk war dance. Instead, Alec repeated the words he had spoken on March 10, 1876: "Mr. Watson, come here, I want you." But in 1915, Thomas Watson had to reply, "It would take me a week to get to you this time."

Alec was still chuckling when another voice came on the line: that of his political hero, President Woodrow Wilson. Speaking from the White House, the president congratulated Alexander Graham Bell on "this notable consummation of your long labors."

The event triggered a burst of pride and patriotic hyperbole. The *San Francisco Examiner* stated that the telephone company men grouped around Thomas Watson were as "jubilant as a lot of boys on an unexpected holiday." *Bell Telephone News* declared that the transcontinental telephone system was "the highest achievement of practical science up to today—no other nation has produced anything like it, nor could any other nation." Oblivious to the years during which the young Alec had lived in Scotland, England, and Canada, the company newsletter insisted that the system was "sui generis, it is gigantic—and it is entirely American."

Chapter 21

THE LAST HURRAH
1915–1923

The outbreak of war in Europe in September 1914 had presented Alexander Graham Bell with a painful dilemma. He was torn between loyalty to the British Empire, which was at war and in which he spent his summers, and to the United States, of which he was a citizen and which remained resolutely neutral during the first years of the European conflict.

However, the Bells did not allow the distant hostilities to disturb their routine of long summers in Cape Breton and winters in Washington. One Sunday afternoon in Cape Breton in July 1917, Alec and Mabel decided to forget "our heavy burden of years, the war, and every other trouble," as Mabel put it, and go for a walk. They were quite a sight: seventy-year-old Alec, his white hair long and uncombed, wore a swimsuit and a Japanese raincoat, while Mabel, aged fifty-nine, sported an old-fashioned ankle-length khaki skirt. But a stranger would quickly realize that this eccentric couple was very close. As usual, they held hands (although the observer wouldn't know that this was so that Alec could spell words into Mabel's hand).

As usual, they often turned to face each other, so Mabel could read her husband's lips.

The Bells started off on the main road from the Point toward Baddeck. But Alec had never been satisfied with the beaten track, and he persuaded Mabel that they should plunge into the woods that covered the slope between the road and the shoreline. There was no path, and the Bells both ended up sliding down to the water's edge on their behinds. Undaunted, they decided to remove their clothes and go for a swim, "the first time in a dozen years I have been in," Mabel confided to Daisy, "and the water was simply delicious." After their dip, they got dressed and walked along the rocky shore for a while, then struck back up the cliff. The going was so steep that they had to clamber up on all fours, grabbing at branches to haul themselves forward. By the time they reappeared at the Point, their clothes were filthy but their expressions were gleeful. "It was simply just such a tramp and scramble as Daddysan loves," Mabel wrote, "but we haven't had it together for a long time." This was a couple for whom thirty-nine years of marriage had done nothing to diminish delight in each other's company.

If there were two things that invigorated Alec, they were a bracing ramble and an embryonic invention. And by 1917, he was convinced he was about to present the world with a new and incredible invention—an innovation that would prove, forty years after the development of the telephone, that Alexander Graham Bell's mind was as creative and original as ever.

Six years earlier, when the Bells and Baldwins had completed their round-the-world trip, Alec and Casey had returned to the challenge of designing their own species of hydrofoil vessels. Alec spent hours at the bench in the long laboratory building (situated between the boathouse and the cottage where Casey and his wife, Kathleen, now lived) overseeing the construction of tin models of hydrofoil hulls. Casey spent his time at the boathouse, fitting foils to existing hulls to see which style provided the most "lift" out of the water. Prototypes ranged from

Alec and Mabel Bell in the garden
at Beinn Bhreagh.

wooden foils of various sizes and shapes to a ladder-like arrangement
of angled steel knife blades attached to the bottom of the hull.

For a while, the experiments went well. Casey labored on, even when
Alec was wintering in Washington or preoccupied with Montessori
education or telephone celebrations. In 1911, a craft called the *HD-
1* (for "hydrodrome," as Alec persisted in calling it), powered by a
fifty-horsepower engine, was ready for testing. It consisted of a rec-
tangular twenty-six-foot-long body with short wings for balance and
an aerial propeller at the stern, and it looked like a stubby seaplane.
Alec and Mabel watched from the wharf as the propeller slowly drove
the chunky construction out into the lake then picked up speed so
that the vessel rose in the water on its foils. Once it was obvious that,
this time, the craft was not going to fall apart, Alec hugged his wife
and announced, "public attention will be riveted on Beinn Bhreagh."
HD-1 was followed by *HD-2* (also known as *Jonah,* since at one point
it was rescued from the deep) and *HD-3,* as Baldwin continued to
refine the HD series design and increase the power of the motor. It
looked, as Mabel told her daughter Elsie, "like [a] grasshopper, having

[a] long slim body with four very long bent legs." By 1913, *HD-3* had achieved speeds of fifty miles an hour. But Alec's hydrofoil boats were still fragile constructions, staying afloat literally on a wing and a prayer. *HD-3* came to a spectacular end when it turned turtle in Baddeck Bay during a special demonstration for the prince of Monaco, whose yacht was moored nearby.

Alec's spirits remained on the upswing, reaching toward the state of euphoria that success always induced. His mind raced with possibilities: he began to talk of building a houseboat with sails, and of propulsion by flapping wing devices. "I believe you can design a boat to beat the world record," he told Casey, whom he had appointed manager of the estate and of the laboratory. He did add, perhaps in recognition of his own tendencies, "[o]therwise we may spend all our days in experiments and improvements and never reach the end." He then encouraged Casey to start another project: a hydrofoil sailboat.

But in 1914, the trials ground to a complete halt: Great Britain declared war on Germany.

In common with most of their fellow North Americans, the Bells were shocked by events in Europe—the continent through which they had traveled so often. "We were all pursuing the even tenor of our lives when suddenly the news was flashed across the ocean," Mabel noted. "It came like a knife cutting sharply and forever our present world from that of yesterday, and left us stunned, bewildered, utterly unable to conceive how this dreadful, this impossible event had happened." As a citizen of the United States, Alec felt that he should remain neutral. So he reluctantly abandoned the hydrofoil vessel trials, feeling that, if successful, they could be used in war. However, during the winter of 1915 he couldn't resist mentioning to Josephus Daniels, the U.S. secretary of the Navy, that *HD-3* had potential as a high-speed submarine chaser. But the secretary of the Navy showed no inclination to rush off in midwinter to a remote and snowbound corner of the British Empire to check out a boat that was unable to reverse direction, invented by an elderly eccentric with an Old Testament beard.

The Bells chafed to do something for the war effort without violating American neutrality. They experimented with drying dandelion leaves and other foodstuffs, in case of food shortages. (They tried rhubarb leaves "and found them very good," Mabel told Daisy, "but Cousin Lily thinks they are not quite safe.") Alec designed a candle-powered heater with which soldiers might dry their clothes in the trenches of France. Their most important initiative was to convert the Beinn Bhreagh laboratories into a boat-building plant, to produce life-boats out of local timber. Mabel bemoaned the impact that this had on the headland: "[W]e are hewing down cherished trees, destroying the best-loved beauty spots on Beinn Bhreagh and erecting in their place huge ugly sheds just in the hope that we may be allowed to build ships for the U.S. government." Nevertheless, in July 1917, an order from the British government for fourteen lifeboats was completed ahead of deadline, and Casey Baldwin delivered them to Sydney.

To celebrate, Alec organized a big dinner at the Point for the forty-four people who had worked on the boats. As a table centerpiece, Mabel had placed little silver paper models of the fourteen boats on mirrors edged with moss. That evening, she wrote to Daisy, "[t]hey sat down at 8.30 and got up somewhere about 11.30 and apparently are still at the piano led by Daddysan who is in high feather, looking his best Santa Claus in white waistcoat and velvet coat and very much alive. He stood talking with old time energy and vim [and] made a long speech on his favorite theme, the laboratory boat building and our fastest motor boat in the world."

Alec's spirits were high because, by the time the lifeboat order was completed, he was back on the hydrofoil warpath, in every sense. On April 6, 1917, the United States had finally declared war on Germany. In common with most of his scientist friends, Alexander Graham Bell had deplored American isolationism. The Bells had been appalled by the bloodshed, brutal conditions, and mounting casualties in France and by the German air raids on London. They shared the widespread outrage in the United States when a German torpedo sank the Cunard liner *Lusitania* off the coast of Ireland in May 1915, with the loss of over 1,000 lives, including those of 124 Americans. "We cannot allow American lives to be endangered in a species of warfare without precedent among civilized nations, and which is a distinct return to the most brutal practices of barbarism," the *Baltimore Sun* had announced, in an editorial that captured the reaction of most of the Bell circle. But it took another two years—and the German decision to begin unrestricted submarine warfare in the Atlantic—before President Wilson led the United States into war.

Alec's regular Wednesday-evening soirée was in full swing when news of the president's decision reached 1331 Connecticut Avenue. "The Graham Bell circle broke loose," according to Dr. L. O. Howard, chief of the United States Bureau of Entomology and one of those present. "Few of [us] realized how heavy a burden of shame [we] were bearing. . . . [Various speakers] gave voice to . . . our enormous joy that at last our country had taken her place on the side of right and justice."

Alec felt there was no time to lose if the hydrofoil was going to be part of the war effort, and he immediately headed for Baddeck. But five days after leaving Washington, he discovered he could not reach his destination. He had crossed the Strait of Canso and traveled as far as Grand Narrows, halfway up Bras d'Or Lake. But the thick ice on the lake had barely begun to break up, and the roads were unpassable. Casey Baldwin had managed to get down to Grand Narrows to meet his boss, and he watched the old man fume with frustration for three days. Finally, as Alec explained to Mabel, Casey had "a brilliant idea." They would take the train along the eastern shore of St. Andrew's Channel, as far as the little fishing village of Shunacadie, and then *row* the twelve miles across the lake to Beinn Bhreagh.

It probably sounded like a merry outing when Casey first aired his plan. In fact, the "brilliant idea" was madness. Soon after the two men set off, a wind sprang up and the water grew increasingly choppy. Alec had anticipated taking his turn at the oars, but Casey refused to contemplate the idea of his boss's 250-pound bulk maneuvering around the little boat. "An upset was an uncomfortable thing to contemplate," Alec admitted. If the boat tipped, there was no prospect of rescue, and they would never survive immersion in the freezing water. Large ice pans knocked against the rowboat, and powerful gusts of wind threatened to blow it too far south. By the time the two men reached the Beinn Bhreagh wharf, nearly five hours later, icicles clung to Alec's beard, and Casey's hands were blistered and raw. But it took only a brandy and hot supper at the Baldwins' house to spur Alec on to planning a new vessel, the *HD-4*, which would be capable of speeds of over fifty miles an hour.

Alec was convinced that a hydrofoil boat could be of incalculable benefit in the Allies' war effort. German U-boats were exacting a vicious toll on Allied shipping, and a hydrofoil, he argued, was the perfect vessel for coastal patrol duty. The minimal area of hydrofoil blades that would be submerged would render the vessel relatively safe in mined areas, and the air propellers would transmit no water vibration that might alert enemy shipping. The U.S. Navy Department was sufficiently intrigued to send two 350-horsepower Liberty motors.

The vessel that Casey and Alec designed that summer was huge. The *HD-4* had a wooden hull that was sixty feet long and shaped like a monster cigar. Outrigger floats projected like fins, one each side at the bow. These carried the engines, and slung on a steel tube that passed through the hull were the two main ladderlike sets of steel hydrofoils. In addition, there were two smaller sets of hydrofoils: one at the bow, to prevent any tendency to dive during liftoff, and one at the stern, to act as a rudder.

The war edged closer to North America. There were even rumors of German submarines in Bras d'Or Lake. On Thursday, December 6, 1917, several of the men working on the *HD-4* heard a huge explosion in the distance and speculated that the Germans had launched an attack nearby. In fact, the noise came not from enemy mortars but from the great Halifax explosion, 250 miles away, the largest man-made disaster

HD-4 *roared down St. Andrew's Channel in 1918.*

in history up to that point. A French munitions ship laden with high explosives had collided with a Belgian relief ship as they navigated the narrow channel between Halifax harbor and Bedford Basin; over two thousand people were killed, and almost ten thousand injured. "I sent all our houseboat blankets to Halifax the very day of the disaster," Mabel informed Bert Grosvenor. "We are told there is not a pane of glass left in Halifax." Alec intensified his efforts on the *HD-4.*

In October of the following year, Alec, Mabel, and the laboratory staff watched anxiously as Casey Baldwin took the *HD-4* on its maiden voyage. The *HD-4* did not let them down. Once Casey had got the craft going at around twenty miles an hour, it lifted smoothly out of the water. Within seconds this monster vessel, weighing close to five and a half tons, was skittering across Bras d'Or Lake on about four square feet of submerged steel blades. The noise was deafening. "If you want to hear anything for the rest of the day," declared the editor of *Motor Boat* magazine after joining Casey in the cockpit for a later trial, "stuff some cotton into your ears. . . . At fifteen knots you feel the machine rising bodily out of the water, and once up and clear of the drag she drives ahead with an acceleration that makes you grip your

An anxious Alec watched Mabel speed across the water.

seat to keep from being left behind. The wind on your face is like the pressure of a giant hand and an occasional dash of fine spray stings like birdshot. . . . Baddeck, a mile away, comes at you with the speed of a railway train."

Alexander Graham Bell himself never rode in this, his latest creation, but his wife had always been more physically intrepid than he was. On November 11, 1919, Alec stood on the wharf, shoulders hunched and fists clenched, as Mabel (accompanied by Casey) went for a spin in *HD-4*, which achieved a speed of thirty-five miles an hour. Mabel was euphoric. "It was a most wonderful trip," she wrote in her husband's notebook. "She felt like a rock, so steady, and kept on an even keel. . . . Really the remarkable thing to me was the feeling of perfect confidence she inspired." As Alec read this, did he notice that his wife had unconsciously described her own role in his life? Probably not, since he rarely explored his own feelings. He was simply overwhelmed with relief that nothing had gone wrong.

With the *HD-4* an unqualified design success, Mabel began to fret about patenting its unique features. She was bitter that Glenn Curtiss had recently made a killing from patents granted to the

Aerial Experiment Association. Curtiss had purchased the AEA patents for a few thousand dollars when he formed his own aircraft company, but he had then sold all his patents, of which the AEA patents were by far the most valuable, to the U.S. government for a reported two million dollars. Curtiss, in Mabel's view, had "been scheming to betray [his AEA partners] in such a way as to make it practically impossible for them to reap their fair share of their mutual benefits." She insisted to her son-in-law David Fairchild that Curtiss owed "the equipping of his shop, the training of his men, and the doing of the experimental work . . . to Mr. Bell's money, his brains, and the doings of his other associates." Had her father, Gardiner Hubbard, been alive, he would never have allowed this to happen. Her husband would have achieved his burning ambition to secure another world-famous invention, and the names and fortunes of Casey Baldwin and Douglas McCurdy would have been made. But none of these three could plot a development strategy like the one that her father, a born promoter as well as a skillful patent attorney and knowledgeable entrepreneur, had devised for the telephone forty years earlier. She appealed to David to find somebody to help Alec and Casey now that they had something else to offer the world.

David Fairchild was unable to find such a person. And Alec had made a lifetime habit of ignoring these kinds of commercial issues. He had no interest in plunging back into patent litigation, especially as he watched Orville and Wilbur Wright conduct the same bitter battles as he had faced with the telephone. (The Wrights' ruthless defense of their 1906 flying-machine patent is estimated to have cost them $150,000, or well over two million dollars in today's terms.) He was far too busy lobbying both the British Admiralty and the U.S. government to take an interest in his new baby. In the summer of 1919, Baddeck pulsated with excitement. The village was full of newspaper men and newsreel photographers waiting for the roar of *HD-4*'s engines. Local residents proudly talked of "our Dr. Bell" as though he were Nova

Scotia born and bred. Small boys hung around the Beinn Bhreagh wharf, staring at the strange craft and hoping to be invited to sit in the cockpit. It was a repeat of the scene at Hammondsport in the summer of 1907, when all those magnificent men in their flying machines had assembled at a disused racetrack to vie for the *Scientific American* trophy. On warm July mornings, the big brown boat would streak across the smooth bay, startling the seagulls into wheeling flight. The roar of its engines would cause everybody around the lakeshore to stop in their tracks and stare out at the torpedo-shaped vessel as it hurtled across the water. The white sails of schooners would flutter in its slipstream, yet it created so little wake that the sailboats would barely stir. To demonstrate its potential as a submarine chaser, *HD-4* carried more than three thousand pounds of extra load in dummy torpedoes, dropping them off one at a time to show that the maneuver would not upset its balance.

Best of all, *HD-4* achieved a speed of 70.86 miles per hour on one test run. This made it the fastest boat in the world. U.S. Navy observers reported that "at high speed, in rough water, the boat is superior to any type of high-speed motor boat or sea sled known." Both Washington and London expressed interest in Alec's hydrofoil work.

But once again, interest fizzled out. What was to blame? This time, it was not Cape Breton's distance from political and financial centers that prevented businessmen and government officials from noticing what was going on there, as had been the case with tetrahedral cell construction. Nor was the problem Alec's lifelong reluctance to hustle on behalf of his inventions. With the hydrofoil, it was timing. The American and British governments had just emerged from a devastating conflict that was already dubbed, with ghastly optimism, "the war to end all wars." The market for new weapons of war had collapsed, and so had any public enthusiasm for spending on naval armaments. It was, wrote Mabel, "the death of all our hopes and high endeavors." A "ship with wings," as it was frequently described in newspapers, seemed more like a rich man's toy in peacetime than a serious naval craft.

Alec and Mabel continued to hope that there might be a private market for the world's fastest boat. Casey's brother-in-law, a Toronto corporate lawyer named Colonel Jack Lash, helped Alec and Casey establish a company to protect their interests, and four patents were issued for original features of the hydrofoil. But Bell-Baldwin Hydrodromes Ltd. was a short-lived affair. In 1923, the company ceased operations and the shell of the amazing *HD-4* was left to rot on the rocky shore of Beinn Bhreagh.

Nevertheless, Alexander Graham Bell's hydrofoil was an extraordinary invention, way ahead of its time. It kept its record as the fastest boat in the world for over a decade. And hydrofoil technology would be rediscovered half a century later.

By the end of the First World War, Alec had begun to show his age. There was no more floating in the lake at night with a lit cigar clamped between his teeth. His energies flagged, and Mabel became increasingly protective of the husband to whom she now wrote as "Darlingest Boysie." In the summer of 1918, Helen Keller asked her great benefactor to appear in a motion picture of her life. Helen's literary style was always lush, but on this occasion she went over the top:

> If I had not had so many proofs of your love and forbearance, I
> should not dare even to consider making the request. . . . Dear
> Dr. Bell, it would be such a happiness to have you beside me
> in my picture travels! . . . Even before my teacher came, you
> held out a warm hand to me in the dark! . . . You have always
> shown a father's joy in my successes and a father's tenderness
> when things have not gone right. . . . You have poured the
> sweet waters of language into the deserts where the ear hears
> not, and you have given might to man's thought, so that on the
> audacious wings of sound it pours over land and sea at his bid-
> ding. Will you not let the thousands who know your name and
> have given you their hearts look upon your face and be glad?

Helen's letter "would move a heart of stone and it has touched me deeply," replied Alec. Although he had the "greatest aversion to appear in a moving picture," he promised to travel down to Boston or Washington if required by the filmmakers. To Helen he wrote, "I can only say that anything you want me to do I will do for your sake." But a few days later, Helen received a further missive from Beinn Bhreagh, this time from Mabel. "Helen dear," she wrote, "my husband is no longer a young man. At any time such a journey . . . would be a great tax on his strength, but just now, in midsummer, the risk would be greater than I am willing he should take." Unlike her husband, Mabel was prepared to say no. "He is dreadfully sorry to refuse any request of yours, and so am I, but I know you will realize how much his life and health mean to me, and that I cannot let him take what I know is a real risk to both."

Alec's health had always been a concern. There were the early brushes with tuberculosis, his abnormal sensitivity to light, severe headaches, frequent insomnia, and breathlessness. Ever since his marriage, Alec had always been a hearty eater, and he was seriously overweight. An exasperated Mabel complained to Daisy in 1914, "Discovered depths of iniquity and deceit in my husband undreamt of during 36 years of marriage. He complains of indigestion. We are all in despair, worried to death over chicken curry. Husband absolutely quiet about surreptitious lunch on ham and bad eggs! Guilt detected by egg spots on waistcoat!"

Moreover, Alec had been diagnosed with diabetes in 1915. The incurable and (before the discovery of insulin) untreatable disease was starting to take its toll. Mabel tried even harder during these years to cut extra calories from her husband's diet. She told Charles to take Alec a lighter cereal for his breakfast. But within minutes of Alec's noticing his breakfast tray, Charles was summoned. "I see you've brought me a dish of shavings for my breakfast," an indignant Alec protested. "Are you playing a prank on me?"

Charles replied, "That is not shavings, sir, it is corn flakes, a new breakfast food." Alec would have none of it. "Then flake it away and bring me my oatmeal and brown sugar," he growled. "Whoever heard of oatmeal hurting a Scotchman?"

For all his bravado, however, Alec felt time pressing on him. He focused his waning energies on his hydrofoils and let other Beinn Bhreagh projects fade away. He decided it was time to get rid of his sheep. "The constant labor of the records was becoming burdensome," according to Alec's secretary, Catherine Mackenzie, "and there were other things to which he wanted to give his energies." Alec was delighted to see his coachman, John McDermid, bidding on some ewes and rams at the auction—he had had no idea that McDermid had an interest in animal husbandry. His delight evaporated when he walked into the sheep barn the following morning and was greeted with a chorus of "baas." Mabel had commissioned McDermid to bid for her at the auction, because she did not want to say goodbye to the sheep. Alec exploded with wrath, telling everyone in earshot, "I thought I was *through* with those damn sheep!" But he swallowed his exasperation because he didn't want to hurt his wife's feelings. "I can see him now," Catherine wrote, "patting Mrs. Bell's hand affectionately and dictating a pleased account of it for 'Home Notes.'"

And yet Alec's fertile brain was as active as ever. He continued to stay abreast of events beyond the Beinn Bhreagh workshops. Although he felt no particular loyalty to any political party, he had become an ardent supporter of Woodrow Wilson after hearing him speak at a dinner in Philadelphia in 1912. He followed avidly the debates at the Paris Peace Conference of 1919, and those on the establishment of League of Nations. Catherine Mackenzie would turn to the report in the *New York Times* while her boss "would pull up the rug around his knees, lay down his glasses, light up his pipe and shout, 'Now! Let's have it, in your best oratorical manner, my dear!' . . . I would declaim the whole speech [with] Mr. Bell shouting, 'Hear! Hear!' 'Yaw! Yaw!' and 'Applause' at intervals." The afternoon would wear on, the fire in the big woodstove in the laboratory office would die down, the pile of letters would remain untouched, and Alec would comfort himself that, thanks to his political hero, the world might achieve a permanent peace.

At the same time, when he wasn't redesigning the angle of hydrofoils, he was exploring other ideas in his notebooks. He returned to the work begun in 1901 on the challenge of condensing drinking water from the sea. His earlier invention consisted of wave-powered bellows pumping fog through sea-cooled bottles, but now he had a new idea: a shallow box holding seawater, with a sloping glass lid on which moisture might condense and then trickle into a container. He began to muse about alternative energy sources, since "coal and oil . . . are strictly limited in quantity," he observed. He urged engineering students to start thinking about ethanol, "a beautifully clean and efficient fuel, which can be produced from vegetable matter of almost any kind." He challenged the orthodoxy of the time, which suggested that atmospheric pollution would mean a cooling climate as the sun's warming rays would be blocked by smog. He even used the phrase "greenhouse effect," as he described how, in his view, a layer of smog would prevent the escape of heat from the earth's surface, just as "a white-washed glass house . . . prevents the escape of heat from the interior." No wonder that a reporter who visited him in December 1921 wrote, "The most remarkable thing about Doctor Bell is that he is younger, in mind, than most men of half his age. Mentally, he seems to have discovered a Fountain of Youth, which keeps him perennially alert and vigorous."

In October 1920, Alec announced he wanted to make a final visit to the land of his birth. The Bells left New York on a White Star liner, accompanied by Catherine Mackenzie and their fifteen-year-old granddaughter Mabel Grosvenor. In Edinburgh, Alec spent afternoons in the city library, checking dusty old city registries for his forebears. So much had changed: tramcars and automobiles had replaced the horse-drawn vehicles of his youth, roads had been widened, a new North Bridge now connected the Old and New Towns, and Princes Street was lined with elaborate new department stores. But there were some familiar landmarks, besides the monument to Sir Walter Scott. Alec took his granddaughter to see the house on Charlotte Street where he was raised, and on the way back, they passed a baker who displayed in his window the kind of mutton pies Alec had enjoyed in his childhood. He gleefully bought a couple for his granddaughter and Catherine and, when the two young women politely declined them, he surreptitiously wolfed them himself.

The 1920 trip was not an entirely happy affair. Although Alec had never been keen on ceremony, he was disappointed, Mabel realized, that there was "no one there to greet him." Old friends were dead and gone. The Bell party drove up to Covesea, where Alec and Mabel had been so happy in 1878. But the wind blew a chilly drizzle in from the North Sea, and the stone cottages on the cliff top were in ruins. At a point in his life when he was ready to be swept by nostalgia, he was instead suffused with a sense of his own irrelevance. On his return to Edinburgh, Alec was accosted by a reporter who asked him about his feelings toward the land of his birth. Alec's spirits were unusually low. He said he felt like a stranger in Scotland. His advice to those tempted to revisit their homeland was, "Don't." Even the sight of a man speaking into a public telephone at the train station couldn't cheer him. The Bell party returned to the Hotel Victoria in London, to catch the next steamer across the Atlantic. He told a reporter from the *Daily Mail*, who was hellbent on a story about the Father of Telephony, that "the telephone has been like the pupil who outgrew his teacher; 35 years ago I realized that my invention was growing too fast for me and abandoned my active work in connection with it. Nobody views each new invention added to it with more amazement than I. [They] make me feel my years."

But then things turned around. The city fathers decided to offer the freedom of Edinburgh to its distinguished native son. Alec immediately recovered his usual exuberance. He personally rebooked the party onto a later sailing, and they took the train north again to Edinburgh. First he spoke at his old school, the Royal High School, and announced that the boys could take the afternoon off. Then he stood in front of seventy-two scarlet-robed councilors in the City Chambers, against a backcloth of sumptuous tapestries and splendid oil paintings, and received a silver casket from the Lord Provost. "I have received many honours in the course of my life," he replied, his voice gruff with emotion, "but none that has so touched my heart as this gift of the freedom of my native city, Edinburgh. . . . I can assure you that I shall always look back upon this scene as the most memorable event in my life." Better still, the following day the *Royal Scotsman* declared that not even Sir Walter Scott had brought more honor to the city.

The following winter, Alec and Mabel Bell toured the Caribbean. In the Bahamas, Alec insisted on being lowered into the ocean in a "Photosphere," an underwater observation tube. In Washington, in the spring of 1922, he acquired one of the latest scientific wonders—a radio. Listeners in Alec's day had to use headphones to hear what was then known as "the wireless set," so he speculated about possible improvements that would allow the wireless set to be heard without earpieces.

But Mabel Bell knew that her husband's physical strength was ebbing away, even as his mind continued to ferment with new ideas. They both returned to their beloved Beinn Bhreagh in June and embarked on all the usual summer routines. Alec made note of the lambs born that year to "Mrs. Bell's Multi-Nippled Twin-Bearing Stock." He watched a new hydrofoil craft developed by Casey Baldwin—a naval towing target with a tetrahedral superstructure—being tested on Bras d'Or Lake's placid waters. But he took more naps than in previous years, and he was white and increasingly listless. He often took a carriage from the Point to his laboratory office a mile away rather than walking. Nevertheless, he would rally to read the newspapers or listen to his nine-year-old granddaughter Nancy Bell Fairchild play the piano. And he entertained his grandchildren by wiggling his toes, even though he had lost sensation in them. None of his family realized that the loss of sensation was an indication that he had pernicious anemia.

The end came fast—too fast for anybody to be ready. Elsie Grosvenor had left for a trip to Rio de Janeiro on National Geographic Society business with her husband, Gilbert. On July 30, the Fairchilds arrived for a prearranged visit to Nova Scotia and were shocked to find a severely diminished Alec, barely able to stir from the chaise lounge on the sunporch to greet them. He rallied at the sight of "dear Daisums" and began to talk about his condition with David Fairchild. David softly suggested that his debility might be electrical in character, and asked his father-in-law if life might not have an electrical basis. "Je ne sais pas, Monsieur," shrugged an exhausted Alec. "Je ne sais pas." The local doctor from Baddeck confirmed that the diabetes had now affected Alec's liver and pancreas.

By the next day, Alec lay breathing heavily in his sleeping porch off his second-floor study. He had no energy to turn his head on his pillow to see Beinn Bhreagh, his beautiful mountain, through the shades. On August 1, he opened his eyes and called for Catherine Mackenzie to take dictation. "Don't hurry," Catherine said, and the old man smiled. "I have to," he replied. Then he laboriously dictated: "I want to say that . . . Mrs. Bell and I have both had a very happy life together, and we couldn't have better daughters than Elsie and Daisy or better sons-in-law than Bert and David, and we couldn't have had finer grandchildren." He mentioned his concern that, since Mabel had legal title to his worldly assets, he had made no financial provisions for his daughters.

Speaking was a supreme effort. Alexander Graham Bell paused, closed his eyes, winced, then continued: "We want to stand by Casey as he has stood by us . . . want to look upon Casey and his wife Kathleen as sort of children . . . ."

He could manage no more: he sank into semiconsciousness. Daisy found herself standing at her father's bedside, watching him struggle for breath. "Daddysan is still here," she wrote to Elsie, in Brazil. "But I have the strange feeling . . . that his spirit is struggling to get loose and is only held to his body by slender cords."

That night, Mabel, Daisy, David Fairchild, Casey, and the doctor clustered around Alec's bed. The sleeping porch was filled with the familiar smells they all loved so dearly—rough soap, tobacco, the rich leathery aroma from the bindings of his *Encyclopaedia Brittanica* (9th edition), the earthy tang of his tweeds, the strong tea he had always loved to sip through his glass tubes. Alec was breathing heavily, lapsing in and out of consciousness and occasionally shifting his weight on the unforgiving horsehair mattress. It was too dark to see his face well; David Fairchild held a flashlight on the dying man's face so that Mabel could watch his lips. Now and again, Alec squeezed Mabel's hand, occasionally opening his eyes and smiling at her.

At midnight, Mabel gently disengaged her hand and went to rest on a sofa in the study; David took her place. But in the early hours of August 2, David felt Alec's pulse slowing. He summoned his mother-in-law, and she was instantly by her husband's side, holding his hand. As she watched Alec's life ebb away, she was suddenly suffused with overwhelming grief. "Don't leave me," she implored, as tears sprang to her eyes. Alec was beyond speech but he managed to spell out "no" into her palm. The intervals between each breath came longer and longer. She held his wrist, willing the pulse to continue. Soon it was imperceptible—yet she could feel his fingers still fluttering weakly as he tried to communicate. Then the fingers were quiet, and he was gone.

"At last," wrote David Fairchild to Bert Grosvenor, "we waited in vain for him to breathe and he didn't. . . . Oh Bert, it was a wonderful going and so perfectly in keeping with the simplicity with which the great soul always surrounded himself. The simple sleeping porch, the lamps, his study, the stars shining in, the stillness of the night. . . . There was no confusion, no noise, but his breathing and [then] the sobs of those around him."

Mabel bowed her head as she took in the enormity of her loss. The man around whom her whole life had revolved since she was seventeen was no longer there. The man who had conquered solitude and brought sound out of silence was now silent himself.

Mabel sat quiet for several minutes, still holding Alec's lifeless hand. Dawn was breaking outside, and daylight seeped through the blinds around Alec's bed. Alec's family quietly wiped away their tears as they waited for his widow to speak. Finally, Mabel took a deep breath. With the fortitude she had displayed all her life, she began to think of what had to be done next. She announced that Alec would be buried on his beloved Beinn Bhreagh, and that the family need not go into mourning. Recalling how Alec had always disliked black, she remarked that if once she started wearing it, she could never bear to put it aside. She began to plan the funeral, allocating responsibilities to all the family and employees. "She goes on just as usual," Daisy wrote to her sister, "makes all the motions, laughs and talks but you never forget for a moment that the heart of everything had gone out of life for her forever."

Within the next few days, a grave was blasted out of the solid rock underneath the tetrahedral tower. The coffin was constructed in the Beinn Bhreagh workshop from local pine, and lined with airplane linen. Alec's body, dressed in his familiar gray jacket and knickerbockers and woolen socks, was placed in the coffin, and his grandchildren gathered balsam branches to lay on it. Meanwhile, the Baddeck telephone switchboard was flooded with messages from all over the world. The telegram from Warren G. Harding, twenty-ninth president of the United States, was especially poignant for Mabel, since she knew how much Alec would have been delighted by Harding's tribute. It spoke of more than the telephone.

> The announcement of your eminent husband's death comes
> as a great shock to me. In common with all of his countrymen,
> I have learned to revere him as one of the great benefactors
> of the race, and one of the foremost Americans of all genera-
> tions. He will be mourned and honored by humankind every-
> where as one who served it greatly, untiringly and usefully.

...as Edison, in West Orange, New Jersey, stepped apart ...me of rivalry with his fellow inventor and acknowledged ...between them, as he told the *New York Times* how he had always regarded [Alexander Graham Bell] very highly, especially his extreme modesty."

Friday, August 4, 1922, dawned damp and gray—the kind of cool, Scottish weather that Alec had always loved. The burial was scheduled for sunset. In the late afternoon, a procession wound its way up the mountain. In front were twenty or thirty bareheaded Baddeck men, next came the coffin on a buckboard wagon, and last came four cars with the immediate family. At the top of the mountain, Mabel Bell, dressed in white, joined the small crowd in the hymn "Bringing in the Sheaves." Next, a local Cape Breton girl sang the first verse of Robert Louis Stevenson's "Requiem":

> Under the wide and starry sky,
> Dig the grave and let me lie.
> Glad did I live and gladly die,
> And I laid me down with a will.
> Home is the sailor, home from sea,
> And the hunter home from the hill.

She was followed, at Mabel's request, by the local Presbyterian minister, who read some verses from Longfellow's "Psalm of Life"—the same poem that, thirty years earlier, twelve-year-old Helen Keller had recited at one of Alec's Wednesday evenings:

> Lives of great men all remind us
> We can make our lives sublime,
> And, departing, leave behind us
> Footprints on the sands of time.

At the end of the little ceremony on Beinn Bhreagh, Daisy's son Sandy and Casey Baldwin's son Bobby raised British and American flags to half-mast on temporary poles.

While Mabel bid her husband farewell, flags stood at half-mast on all Bell buildings in North America. Every AT&T telephone on the continent was silent for one minute to honor the passing of Alexander Graham Bell.

In the next few months, Mabel Bell, who was Alec's sole heir, kept herself busy as she set about completing her husband's projects. She insisted that the whole family should remain at Beinn Bhreagh. She made the necessary financial provisions for Casey Baldwin to continue the hydrofoil research. She kept an eye on the sheep. She urged her children to stay connected with the work on behalf of the deaf that both their father and their Grandfather Hubbard had pursued. She began discussions with Bert Grosvenor on finding someone to write a biography of her husband, but discouraged him from commissioning Lytton Strachey, renowned for his recent biography of Queen Victoria, because she wanted a biographer who would write "from his own feeling of fascination." And she wanted the biography to capture Alexander Graham Bell in all his complexity: "You must see that the biography does not picture Father as a perfect man. He was a very clever man and a good man, but he had his faults, just like every other human being. And I loved him for his faults. . . . I want people to realize that he was a human being and no saint. I don't mean that he ever did anything morally wrong. Sometimes he was very inconsiderate of me, but I loved him for it."

But Mabel herself was starting to flag. "I have my hands full of things I want to do for Father and I have not much energy, it takes a long time to get anything done," she confided to Bert Grosvenor. Her world was shrinking in every way. She was having great difficulty in making out what people were saying, because her sight had started to deteriorate. Neither of her daughters or anybody else in the family knew the manual language that Alec had sometimes used with her.

On Beinn Bhreagh's mountaintop, a brass memorial tablet marks Alec's and Mabel's graves.

As the reality of isolation bore down on her, the darkness created by Alec's death became more and more oppressive. What was left for her to live for? She missed her husband dreadfully; she wept copiously, wretchedly, miserably. Her daughters realized that their mother was suffering more than normal grief.

When Mabel returned to Washington in December, she finally saw a physician. He told her that she had a terminal form of pancreatic cancer. She was almost relieved to hear the diagnosis. "Wasn't I clever," she remarked to Daisy, "not to get ill until Daddysan didn't need me anymore?" She died only days later, on January 3, 1923, five months after Alec. Below the headline "Widow of Inventor Is Dead," the *New York Times* reported that "[t]he recent death of her famous husband affected Mrs. Bell's health. Since the day the inventor was buried in Nova Scotia, his widow has been grieving. She never recovered from the shock, it was said today." The Associated Press report, widely republished, was blunter: "Mrs. Bell passed away after a long illness beginning with a breakdown suffered at the time of Dr. Bell's death last August."

Mabel Bell's funeral took place at Twin Oaks, once her parents' home, where her last surviving sister Grace now lived with her husband (and Alec's cousin) Charlie Bell. The following summer, exactly one year after the funeral of Alexander Graham Bell, Mabel's son-in-law Gilbert Grosvenor placed her ashes in the grave beside him on Beinn Bhreagh.

A large rock, with a brass memorial tablet on it, now marks the final resting place of both Bells. The symmetrical inscriptions on the memorial reflect the balance and harmony of their relationship. On the left-hand side, the tablet reads, "Alexander Graham Bell, Inventor Teacher, Born Edinburgh March 3 1847, Died a Citizen of the U.S.A., August 2, 1922." On the right-hand side, it reads, "Mabel Hubbard Bell, His Beloved Wife, Born Cambridge Mass, November 23 1857, Died Washington D.C., January 3 1923."

Nearly a century after the deaths of Alexander Graham Bell and his wife, the headland they loved so dearly remains undisturbed—still owned and visited every year by their many descendants. Today, in front of their grave, Bras d'Or Lake sparkles in summer sunshine or glitters in winter frosts. The Cape Breton bald eagles soar above it, riding the air currents with effortless majesty.

"An inventor," Bell once remarked, "is a man who looks around the world and is not content with things the way they are; he wants to improve what he sees; he wants to benefit the world." Alexander Graham Bell was such a man.

Epilogue

THE LEGACIES OF ALEXANDER GRAHAM BELL

Alexander Graham Bell's greatest invention changed the world forever. If Samuel Morse's telegraph shrank geography, by compressing distance, Bell's telephone liberated the individual because it allowed the transmission of a human voice. Almost all the inventions and technologies of the Industrial Revolution—steam engines, machine tools, mining equipment, microscopes, textile machinery—required special training in their use, and in some cases could be dangerous to their operators. But the telephone was easy and safe to use, and, unlike the telegraph, anybody could use it. It had a profound impact on both personal relationships and the social fabric of society. No wonder Stalin vetoed the idea of a modern telephone system in Russia after the revolution, according to Trotsky's *Life of Stalin*. "It will unmake our work," said the dictator. "No greater instrument for counterrevolution and conspiracy can be imagined."

The effect of Bell's invention on business, commerce, industry, politics, and warfare in the early twentieth century was instant and quickly self-evident. Thanks to the telephone, generals could stay in contact with field officers, an architect could confer with a foreman straddling a girder hundreds of feet above him, politicians could rally supporters during elections. The telephone has been blamed for urban sprawl (buildings no longer had to be within walking distance of each other for ease of communication) and credited with suppressing crime (police officers could summon help). In rural areas, the early party-line system created a virtual community, as farmers gathered by their phones each evening to exchange news and gossip. At Beinn Bhreagh, the grandchildren and great-grandchildren of the telephone's inventor quickly learned to be discreet on the party line that Alec Bell himself had rigged up between various houses and offices on the estate. "Aunt Daisy loved to listen in to our conversations," recalled Dr. Mabel Grosvenor, aged one hundred in 2005 and the Bells' last surviving grandchild. "If the conversation got too personal, we'd hear her voice: 'Don't forget, I'm here!'"

By the early decades of the twentieth century, the telephone was entrenched in popular culture and literature. Songs with titles like "Hello, Central, Give Me Heaven" and "All Alone by the Telephone" became parlor favorites in an era when every home had a piano. Telephone calls crop up in two of the masterpieces of modern literature: James Joyce's *Ulysses,* first published in 1922, and Marcel Proust's *Remembrance of Things Past,* published from 1912 to 1922. The telephone came to loom so large in stage plays that, as writer John Brooks has pointed out, "the telephone onstage became the leading cliché of Broadway." And we can all list movies in which the telephone has a starring role, from the 1950s classics like *Dial M for Murder* and *Pillow Talk* to more recent releases such as *Legally Blonde,* in which cellphones seem to be glued to the ears of the main characters.

The telephone challenged existing technologies (local mail service lost business), existing hierarchies (France was slow to develop a phone system because the state controlled access), and existing employment patterns. The demand for "hello girls," as switchboard operators were known, opened a vast new field of white-collar employment for women.

Genius at work.

By the time Bell himself died, a continental web of telephone wires, suspended from telephone poles for which whole forests had been razed, covered North America and Europe. And that was all before technology took another leap forward, with the introduction of wireless service. Today, a world without instant voice communication to any point on the globe is, for most of us, inconceivable.

Alexander Graham Bell is far more than his telephone. But this one device says so much about his genius—it evolved from his determination to help the deaf; a process of intuition, rather than calculation, prompted the crucial conceptual breakthrough that allowed him to transform electric impulses into sound; and once he had made that breakthrough, he was eager to move on to other things. Bell's gift as an inventor was creative leaps of the imagination rather than the rigorous research that characterized such contemporaries as Thomas Edison. Wealth and fame were important to him only insofar as they freed him up to *think*. For the rest of his life, he was a spectator on the industry he had spawned, as others transformed the telephone into an effective instrument by means of marketing and technological improvements such as automatic switchboards and multiplexing systems that can send many conversations simultaneously over a single pair of wires. His patents and his family's stock in the Bell Telephone Company made him rich, but he showed little interest in subsequent advances in the technology.

But Bell ached to be more than a one-hit wonder. What of his other inventions? What about the patents he received for the photophone, the graphophone, tetrahedral cells, flying machines, and hydrodromes? Did any of these leave a lasting mark on the world, or make him any money?

The patents that Bell and Sumner Tainter received for the graphophone yielded some capital for the Bells, when Thomas Edison purchased the patents in order to work on a device that recorded sound. The Edison phonographs, as early record players were known, owed some of their technology to Bell's work, but it was Edison rather than Bell who got rich from them. The technical breakthroughs achieved by the Aerial Experiment Association were incorporated into subsequent flying machines, but the Wright brothers had done most of the groundbreaking work on early airplanes. The AEA patents did have a small payoff to Alexander Graham Bell, when Glenn Curtiss purchased them for $5,899.49 in cash and $50,000 in Curtiss stock in 1917, but Mabel Bell always felt that Curtiss had cheated them, because Curtiss turned around and sold all his patents, of which the AEA ones were considered the most valuable, to the United States government for a reported $2 million, and kept all the proceeds.

The rest of the patents granted to Alexander Graham Bell and his associates—the patents for photophones, tetrahedrals, and hydrodromes—expired before anyone showed much interest in them. Yet the technologies they described would all be exploited in subsequent years. In 1957, Charles Townes and Arthur Schawlow developed the laser for Bell Laboratories, and in 1977 the Bell Corporation installed under the streets of Chicago a fiber-optic system that carried digital data and voice and video signals on pulses of light. Tetrahedral kites never justified their inventor's faith in them as manned flying machines, but the American architect, engineer, and poet Buckminster Fuller popularized tetrahedral construction in 1957 when he constructed his first geodesic dome. Today, several large structures employ the technology—the retractable roof of Toronto's Rogers Centre stadium, for one, relies on tetrahedral construction. And the research done by Alexander Graham Bell and Casey Baldwin on hydrofoils was the foundation for the development of various naval prototypes after the Second World War. Canada's Defence Research Board built a spectacular vessel and named it the *Bras d'Or* in memory of Bell. These days, commercial hydrofoils operate on inland and coastal waterways around the world, particularly in Japan and Russia.

In 1997, when I first visited the Alexander Graham Bell National Historic Site in Baddeck, Nova Scotia, I was immediately intrigued by both Bell and his wife. As I entered the building, I was confronted first by a huge photo of Dr. Bell, bearded and benevolent, and next by a picture of a long-skirted Mabel Bell, with an expression that was both girlish and maternal. I was hooked. I wanted to understand how he invented the telephone, but I also wanted to find out why he had left Scotland, what had propelled him to marry a woman so different in age and background, whether she had played an important role in his life, and how he came to build a research laboratory in one of North America's most remote corners. When I set out to write about Alexander Graham Bell a couple of years later, I was determined to write more than a biography of an inventor. I wanted to explore his life within the context of his family relationships and of the ferment of the late nineteenth century.

During the years that I have spent in the company of this man, I have discovered a much more neurotic and unconventional individual than I had expected. I have come to realize his incredible dependence on his wife, Mabel Hubbard Bell. Brilliantly intuitive in his research,

Bell could be demanding and insensitive to those he loved. Mabel Hubbard was a warm-hearted, clever woman who had to make all the compromises and sacrifices in their marriage. Yet, as I traced the Bells' relationship through their journals and letters, I saw a wonderful love affair, in which each partner supplied what the other most needed. Mabel gave Alec stability and a safe haven in which he could pursue his obsessions. Alec ensured that Mabel, far from being shoved to the margins of speaking society like most deaf people in her day, had an exhilarating and challenging life. When Alec predeceased his wife, her world collapsed. Maybe cancer was the word on Mabel Bell's death certificate, but those around her all knew that she died of a broken heart. As the Bells' granddaughter Dr. Mabel Grosvenor observes, "[s]he was the center of the family for the rest of us, but everything she did was for him."

Early success was both a blessing and a curse for Alexander Graham Bell. It proved his genius and made him wealthy, but it also reinforced his reluctance to move his ideas from the drawing board to the marketplace. Born in 1847 into a world without electric lights, automobiles, planes, radios, or refrigeration, he died when television, space travel, computers, and iPods were still way beyond most people's imaginations. Today, he is remembered mainly as the avuncular figure captured in photographs taken at the end of his life, when he was revered as, in the endlessly repeated cliché, "the Father of Telephony."

Yet he *was* a genius, even if he was always reluctant to take his inventions from the laboratory bench to the commercial world. Studying his notebooks, I've become convinced that he probably envisioned not only the technologies that we take for granted today but even some of those that still await invention. His imagination, like his spirit, knew no bounds.

Appendix

PATENTS ISSUED BY THE U.S. PATENT OFFICE
TO ALEXANDER GRAHAM BELL AND HIS ASSOCIATES

Patents Issued for Telephone (13)

No. 161,739	Apr. 6, 1875	Transmitter Receiver Electric Telegraph
No. 174,465	Mar. 7, 1876	Telegraphy
No. 178,399	June 6, 1876	Telephonic Telegraphic Receiver
No. 181,553	Aug. 29, 1876	Generating Electric Currents
No. 186,787	Jan. 30, 1877	Electric Telegraphy
No. 201,488	Mar. 19, 1878	Speaking Telephone
No. 213,090	Mar. 11, 1879	Electric Speaking Telephone
No. 220,791	Oct. 21, 1879	Telephone Circuit
No. 228,507	June 8, 1880	Electric Telephone Transmitter
No. 230,168	July 20, 1880	Automatic Short Circuit for Telephone
No. 238,833	Mar. 15, 1881	Electric Call-Bell
No. 241,184	May 10, 1881	Telephonic Receiver
No. 244,426	July 19, 1881	Telephone Circuit

Patents Issued to Bell and/or S. Tainter for Photophone (6)

No. 235,199	Dec. 7, 1880	Apparatus for Signaling and Communicating—Photophone
No. 235,496	Dec. 14, 1880	Photophone Transmitter
No. 235,497	Dec. 14, 1880	Selenium Cells
No. 235,590	Dec. 14, 1880	Selenium Cell
No. 235,616	Dec. 21, 1880	Process of Treating Selenium to Increase Its Electric Conductivity
No. 241,909	May 24, 1881	Photophonic Receiver

Patents Issued to Bell, S. Tainter, and C. Bell for Graphophone (2)

No. 341,212	May 4, 1886	Reproducing Sounds from Phonograph Records
No. 341,213	May 4, 1886	Transmitting and Recording Sounds by Radiant Energy

Patents Issued to Bell and/or Hector McNeil for Tetrahedral (3)

No. 757,012	Apr. 12, 1904	Aerial Vehicle
No. 770,626	Sept. 20, 1904	Aerial Vehicle or Other Structure
No. 856,838	June 11, 1907	Connection Device for the Frames of Aerial Vehicles and Other Structures

Patents Issued to AEA and/or Bell and F. W. Baldwin for Flying Machine (3)

No. 1,010,842	Dec. 5, 1911	Flying Machine
No. 1,011,106	Dec. 5, 1911	Flying Machine
No. 1,050,601	Jan. 4, 1913	Flying Machine

Patents Issued to Bell and F. W. Baldwin and/or S. S. Breese for Hydrodromes (4)

No. 1,410,874	Mar. 28, 1922	Hydrodrome, Hydroaeroplane, and the Like
No. 1,410,875	Mar. 28, 1922	Hydrodrome, Hydroaeroplane, and the Like
No. 1,410,876	Mar. 28, 1922	Hydrodrome, Hydroaeroplane, and the Like
No. 1,410,877	Mar. 28, 1922	Hydrodrome, Hydroaeroplane, and the Like

Sources

A wealth of primary material is a treat for a biographer. For this book, I was blessed, if not almost overwhelmed. The Alexander Graham Bell Family Papers, which cover the years from 1834 to 1974, constitute one of the most extensive family collections I have ever seen. The collection contains correspondence, diaries, journals, laboratory notebooks, patent records, speeches, writings, subject files, genealogical records, printed material, and other papers pertaining to Bell's work in a wide range of scientific and technological fields, including communication, aviation, eugenics, and marine engineering. It also includes material documenting his contributions to the education of the deaf, and correspondence with individuals, ranging from Helen Keller to Theodore Roosevelt, Guglielmo Marconi to Woodrow Wilson. The originals of the papers are housed in Washington, at the Library of Congress; the collection there consists of about 147,700 items in 446 containers plus 8 oversize boxes. There are duplicates of most items at the Alexander Graham Bell National Historic Site (AGBNHS) in Baddeck, Nova Scotia, in 180 three-ring binders of personal letters, which were transcribed in the 1920s in preparation for an earlier biography. In Baddeck, there are also 135 volumes of "Home Notes" (handwritten accounts of life in the Bell household, including visitors received, stray thoughts, and weather reports), 37 volumes of "Lab. Notes" (handwritten notes and hand-drawn diagrams of experiments), 25 volumes of the *Beinn Bhreagh Recorder* (the informal and irregular magazine that Bell enjoyed compiling for family and friends from 1910 onward), 5 volumes of "Dictated Notes," and 7 volumes of the *Aerial Experiment Association Bulletin.*

A substantial proportion of these papers consists of letters exchanged between various family members, and in particular between Bell himself and his wife, Mabel. These were my raw material for this biography. Vivid, intimate, colorful, poignant—there was so much to work with in this extraordinarily rich resource. Every piece of dialog in this book, every sentence in quotation marks, is taken word for word from letters and diaries in the collections. I have not invented anything. I have not footnoted the origin of each quotation, as my text usually makes obvious the source within those dozens of letters, chronologically cataloged. However, I note in this section those letters on which I have drawn that come out of sequence.

The passages in italics that I quote from these letters and journals were underlined in the originals. Throughout his life, Bell used English spelling, while his Boston-raised wife used American spelling: their different styles have been preserved in the transcripts in the Bell archive by earlier authors and by me.

I was far from being the first visitor to this gold mine, and I was fortunate to be able to draw on previous biographies of the Bells. The most authoritative and traditional treatment of Alexander Graham Bell, particularly during his early career, is Robert V. Bruce's *Bell: Alexander Graham Bell and the Conquest of Solitude* (Boston: Little, Brown, 1973). Professor Bruce's book is detailed, insightful, and fascinating, written by a scientist and assuming a degree of scientific knowledge. Professor Bruce taught me a lot. I also learned much from the large-format and more recent *Alexander Graham Bell: The Life and Times of the Man Who Invented the Telephone* (New York: Abrams, 1997), co-authored by Bell's great-grandson Edwin S. Grosvenor and Morgan Wesson, which contains many previously unpublished photographs and an interesting exploration of the telephone industry.

I turned to four books for personal reminiscences of the Bell family. These were *Alexander Graham Bell: The Man Who Contracted Space* (Boston: Grosset and Dunlap, 1928), by Catherine Mackenzie, Bell's secretary in the last eight years of his life; *The World Was My Garden: Travels of a Plant Explorer* (New York: Scribner, 1938), by the Bells' son-in-law David Fairchild; *Chord of Steel* (New York: Doubleday, 1960), by Thomas Costain, a Brantford-born Canadian author who interviewed Brantford residents about their memories of the Bells; and *Genius at Work: Images of Alexander Graham Bell* (Toronto: McClelland and Stewart, 1982), by Dorothy Harley Eber, who recorded many elderly Baddeck locals as they reminisced about their illustrious American neighbors.

The first book to cover the lives of both Alexander Graham Bell and Mabel Hubbard was Helen E. Waite's *Make a Joyful Sound: The Romance of Mabel Hubbard and Alexander Graham Bell* (Philadelphia: MacRae Smith, 1961). Waite was able to talk to the Bells' two daughters, to whom the book is dedicated. Although dated in style, the book is modern in perspective: it endeavors to give Mabel Bell credit for her contributions to her husband's achievements. Similarly, Lilias M. Toward's *Mabel Bell: Alexander's Silent Partner* (Toronto: Methuen, 1984; published in large print, Wreck Coves, N.S.: Breton Books, 1996) traces Mabel Bell's role in the Bell story.

I also looked at *Sounds Out of Silence: A Life of Alexander Graham Bell* (Edinburgh: Mainstream Publishing, 1997), by James Mackay, which contains many new details of Bell's childhood in Scotland and of his visits to Britain.

Lastly, I felt a very direct connection with some of the people mentioned in this book when I listened to *Voices of History 2*, recordings from the British Library Sound Archive (British Library Board, 2005) at www.bl.uk/soundarchive. The two-CD set includes short speeches by Thomas Edison, Thomas Watson, and Lord Kelvin. No recording of Bell's voice is known to survive.

CHAPTERS 1, 2, AND 3

For descriptions of Bell's childhood in Scotland and England, I drew on accounts recorded by Bell himself, plus personal reminiscences solicited immediately after his death by Fred DeLand, the superintendent of the Volta Bureau who was given the task of arranging and cataloging the family's archives for the newly established Bell Room at the National Geographic Society's headquarters in Washington. (This is the collection that was subsequently transferred to the Library of Congress.) These reminiscences are in Volumes 112–113 at AGBNHS.

For background on the look and feel of Edinburgh during the mid-nineteenth century, I consulted Arthur Herman's *How the Scots Invented the Modern World* (New York: Three Rivers Press, 2001) and Michael T. R. B. Turnbull's *Curious Edinburgh* (Stroud, U.K.: Sutton Publishing, 2005). My own biography of poet Pauline Johnson, *Flint & Feather: The Life and Times of E. Pauline Johnson, Tekahionwake* (Toronto: HarperCollins, 2002), describes nineteenth-century Canada, and in particular the little Ontario town of Brantford. For Bell's early Boston years, I relied on Thomas H. O'Connor's *The Hub: Boston Past and Present* (Boston: Northeastern University Press, 2001) as an introduction to that city. I also used *Boston University* (Charleston, S.C.: Arcadia Publishing, 2002), by Sally Ann Kydd.

Kenneth Silverman's *Lightning Man: The Accursed Life of Samuel F. B. Morse* (New York: Alfred A. Knopf, 2003) is an excellent account of the birth of the telegraph, and Neil Baldwin's *Edison: Inventing the Century* (New York: Hyperion, 1995) explores the frenetic scientific activity when Bell was a youth. *They Made America,* by Harold Evans (New York: Little, Brown, 2004), was a useful source on the climate of innovation and on individual inventors in this era.

For discussions of different theories of education for the deaf and hard of hearing in this and subsequent chapters, I did a thorough literature search and learned much from Jonathan Rée's *I See a Voice* (London: HarperCollins, 1999) and Harlan Lane's *When the Mind Hears: A History of the Deaf* (New York: Random House, 1984).

CHAPTER 4

Mabel Hubbard Bell described her education to Carolyn Yale for the latter's article "Mabel Hubbard Bell, 1859–1923," *Volta Review* 25 (1923): 107–110. She also sporadically kept a journal during the years 1870 to 1873 (vols. 103 and 104, AGBNHS). The story of Laura Bridgman, the little deaf and blind girl who was taught by Dr. Samuel Gridley Howe, is told in two recent biographies: *The Education of Laura Bridgman,* by Ernest Freeberg (Cambridge, Mass.: Harvard University Press, 2001), and Elizabeth Gitter's *The Imprisoned Guest: Samuel Howe and Laura Bridgman, the Original Deaf-Blind Girl* (New York: Farrar, Strauss and Giroux, 2001).

CHAPTERS 5 AND 6

The best description of the research leading to the invention of the telephone appears in Robert V. Bruce's *Bell.* Thomas Watson described his work with the inventor in *Exploring Life: The Autobiography of Thomas A. Watson* (New York: D. Appleton, 1926). The development of the telephone is also explained well in several books for younger readers, most notably Naomia Pasachoff's *Alexander Graham Bell: Making Connections* (New York: Oxford University Press, 1996) and A. Roy Petrie's *Alexander Graham Bell* (Don Mills, Ont.: Fitzhenry and Whiteside, 1975).

CHAPTER 7

My account of Alexander Graham Bell's courtship of Mabel Hubbard is drawn from Bell's own account in his letters and in a journal he kept at this period (vol. 123, AGBNHS). For background color, I drew on Henry James's novels, particularly *The Bostonians* (New York, 1886).

CHAPTER 8

There are many accounts of the 1876 Philadelphia Exhibition; one of the best is available at http://www.nga.gov/resources/expo1876. htm. My picture of Dom Pedro is drawn from Lilia Moritz Schwarcz's biography *The Emperor's Beard: Dom Pedro II and His Tropical Monarchy in Brazil,* translated by John Gledson (New York: Hill and Wang, 2003).

The description of Mabel and Alec Bell's conversation at Boston Station is taken from the article by their daughter Elsie Grosvenor, "Mrs. Alexander Graham Bell—A Reminiscence," in the *Volta Review* 59 (1957): 209–305, and "Notes on Dr. Alexander Graham Bell by Mrs. Elsie May Bell Grosvenor" (vol. 175, AGBNHS).

CHAPTER 9

Background on the fight for women's equality in America during these years comes from Linda K. Kerber and Jane Sherron de Hart, eds., *Women's America: Refocusing the Past,* 4th ed. (New York: Oxford University Press, 1995).

CHAPTER 10

Most of the descriptions of the Bells' sojourn in Britain come from Mabel's letters to her mother. I drew on Asa Briggs's *Victorian Things,* 4th ed. (Stroud, U.K.: Sutton Publishing, 2003) and Judith Flanders's *The Victorian House* (London: HarperCollins, 2003) for further domestic details. Queen Victoria's reaction to the telephone can be found in her edited journal.

CHAPTER 11

James M. Goode has resurrected many of the streetscapes and mansions that the Bells knew in Washington in his monumental *Capital Losses: A Cultural History of Washington's Destroyed Buildings,* 2nd ed. (Washington, D.C.: Smithsonian Books, 2003). For information about big business in the late nineteenth century, I consulted *The*

Gilded Age, edited by H. Wayne Morgan (Syracuse, N.Y.: Syracuse University Press, 1970), and *Essays on the Gilded Age,* by Carter E. Boren, Robert W. Amsler, Audra L. Prewitt, and H. Wayne Morgan (Austin: University of Texas Press, 1973). Gertrude Hubbard's remarks to Mabel Bell about the conduct of Western Union during the copyright case were quoted by Mabel in a letter to her mother, written on November 22, 1898, after her father's death (vol. 88, AGBNHS).

CHAPTER 12

Bell himself gave a full description of his treatment of President Garfield, in his correspondence with his wife. Other details appear in Catherine Mackenzie's *Alexander Graham Bell.* Helen Waite supplied information about Bell's attitude to his American citizenship in *Make a Joyful Sound.*

CHAPTERS 13 AND 14

Charles Dudley Warner's travel essay, *Baddeck, and That Sort of Thing: Notes of a Sunny Fortnight in the Provinces,* was originally published in Boston in 1874. The full text is now available online at http://cdl. library.cornell.edu/moa.browse.author/w.29.html.

The Bells' first visit to Cape Breton is the stuff of local legend. The story of Alec Bell's repair of the local telephone was included in an address given by J. A. D. McCurdy to the Canadian Society of Cost Accountants and Industrial Engineers, Hamilton, Ontario, December 10, 1941 (vol. 174, AGBNHS). Maud Dunlop's memoir appeared in the *Sunday Leader,* March 29, 1925. Marian H. Bell Fairchild wrote down her earliest memories of family holidays in Baddeck in 1944 (vol. 174, AGBNHS).

For my account of the demography and culture of Cape Breton, I relied on *Cape Breton Over,* by Clara Dennis (Toronto: Ryerson Press, 1942); *Beyond the Hebrides,* by Donald A. Fergusson (Halifax, N.S.: Lawson Graphics Atlantic, 1977), and Marcus Tanner's *The Last of the Celts* (New Haven, Conn.: Yale University Press, 2004).

One of the most valuable sources on the routines and dynamics of the Bell household was Charles F. Thompson, who worked for the family for over thirty years. The Bell archives contain two accounts by Charles, written on February 23, 1923, and March 20, 1924 (vol. 113, AGBNHS). Bell's Wednesday soirées are described in David Fairchild's *The World Was My Garden*.

CHAPTER 15

Alec Bell's relationship with Helen Keller is part of a much larger story: the extraordinary life and career of Helen Keller herself. The classic biography of Keller is Joseph P. Lash's *Helen and Teacher: The Story of Helen Keller and Anne Sullivan Macy* (New York: Delacorte Press, 1980). Keller herself wrote movingly of both her disability and her debt to Bell in *The Story of My Life* (New York: Doubleday, 1902), *Midstream: My Later Life* (New York: Doubleday, Doran, 1930), and the recently republished *The World I Live In*, edited by Roger Shattuck (1903; New York: New York Review Books, 2003).

Richard Winefield describes the clash between Bell and Gallaudet in *Never the Twain Shall Meet: Bell, Gallaudet, and the Communications Debate* (Washington, D.C.: Gallaudet University Press, 1987). In *The Mask of Benevolence: Disabling the Deaf Community*, Harlan Lane discusses Bell's interest in the genetic origins of deafness (New York: Alfred A. Knopf, 1992).

Mabel Bell made her admission about her coolness toward deaf people in a letter to her son-in-law Gilbert Grosvenor, dated October 11, 1921 (vol. 99, AGBNHS). She herself wrote a paper entitled "World of Silence," which is in the Bell Papers and is quoted in Winefield, *Never the Twain Shall Meet*, 76–77.

CHAPTERS 16 AND 17

Much of the information about the Bells' life in Baddeck comes from the reminiscences of Charles Thompson and from letters written by Cousin Mary Blatchford describing her visits in 1891 and 1911; these letters appeared in the *Beinn Bhreagh Recorder,* vol. 23, October 14, 1919. The *Halifax Chronicle's* description of Beinn Bhreagh comes from Robert V. Bruce's *Bell.* The letter in which Alec Bell describes his experiments to his wife is dated December 14, 1893.

Social life in Washington is captured in James M. Goode's *Capital Losses* and Julia B. Foraker's lively (if autocratic) memoir, *I Would Live It Again: Memories of a Vivid Life* (New York and London: Harper and Brothers, 1922). Robert M. Poole describes the founding role played by the Grosvenor and Bell families in the National Geographic Society and its magazine in *Explorers House: National Geographic and the World It Made* (New York: Penguin, 2004).

The growing interest in flying machines in the late nineteenth century has been explored by many authors and experts. Two sources I found particularly useful were *A Dream of Wings: Americans and the Airplane, 1875–1905* by Tom. D. Crouch (New York: Norton, 1981) and the article "Samuel Langley's Steam-Powered Flying Machines," by C. David Gierke, which appeared in the magazine *Aviation History* in July 1998.

CHAPTERS 18 AND 19

Statistics about Rochester's streets appear in *Essays on the Gilded Age* by Carter E. Boren, et al., 99. The information about the Bells' trip to Genoa for Smithson's remains comes from Mabel's journal and Charles Thompson's memoirs. The incident is also described in *The Stranger and the Statesman: James Smithson, John Quincy Adams, and the Making of America's Greatest Museum: The Smithsonian,* by Nina Burleigh (New York: William Morrow, 2003).

Bells' Boys have been described in many volumes on aviation history, including those cited above. The most exhaustive treatment of them is found in *Bell and Baldwin: Their Development of Aerodromes and Hydrodromes at Baddeck, Nova Scotia,* by J. H. Parkin (Toronto: University of Toronto Press, 1964). Additional sources for this chapter included the article "The Tetrahedral Principle in Kite Structure," which appeared in the *National Geographic Magazine*, 14 (1903), and the article "A Few Notes of Progress in the Construction of an Aerodrome," by Bell himself, which appeared in the *Beinn Bhreagh Recorder* of October 12, 1910. In *High Flight: Aviation and the Canadian Imagination*, Jonathan F. Vance explores the obsession with being airborne that characterizes Canada, as well as Alec Bell. The contributions to aviation technology made by Bells' Boys are described in *Ideas in Exile: A History of Canadian Invention*, by J. J. Brown (Toronto: McClelland and Stewart, 1967).

For information about the Wright brothers, I turned to *The Bishop's Boys: A Life of Wilbur and Orville Wright*, by Tom D. Crouch (New York: Norton, 1989), and *On Great White Wings: The Wright Brothers and the Race for Flight*, by Fred E. C. Culick and Spencer Dunmore (Toronto: McArthur and Company, 2001). Seth Shulman's *Unlocking the Skies* (New York: HarperCollins, 2002) is a lively account of the career of Glenn Hammond Curtiss.

CHAPTERS 20 AND 21

Catherine Mackenzie wrote "Some Notes about Dr. Alexander Graham Bell," covering her former boss's work routines, for Fred DeLand in the 1920s (vol. 112, AGBNHS). Much of the information about Bell with his grandchildren, as well as the Mary Pickford anecdote, appears in a delightful unpublished memoir by Melville Bell Grosvenor, entitled "Life with Grandfather." I also gathered additional information from Dr. Mabel Grosvenor, the Bells' granddaughter, who now lives in a Washington retirement home, during interviews in 2003, 2004, and 2005. Bell's work with hydrofoil craft is covered in John Boileau's fascinating *Fastest in the World: The Saga of Canada's Revolutionary Hydrofoils* (Halifax, N.S.: Formac, 2004). The description of Bell and Baldwin's voyage in a rowboat across Bras d'Or Lake in April 1917 appeared in the *Beinn Bhreagh Recorder*, vol. 21, 86–91.

EPILOGUE

I drew on Robert V. Bruce's *Bell* for much of the information about
the subsequent development of Bell's patented inventions. For the
impact of Bell's most important invention, I turned to *The Social
Impact of the Telephone,* edited by Ithiel de Sola Pool (Cambridge,
Mass.: MIT Press, 1977), and *The Creation of the Media: Political
Origins of Modern Communications,* by Paul Starr (New York: Basic
Books, 2004).

Photo Sources

All photos and illustrations are from the photo archives at the Alexander Graham Bell National Historic Site (AGBNHS), Baddeck, N.S., except for the following images:

p. iv: Portrait of Alexander Graham Bell by Timoléon Lobrichon, (artist), from the Gilbert H. Grosvenor Collection of Alexander Graham Bell photographs, Prints and Photographs Division, Library of Congress (inset portrait of Mabel Hubbard from the AGBNHS).

p. 23:	Library and Archives Canada	PA26916
p. 34:	Library of Congress	USZ62 102341
p. 53:	Library of Congress	USZ6 2008
p. 54:	Library of Congress	USZ6 2015
p. 86:	Library of Congress	USZ62 108154
p. 100:	Library of Congress	USZ62 117584
p. 106:	Library of Congress	USZ62 71316
p. 130:	Library of Congress	USZ62 106472
p. 131:	Library of Congress	USZ62 57385
p. 133:	Library of Congress	USZ62 96109
p. 219:	Library of Congress	USZ62 134586

All maps (pages 50, 116 and 316) are by Dawn Huck.

Acknowledgments

I must begin my acknowledgments with heartfelt thanks to Dr. Mabel Grosvenor, the formidable 101-year-old former pediatrician who is the Bells' last surviving grandchild. When I first visited "Dr. Mabel," as she is always known, in Washington in 2003, she welcomed me and gave me the kind of information about her grandfather that only someone with personal memories would have. She told me, for example, that Alexander Graham Bell spoke English with a standard middle-class English accent, rather than a Scottish, Canadian, or American accent. She also talked to me about the crucial importance of her grandmother, Mabel Grosvenor Bell: without his wife, Dr. Mabel suggested, this brilliant man "would have been lost." Dr. Mabel subsequently made arrangements for me to visit the Point, the Bells' home near Baddeck, Nova Scotia, which is still in the family and has altered very little since Bell passed away there in August 1922. It was important to me to have the imprimatur of Dr. Mabel for my efforts, and I am grateful to her.

Similarly, other members of the Bell family, especially Grosvenor

Blair, Joe Blair, Hugh and Jeanne Muller, and Elsie Myers Martin, were helpful and encouraging.

When I embarked on this book, I assumed that I would do most of my research at the Library of Congress in Washington. But I decided to start with a return visit to the Alexander Graham Bell National Historic Site (AGBNHS) in Baddeck, which I first saw in July 1997. It was a challenge to get there from my home in Ottawa. In December 2003, I flew to Halifax and then drove a rented car across mainland Nova Scotia to the Canso Causeway that has linked the mainland to Cape Breton Island since 1955. After a four-and-a-half-hour journey through blinding snow, I finally arrived. There I discovered that the Bell museum, as the site is colloquially known, had copies of most items that reside in the Bell archives in Washington. I realized I would be crazy to pursue my research anywhere other than Baddeck. I cannot thank enough the staff at the Alexander Graham Bell National Historic Site, including Joyce Hill, Jess Fraser, Sharon Morrow, Valerie Mason, and Rosalynd Ingraham, who welcomed me and allowed me to monopolize the photocopier. Most of all, I am deeply indebted to the manager of the AGBNHS, Aynsley MacFarlane, who made my six research trips feel more like vacations, introduced me to the magic of Cape Breton on our frequent drives together, and corrected many mistakes in my manuscript.

Many people helped with the research and writing of this book. Dr. Duncan McDowall of Carleton University suggested background reading for historical context. Dr. Christina Cameron, then director of National Historic Sites in the Department of Canadian Heritage, and Judith Tulloch in Halifax facilitated my research. John Wesley Chisholm, the Halifax filmmaker who made a superb documentary about the Bell hydrofoils, gave me useful advice. The staff at the Cambridge Historical Society and the Bostonian Society were resources on the Bells' Boston years. Maureen Boyd shepherded me around Washington, helping me locate Bell homes and sources of information about the capital in the late nineteenth century. Ryan Shepard, collections librarian at the

Kiplinger Library at the Historical Society of Washington, D.C., talked to me about the capital's late-nineteenth-century architecture. Patricia Potts, in the United Kingdom, enlarged my understanding of the politics of deaf education. Brian Wood, curator of the Bell Homestead National Historic Site, supplied information on the Bells' years in Brantford, Ontario, when they lived in the house now called Tutela Heights but known in their day as either Melville House or Tutelo Heights. Roddy McFall showed me copies of Bell patents in Library and Archives Canada. John Boileau generously checked my version of the development of Bell's hydrofoils. Dr. Sandy Campbell, Sheila Williams, Ernest Hillen, and Marta Tomins read all or parts of the manuscript and gave helpful critiques. I owe Morgan Wesson a debt of gratitude for his help on picture research. Thanks to C. M. Townsend for information about the first phone call from Brantford to Paris.

I also owe thanks to Marie Wood, Alan Rayburn, Shona Cook, Mona and Leo Asaph, Monic Charlebois, and Mike Lavergne. I am very grateful to the posse of friends who have given me support, and asked the kind of questions that prompt new insights, during my journey with the Bells: Judith Moses, Maureen Boyd, Wendy Bryans, and Cathy Beehan. I owe my deepest gratitude to my husband, George Anderson, on whose sound judgment, balanced advice, and warm encouragement I depend, and to my sons, Alexander, Nicholas, and Oliver, who are smart, funny, and a reminder that there is life in the twenty-first century too.

It was a pleasure to be working again with Phyllis Bruce at HarperCollins Canada and the formidable team of professionals there, including managing editor Noelle Zitzer, copy editor and indexer Stephanie Fysh, type designer Sharon Kish, jacket designer Alan Jones, proofreader Curtis Fahey, science reader David Peebles, and publicists Lisa Zaritzky and Melanie Storoschuk. I have enjoyed working with Cal Barksdale at Arcade Publishing in New York. I would also like to thank my agents, Jackie Kaiser and John Pearce, at Westwood Creative Artists, for all their efforts on my behalf.

Finally, I am grateful to the Office of Cultural Affairs at the City of Ottawa, and to the Canada Council, for financial assistance and their continued support of Canadian writers.

Index